Qp 86 .H343

101875

Davis, Robert L.
Handbook of Geriatric
Nutrition

East Texas Baptist College
Library
Marshall, Texas

HANDBOOK OF GERIATRIC NUTRITION

HANDBOOK OF GERIATRIC NUTRITION

Principles and Applications for Nutrition and Diet in Aging

Edited by

JENG M. HSU, D.V.M., Ph.D.

and

ROBERT L. DAVIS, Sc.D.

Veterans Administration Medical Center
Bay Pines, Florida

NOYES PUBLICATIONS
Park Ridge, New Jersey, U.S.A.

Copyright © 1981 by Noyes Publications
No part of this book may be reproduced in any form
without permission in writing from the Publisher.
Library of Congress Catalog Card Number: 81-16853
ISBN: 0-8155-0880-8
Printed in the United States

Published in the United States of America by
Noyes Publications
Mill Road, Park Ridge, New Jersey 07656

Library of Congress Cataloging in Publication Data
Main entry under title:

Handbook of geriatric nutrition.

Bibliography: p.
Includes index.
1. Aging--Nutritional aspects. 2. Aged--
Nutritional aspects. 3. Geriatrics. I. Hsu,
Jeng M. II. Davis, Robert L. [DNLM: 1. Diet--
In old age. 2. Nutrition--In old age. QU 145
H235]
QP86.H343   613.2'0880565   81-16853
ISBN 0-8155-0880-8   AACR2

Foreword

For far too long, nutrition has been relegated to the realm of fadism and folklore. No single group deserves all the blame; we share it—health professionals, social service workers, educators, and policymakers.

Happily, and not a minute too soon, the role of nutrition in general well-being is gaining some ascendancy. And its role in contributing to, or preventing, serious illness is finally getting the attention it deserves in the biomedical research community.

Many Americans finally are discovering that perhaps mother was right about the importance of a balanced diet. But we must not rely on conversion of our people one-by-one. Our national consciousness, and national policies, must begin to reflect what we know and what we are beginning to learn about nutrition.

Some might assume that the elderly—the Nation's grandmothers and grandfathers and great-grandparents—have better eating habits than their progeny. On the contrary, many are woefully ignorant of what they should know in order to lead healthful lives. And at no time in life is sound nutrition more important than in later years, that time when the body needs special attention, when the system is called on to make different kinds of accommodations.

It is in this area that the contributors to this book have done us a great favor. By calling attention to the unique characteristics of older people they have added in a very significant way to the growing store of information about those things that make aging people special. I commend to older persons themselves, and any who minister to them, this effort and others which surely will follow to understand the aging body and its nutritional needs.

Claude Pepper
Chairman, Select Committee on Aging
United States House of Representatives

Contributors

Anthony A. Albanese, Ph.D.
Nutrition and Metabolic Research Division
Burke Rehabilitation Center
White Plains, New York

John F. Aloia, M.D.
Department of Medicine
Nassau Hospital
Mineola, New York

Charles H. Barrows, Sc.D.
Gerontology Research Center
National Institute on Aging
Baltimore City Hospitals
Baltimore, Maryland

Robert F. Borgman, D.V.M., Ph.D.
Clemson University
Clemson, South Carolina

Linda H. Chen, Ph.D.
Department of Nutrition and Food Science
University of Kentucky
Lexington, Kentucky

Audrey K. Davis, R.D.
Nutrition Therapy, Education, and Research Section
Veterans Administration
Bay Pines, Florida

Contributors

Robert L. Davis, Sc.D.
Medical Research
Veterans Administration
Bay Pines, Florida

Mary Jean Etten, R.N., Ed.D.
Department of Nursing and Gerontology
St. Petersburg Junior College
St. Petersburg, Florida

Judith Hallfrisch, M.S.
Beltsville Human Nutrition Research Center
United States Department of Agriculture
Beltsville, Maryland

H. Steve Hsieh, Ph.D.
Department of Nutrition Sciences
School of Dentistry and Medicine
University of Alabama in Birmingham
Birmingham, Alabama

Jeng M. Hsu, D.V.M., Ph.D.
Veterans Administration Medical Center
Bay Pines, Florida

Kung-Ying Tang Kao, M.D., Ph.D.
Geriatrics Research Laboratory
Veterans Administration Medical Center
Martinsburg, West Virginia

Mushtaq Ahmad Khan, Ph.D.
Food and Drug Administration
Department of Health and Human Services
Washington, D.C.

Gertrude C. Kokkonen, B.S.
Gerontology Research Center
National Institute on Aging
Baltimore City Hospitals
Baltimore, Maryland

Florence Lazicki Lakshmanan, Ph.D.
Nutrition Institute
Science and Education Administration
Beltsville, Maryland

Alfred H. Lawton, M.D., Ph.D.
College of Medicine
University of Florida
Gainesville, Florida

Menahem Lender, M.D.
Metabolic Section
Veterans Administration Medical Center
Hines, Illinois

Alif M. Manejwala, M.D.
Laurel, Maryland

Dace Osis
Veterans Administration Medical Center
Hines, Illinois

Claude Pepper
Chairman, Select Committee on Aging
United States House of Representatives
Washington, D.C.

Sheldon Reiser, Ph.D.
Carbohydrate Nutrition Laboratory
Beltsville Human Nutrition Research Center
United States Department of Agriculture
Beltsville, Maryland

Sue V. Saxon, Ph.D.
Department of Gerontology
University of South Florida
Tampa, Florida

Herta Spencer, M.D.
Metabolic Section
Veterans Administration Medical Center
Hines, Illinois

Donald M. Watkin, M.D., M.P.H.
Office of Aviation Medicine
Federal Aviation Administration
Washington, D.C.

Albert J.E. Wilson, III, Ph.D.
University of Tampa
Tampa, Florida

Contents

Foreword v
Contributors vii
1. **Introduction:** *Donald M. Watkin* 1
   Lifelong Attention 1
   Focus on Research 3
   Service Today 5
   References 5
2. **Maternal, Early Neonatal and Subsequent Nutrition Influence on Aging:** *Robert L. Davis* 8
   Introduction 8
   Influences of Maternal and Postnatal Nutrition on Aging 8
   Selected Human Studies 13
   Conclusion 14
   References 15
3. **Psychological Aspects of Nutrition in Aging:** *Sue V. Saxon and Mary Jean Etten* 19
   Introduction 19
   Psychosocial Development 19
   Meaning of Food 21
   Appetite and Hunger 24
   Overnutrition and Undernutrition 26
   Physical Changes 29
   Sensory Systems 30
   Psychodynamics of Aging and Nutrition 34
   References 37
4. **Sociological Aspects of Nutrition and Aging:** *Albert J.E. Wilson III* 43
   Introduction 43
   Functions of Food 44
   Relationship of Studies of Early Life Nutrition to Aging 44
   Nutritional Status of Older Americans 45
   Sociological Factors as Independent Variables 46
   Nutritional Status of the Aging as an Independent Variable 51
   Summary 53

References 54

5. **Protein Nutrition and Aging:** *Kung-Ying Tang Kao and Florence Lazicki Lakshmanan* 56
   Protein Metabolism and Aging..56
   Aging-Related Diseases and Protein Metabolism 66
   Protein Requirements and Aging 71
   Concluding Remarks and Summary 77
   References 78

6. **Cholesterol Metabolism and Atherosclerosis in Aging:** *Mushtaq Ahmad Khan and Alif M. Manejwala*..88
   Introduction..88
   Age-Related Changes in the Arteries....................................89
   Evolution of Atherosclerotic Lesion......................................91
   Risk Factors Associated with Atherogenesis 93
   References 104

7. **Carbohydrate Nutrition and Aging:** *Sheldon Reiser and Judith Hallfrisch*..110
   Introduction..110
   Animal Studies..110
   Human Studies..118
   Summary and Conclusion...123
   References 124

8. **The Current Status of Ascorbic Acid, Vitamin B_6, Folic Acid and Vitamin B_{12} in the Elderly:** *Jeng M. Hsu*.........................128
   Introduction..128
   Ascorbic Acid..129
   Vitamin B_6...133
   Folic Acid..136
   Vitamin B_{12} 140
   Concluding Remarks...145
   References 146

9. **Vitamin D Metabolism in Aging:** *John F. Aloia* 157
   Introduction..157
   Background 157
   Aging and $25OHD_3$...164
   Hormones that Interact with Vitamin D 166
   Drugs and Illnesses in the Elderly Which Affect Vitamin D Metabolism..167
   Involutional Osteopenia 168
   Summary 169
   References 170

10. **Vitamin E and Aging:** *Linda H. Chen* 176
    Introduction..176
    Chemistry, Physiology, Food Sources and Daily Allowance of Vitamin E 177
    Vitamin E Deficiencies in Man and Animals 178
    Vitamin E and Aging...178
    Vitamin E Controversy...182
    Megadoses of Vitamin E 182

Concluding Remarks 183
References 183

11. **The Current Status of Zinc, Copper, Selenium and Chromium in Aging:** *Jeng M. Hsu and H. Steve Hsieh* 188
    Introduction 188
    The Status of Zinc in Human Nutrition 188
    The Status of Copper in Human Nutrition 195
    The Status of Selenium in Human Nutrition 201
    The Status of Chromium in Human Nutrition 203
    Conclusion 206
    References 207

12. **Magnesium, Phosphorus and Calcium Needs for Bone Health:** *Anthony A. Albanese* 219
    Magnesium 219
    Phosphorus 221
    Calcium 223
    Sources of Calcium 224
    Bioavailability 224
    Interaction of Dietary Constituents 227
    Skeletal Bone Loss 235
    Alveolar Bone 244
    Summary 247
    References 248

13. **Fluoride Metabolism and Aging:** *Herta Spencer, Dace Osis and Menahem Lender* 250
    Introduction 250
    Methodology 251
    Results 252
    Discussion 259
    References 261

14. **Nutrition Related Diseases of the Aged:** *Alfred H. Lawton* 266
    Introduction 266
    Diabetes Mellitus as a Representative Disease 266
    Chronic Disease, Age, and Nutritional Factors 268
    Concluding Remarks 275
    References 276

15. **Dietary Fiber and the Aging Processes:** *Robert F. Borgman* 279
    Introduction 279
    Chemical and Physiological Properties of Dietary Fiber 280
    Role of Dietary Fiber in Prevention of Diseases Affecting
      the Intestinal Tract 282
    Diseases Associated with Straining During Defecation
      (Hernias and Varicose Veins) 287
    Diseases Associated with Atherosclerosis (Ischemic Heart
      Disease, Stroke, etc.) 289
    Cholelithiasis 291
    Obesity 291
    Diabetes Mellitus 292
    Possible Problems with Increasing Dietary Fiber 292
    Sources of Dietary Fiber 292

Prospectus on the Use of Dietary Fiber to Prevent Premature
Aging 293
References 293

16. **Nutritional Hazards of Retirement:** *Audrey K. Davis* 296
    Introduction 296
    Low Income 297
    Poor Dentition 298
    Boredom 299
    Erratic Food Consumption Patterns 300
    Retiree-Spouse Relationships 301
    Influence of Advertising 302
    Obesity 304
    Lack of Exercise 305
    Need for Socialization 306
    Nutritional Assessment and Counseling 308
    Summary 308
    References 309

17. **Modern Nutrition for Those Who Are Already Old:** *Donald M. Watkin* 312
    Introduction 312
    Futility of Conventional Prevention by Nutritional Means
    Among the Elderly 313
    Nutritional Management of the Genetically Nonelite 314
    Nutritional Management of Acute Illnesses 317
    Nutritional Management of Trauma 318
    Nutritional Management of Older Persons Recovering from
    Acute Illness or Trauma 320
    Nutritional Management of Older Persons not Afflicted by
    Acute Illness or Trauma 321
    Conclusion 323
    References 324

18. **Food Facts, Fads, Fallacies and Folklore of the Elderly:** *Audrey K. Davis and Robert L. Davis* 328
    Introduction 328
    Sources of Nutrition Misinformation 329
    Food Faddism and Quackery 330
    Why Misinformation is Dangerous 333
    Vulnerability of the Elderly 333
    Nutrition Facts and Fallacies 334
    Types of Nutrition Nonsense 344
    Realities of Advertising 345
    Personal and Public Responsibility 346
    References 347

19. **Research Needs:** *Charles H. Barrows and Gertrude C. Kokkonen* 350
    General Considerations 350
    Effect of Age on Adult Nutritional Requirements 351
    Dietary Restriction and Life Extension 353
    References 358

Index 361

–1–
Introduction

Donald M. Watkin

The editors have assembled contributions from authors who have had experience in the basic science and applied aspects of aging as a biologic, psychologic, economic and sociologic process and in educating professionals and the public in measures important in the successful management of that process. With reason, therefore, they have focused this handbook on both the scientific principles and the applied aspects of nutrition and aging. Only by combining scientific with medical, public health, sociologic and economic considerations can rational approaches to the triad of nutrition, health and aging, each component of which is an agglomerate of infinite dimensions interacting with the other two, be conceptualized today.[25] This handbook provides the presently known ingredients needed to approach that objective in the decade of the eighties.

LIFELONG ATTENTION

Chapter 2 introduces a concept on which all considerations of nutrition and aging must be founded. Aging is a process beginning no later than conception; nutrition begins to influence that process from its inception.

Nutrition's influence on the process of aging is greater *in utero* and during the first two years of life. However, as the major environmental factors directly under the control of the parent of a child or of the adolescent or mature adult, nutrition has a direct bearing on the physical and mental health and on the behavior of all aging human beings. This influence prevails throughout life. Hence, in considering nutrition and aging, the scientist and the practitioner must scrupulously avoid regarding aging as a process occupying only the latter years of life.

In regard to the influence of nutrition on mature adults, Belloc and Breslow[3] have clearly demonstrated the desirable effect of lifelong devotion to seven good health practices. In paraphrased form, these are: avoid gluttony, tobacco and excessive alcohol consumption; eat nutritionally adequate regularly scheduled meals; and incorporate into one's life style regularly scheduled hours of sleep, exercise and rest and relaxation. Four of these are definitely nutrition-oriented; avoidance of tobacco and scheduling adequate rest and relaxation have many nutrition-related connotations. The Belloc and Breslow study showed that the health status (measured in an arbitrary unit called a "ridit" [relative to an identified distribution], low indicating good health and high indicating the proximity of death) at any age was inversely proportional to the number of health practices adhered to throughout life. They found persons 85 years of age who had followed six to seven of the life styles with ridits identical to 35-year-olds who had followed from none to two.

Although virtually all authorities agree with the Belloc and Breslow concept, most authorities acknowledge that few persons are fortunate enough to experience the ideal environment and life style from conception to death. The population cohorts with the greatest number of health problems in today's societies are those for whom the values of appropriate environment (including nutrition) and desirable life styles were not recognized at critical points in their aging process. Hence, over half of today's elderly are persons who have one to many lifelong illnesses and disabilities which, since they are *in situ*, can not be prevented but still must be managed in manners such that the optimum health status possible in the face of specific pathologies may be maintained for the remainder of life.

This need for attention to health issues among the elderly may last for as long as 50 years, assuming one becomes legally elderly at age 60 and may survive until 110. Nutrition's role in maintaining optimum health among the elderly is far different from the one it plays among younger and healthier age cohorts.

Among the elderly, nutritional regimens must be modified to counter the impacts of medications and/or surgery. They must participate with drugs and surgery in the management of acute illnesses and accidents which afflict the elderly at rates far higher than those among younger age cohorts. They must also be modified to meet the economic, psychologic and sociologic problems faced by the elderly and described in detail in Chapters 3, 4 and 16.

Prevention of disease through nutritional modifications—a valid consideration among fetuses, children and young adults—has not been demonstrated to apply to elderly persons. Any claims implying prevention of disease through modifications in dietary regimens should be viewed with great suspicion. However, as stressed above, management of diseases and disabilities by appropriate nutritional regimens is vital to assuring optimum possible health status among older persons.

In summary, nutrition influences biologic aging throughout life but particularly in the very early years of life. In mature adults, lifelong applications of nutrition-related life styles have been shown to be related to health status. Among those who are already old and who were raised at times when and in places where present knowledge of nutrition, health and aging was unknown, nutrition combined with appropriate health care can optimize health status even in the face of existing pathologies, acute illnesses, indicated surgery and serious accidents.

FOCUS ON RESEARCH

Chapters 5 through 13 and 15 detail present knowledge of major nutrients and other food components and their relation to aging as a lifelong process, to aging as the state of being old, and to aging as a factor in the management of diseases and disabilities which will remain with their hosts throughout life. This present knowledge is fragmentary, often contradictory and based on studies of very small numbers of human beings. Since the variance of any measurable parameter increases with advancing age, a serious problem obviously lies in the small numbers of persons who have been observed in various research efforts.

A related problem lies in the very small numbers of persons who have been studied during the last quarter of the technical life span of man (*viz.*, persons 86 years of age or older). Studies of persons aged 60 through 85 have been somewhat more numerous but include large numbers of the so-called frisky elderly who may have been equivalent on a functional scale to persons up to two decades younger. Until studies are conducted on large numbers of clinically healthy persons 86 or over, little solid evidence will be available in regard to the nutritional status or the nutrient requirements as they are related to old age *per se*.

A third problem is the lack of acceptable scientific evidence that altering nutritional regimens will influence the health and longevity of human beings. Prospective, controlled studies in large numbers of persons of any age have not been performed. The appalling lack of knowledge on the effects of change can be illustrated by three examples.

First, the National Heart, Lung and Blood Institute (NHLBI) has been conducting since 1973 a 10-year study[16], the Coronary Primary Prevention Trial (CPPT), to test the hypothesis that reducing the serum cholesterol level will reduce the number of cardiovascular endpoints (*i.e.*, indicators of serious pathology caused by atherosclerosis). In this case, serum cholesterol is being reduced by a drug—cholestyramine—combined with a diet which is capable of producing only modest reductions in serum cholesterol concentration but is not a cholesterol-lowering regimen *per se*. The point deserving emphasis here is that the lipid hypothesis itself previously has not been tested scientifically in man. In other words, in spite of the concern over the cholesterol contents of diets expressed since the first decade of this century, there is no evidence apart from epidemiologic associations indicating that lowering the serum cholesterol in man by dietary modifications or by drug therapy will diminish the number of cardiovascular endpoints. The NHLBI's CPPT data will not be available until 1983 at the earliest.

Second, since 1959, the principal standards of weight for height in the United States have been the desirable weight tables prepared by the Metropolitan Life Insurance Company[15] based on actuarial experience of United States insurance companies reported in the Build and Blood Pressure Study (BBPS).[17] The BBPS data imply a direct linear relationship (except at the very low end of the scale) between weight and mortality. In addition, the Ninth Edition (1980) of the Recommended Dietary Allowances (RDA's)[11] suggest guidelines for desirable weight lower than those of the Eighth Edition (1974).[10] These recommendations have appeared in many publications.[4,5,20]

In 1980, however, an analysis of data from the Framingham Study,[18,21] using the same definitions of body build and similar follow-up periods for mortality as were used in the BBPS, suggests that mortality is lowest at average (not ideal or desirable) body weights and that mortality increases as body weight is lowered as well as when it increases. This excess mortality in persons below average weight obtained even when all persons with definite clinical evidence of potentially lethal diseases and when smokers were excluded from the analysis. Similar findings have been reported from an analysis of 14 years of follow-up in the Chicago Peoples Gas Company study.[8] Andres[1] has reviewed Belloc's study[2] of Alemeda County, California; Cole's study[6] in Derbyshire in the United Kingdom; Libow's[13] Human Aging Study; and the Baltimore Longitudinal Study and concluded that overweight persons in the range 110 to 130% of desirable weight have lower mortality than persons in the 80 to 110% range.

Third, a potentially very important factor influencing energy metabolism of human beings has begun to surface in the clinical literature,[9,19] This is the observation that brown fat under the influence of norepinephrine is capable of converting food energy to heat at rates sufficient to enable some persons endowed with copious quantities of brown fat to consume dietaries containing relatively speaking enormous amounts of energy without gaining weight. Changes in the quantity and activity of brown fat with age are unknown. The role of brown fat in determining the energy requirements of older persons has never been explored.

Clearly, these three examples suggest that far more research in human beings must be done to clarify even the most basic relationships of nutrition and aging. The well-known 1980 conflict[14] arising over the DHEW-USDA *Dietary Guidelines for Americans*[20] and the Food and Nutrition Board's *Toward Healthful Diets*[12] is clear confirmation of this need. While the cost of such studies would be great in terms of personnel, time and money,[7] the cost in terms of human misery and anxiety and of money spent in implementing unproven hypotheses suggests that cost-effectiveness favors additional carefully planned research.

SERVICE TODAY

Chapters 14, 17, 18 and 19 are dedicated to what can be done particularly for those who are already old based on principles derived from the limited data now available. Essential as it is to acquire new and better data through more research, it is even more important to maintain the confidence of those who are already old by applying principles based on the scientific method to assist the aged of today.[23] Fallacies, fads and quackery (*v.i.*, Chapter 18) thrive because the elderly themselves and even many who are devoted to serving them lack the knowledge now extant which would devastate the faddists and quacks were it applied.[22] Distribution of that knowledge to professionals working in the nutrition, health and aging fields is a major objective of this book. Its application is dependent on the same knowledge being translated into terms understandable especially to those who are already old.[24] If the older Americans of today accept nutrition's critical role in all phases of aging as a process, they can become a vital force, educating younger members of society in utilizing nutrition throughout life to minimize in the future the prevalence of the pathology seen among the aged of today.[24]

Education is best realized through example. Application of modern nutrition principles to alleviate the problems of those already old can create lasting impressions among those assisted. They in turn can pass the principles as applied on to their peers and to younger members of their families, their friends and their communities.

The editors of and contributors to this work are unanimous in their conviction that nutrition principles—present and those which will be scientifically confirmed in the future—can be applied successfully for the betterment of all who are aging.

REFERENCES

(1) Andres, R. 1980. Influence of obesity on longevity in the aged. In C. Borek (ed.) *Aging, Cancer and Cell Membranes*. Volume VII, Advances in Cancer Biology Series. Stratton Intercontinental Medical Book Corporation, New York.

(2) Belloc, N.B. 1973. Relationship of health practices and mortality. *Prev. Med.* 2:67-81.

(3) Belloc, N.B. and Breslow, L. 1972. Relationship of physical health status and health practices. *Prev. Med.* 1:409-421.
(4) Bray, G.A., Editor. 1976. *Obesity in Perspective, Parts I and II.* Fogarty International Center Series on Preventive Medicine, Volume 2. DHEW Publication No. (NIH) 76-852. U.S. Government Printing Office, Washington, D.C.
(5) Bray, G.A., Editor. 1979. *Obesity in America: Proceedings of a Conference at the National Institutes of Health, October, 1977.* NIH Publication No. 79-359. U.S. Government Printing Office, Washington, D.C.
(6) Cole, T.G., Gibson, J.C. and Olsen, H.C. 1974. Smoking and obesity in an English and Danish town. Male deaths after a 10-year follow-up. *Bull. Physio-pathologie Respiratoire* 10:657-679.
(7) Diet-Heart Panel of the National Heart Institute, E.H. Ahrens, Jr., Chairman. 1969. Mass field trials and the diet-heart question: their significance, timeliness, feasibility and applicability. American Heart Association Monograph No. 28. American Heart Association, Inc. New York.
(8) Dyer, A.R., Stamler, J., Berkson, D.M., and Lindberg, H.A. 1975. Relationship of relative weight and body mass index to 14-year mortality in the Chicago Peoples Gas Company Study. *J. Chronic Dis.* 28:109-123.
(9) Elliott, J. 1980. Blame it all on brown fat now. *J. Amer. Med. Assoc.* 243:1983-1985.
(10) Food and Nutrition Board, Committee on Dietary Allowances, A.E. Harper, Chairman. 1974. *Recommended Dietary Allowances, Eighth Revised Edition.* National Research Council. National Academy of Sciences. Washington, D.C.
(11) Food and Nutrition Board, Committee on Dietary Allowances, H.N. Munro, Chairman. 1980. *Recommended Dietary Allowances, Ninth Revised Edition.* National Research Council. National Academy of Sciences. Washington, D.C.
(12) Food and Nutrition Board, Task Force on Guidelines toward Healthful Diets, R.E. Olson, Chairman. 1980. *Toward Healthful Diets.* National Research Council. National Academy of Sciences. Washington, D.C.
(13) Libow, L.S. 1974. Interaction of medical, biologic and behavioral factors on aging adaptation and survival. An 11-year longitudinal study. *Geriatrics* 29:75-88.
(14) MacNeil, R., Lehrer, J., Goodman, D., Calloway, W., and Olson, R. The cholesterol question. *The MacNeil/Lehrer Report, May 28, 1980.* Library #1218, show #5238. Journal Graphics, Inc., New York.
(15) Metropolitan Life Insurance Company. 1959. New weight standards for men and women. *Statistical Bull.* 40:1-4.
(16) Rifkin, B.M., Chief, Program Office. 1979. The Coronary Primary Prevention Trial: design and implementation. The Lipid Research Clinics Program. *J. Chron. Dis.* 32:609-631.
(17) Society of Actuaries. 1960. *Build and Blood Pressure Study, 1959. Vol. I.* Chicago.
(18) Sorlie, P., Gordon, T., and Kannel, W.B. 1980. Body build and mortality: The Framingham Study. *J. Amer. Med. Assoc.* 243:1828-1831.
(19) Stock, M.J. 1981. Diet-induced thermogenesis: a role for brown adipose tissue. In E.J. Bassett and R. Beers, Jr. (eds.), *Nutrition Factors: Modulating Effects on Metabolic Processes.* Proceedings of the 13th Miles

International Symposium, Baltimore, June 18-20, 1980. Raven Press, New York (In press).
(20) U.S. Department of Agriculture and U.S. Department of Health, Education and Welfare, 1980. *Nutrition and Your Health: Dietary Guidelines for Americans.* U.S. Government Printing Office. Washington, D.C.
(21) Vaisrub, S. 1980. Editorial. Beware of the lean and hungry look. *J. Amer. Med. Assoc.* 243:1844.
(22) Watkin, D.M. 1980. Better nutrition for those already old: the challenges of the eighties. *Aging.* Sept.-Oct. 1980. Nos. 311-312.
(23) Watkin, D.M. 1981. Modern nutrition for those who are already old. This volume.
(24) Watkin, D.M. 1981. Certain aspects of the effect of age on the acquisition of nutrients. In D. Harman (ed.). *Proceedings of a Symposium of Aging and Nutrition at the Ninth Annual National Meeting of the American Aging Association (AGE). Washington, D.C.* September 20-22, 1979. Raven Press, New York (In Press).
(25) Watkin, D.M. 1981. *The Nutrition-Health-Aging Triad: Integrating the Sciences, Medicine and Public Health.* Raven Press, New York (In press).

–2–
Maternal, Early Neonatal and Subsequent Nutrition Influence on Aging

Robert L. Davis

INTRODUCTION

The concept that there exist certain critical periods in the development of the animal including man is not novel. It is well known that inadequacies during early nutrition may cause a variety of organic, functional deficiencies and metabolic derangements, and that many of these abnormalities thus inflicted may persist throughout the life span of the animal, regardless of the adequacy of subsequent dietary intake.[8,15,24,35,44,45]

It is becoming more evident that early nutritional influences have a pronounced effect upon anatomic, metabolic and behavioral characteristics of man and animal. The small stature of the Japanese population was attributed earlier to genetic endowment. Currently, teenagers in Japan are approximately as tall as their American counterparts. This marked change certainly cannot be attributed to genetic change but to post-war changes in life style, especially nutritional, maternal in particular. In Israel, a similar trend has taken place.

Barrows[7] proposes "that the orderly sequence of events that takes place throughout the life span of an organism is a genetically programmed phenomenon. However, the rate of expression of the information of the program is subject to environmental influences such as nutrition."

INFLUENCES OF MATERNAL AND POSTNATAL NUTRITION ON AGING

Life Span

The discovery of McCay[36] that the life span of rodents could be significantly extended by dietary means was most important. McCay observed that dietary restriction, sufficient to retard growth, could extend the life span. He also demonstrated that growth retardation by the feeding of a low protein diet beginning early in life resulted in a longer life span. The most striking finding was a significant delay in the development of diseases of the lung, kidneys and tumors of all types.

Ross and associates[47-50] carried out extensive studies on the effect of different diets upon the life span and incidence of disease of animals. The effect depended in part when during the life span the diets were fed. Their findings indicate that an increase in protein early in life and a decrease later in life were associated with an increase in life span. Ross[50] also observed that animals could voluntarily alter their selection of diets which could extend their life span.

Metabolism

Many investigators[15,29,31,32,44,45,63,65,66] have provided ample evidence that alterations in the diet of animals during gestation, lactation and infancy may induce metabolic abnormalities of the offspring as well as psychosocial changes. It has also been reported that the composition of the maternal and early postnatal diet can have profound effects on the development of the progeny, even when they are fed an adequate diet on an unlimited basis after weaning.[15,33,44,45]

Quantitative restriction or restriction in protein and or energy intake[2,6,31,40,44,51,52,58] of the maternal diet of the rat and other animals results in progeny with reduced feed efficiency,[14,15,33,45] disorders in nitrogen utilization,[6,44,45] abnormal behavioral and mental development,[56,57,63,64] and irreversible stunting of growth.[15,16,31,32] These metabolic disturbances appear when maternal restriction occurs during gestation but not during lactation period alone if feed efficiency is used as a basis.[24,45] Even when a diet is 50% deficient in calories, protein has a unique role in promoting behavioral and emotional stability and lessening neonatal mortality of rats.[44] Behavioral abnormalities such as tremors, convulsions, bizarre stance and hyperexcitability commonly observed in progeny of mothers subjected to both protein and calorie dietary restriction have not been observed in the progeny of calorie-restricted dams when adequate protein was available in the maternal diet.

Guthrie and Brown[20] observed that children who have experienced a severe protein calorie malnutrition in early infancy had brain size and intellectual development damage refractory to subsequent rehabilitation. Maternal diets restricted in both protein and energy intake also produce reduced liver and brain weight and total organ DNA, RNA and protein in offspring at weaning.[58] Naege[42] reported that substantial maternal protein undernutrition during pregnancy leads to low birth weight babies and to an increase in the perinatal death rate. Low birth weight infants have reduced cells which are small in size compared to those of infants of normal weight. Undernutrition also results in smaller thymus, spleen, liver and adrenal glands.

Mice neonatally infected with an enterovirus or malnourished during early life showed evidence of lasting depression of body weight and decreased biosynthesis of protein and ribonucleic acid in various tissues and subcellular fractions of brain.[32,33] The depression of body weight in neonatally infected or malnourished mice was partially corrected by combined administration of growth hormone and insulin. Lee[32] reported that early disturbances caused by neonatal infection or under nutrition can result in changes in the metabolism and structure of brain protein.

The inclusion of high levels of sucrose in the diet of mice during gestation and lactation result in progeny significantly heavier than control progeny fed laboratory chow.[15] Many litters of the progeny of sucrose-fed mothers were permanently obese and showed a markedly different disposition of a glucose test dose than controls. Significant differences in body composition between the two groups were also observed.[15]

Turner[60] suggests that the state of nutrition during the growth period may be more important in determining adult function than adult diet. This is not only in respect of reproductive performance, but also of diabetes mellitus and atherosclerosis.

Psychosocial Aspects

Many investigators have reported that undernutrition or faulty diet during prenatal and early life results in lasting effects on physical activity, as well as learning and social behavior of man and animal.[8,11,22,39,43,53,56,57,63] The effects of malnutrition vary in accordance with the time in the animal's life at which it is experienced. In some animals the effects are most severe if the nutritional stress occurs in the prenatal period, in others during early postnatal life.[8] Behavioral and psychosocial measurements reflect the integration of many complex processes in the organism and have been utilized to assess age-associated changes. They provide a useful means of monitoring rate of aging in animals subjected to various nutritional deprivations during pregnancy.

Early postnatal malnutrition in children may result in greater fatigue and lesser ability to sustain either prolonged physical or mental effort. Nutritional inadequacy increases risk of infection, interferes with immune mechanisms, and results in illness. The combination of subnutrition and disease reduces time available for learning and disrupts orderly acquisition of intellectual growth.[8,39]

However, Frisch[18] states that the supposition that malnutrition causes mental retardation in humans has little conclusive scientific information. It is difficult to apply conclusions from animal experiments to human beings since critical periods of brain growth in most test animals (rat, dog) is very different from man. The pig has a period of rapid brain growth most similar to man. The results of experimental malnutrition in animals, while suggestive, cannot be considered conclusive evidence for human beings. The social-environmental factors, particularly the role of the mother, can be most important in the influence of mental functioning and development of children.

The severely protein-calorie malnourished child exhibits behavioral abnormalities. Nutritional deprivation in early life may have a permanent retarding influence on mental development.[4,5,11,39] Nutrition deficiency in early life affects later feeding behavior.[5,6] Comparisons of children previously malnourished in infancy with their siblings as well as with children of similar social background indicate the malnourished group showed significant differences in verbal performance and intelligence quotient (IQ).[8]

Assessing the available data, Monckeberg[39] reports that severe early malnutrition produces a long-term persistent effect, not only on intelligence, but also upon basic learning academic skills. Survivors of severe early malnutrition are physically and behaviorally different from normal children. Further research[39] of longitudinal perinatal studies and studies of twins, gives new insight as to the significance of fetal nutrition, suggesting that small for age infants are not only developmentally retarded in utero, but also persistently physically and mentally retarded throughout life. Intellectual performance of children malnourished in infancy was significantly inferior to that of adequately nourished children from an otherwise almost identical environment. Mental retardation was also correlated with retarded growth. Growth retardation and a deficit in physical growth were observed to be significantly correlated to IQ.

What is the evidence that a reduced number of brain cells in a malnourished animal is a factor in its behavior?[64] Chase et al[13] reported alterations in human brain biochemistry following intrauterine growth retardation. Reduction in brain cell number has been demonstrated in human brain in infants who have died of severe malnutrition.[64] Enzymatic maturation and development in brain is also affected with defective enzyme organization in the brains of malnourished animals.[64] Brain underdevelopment caused by prenatal malnutrition in females of one generation can be transmitted to the next generation, even in absence of postnatal malnutrition of first generation.[66] Despite the many observations implicating undernutrition with learning, behavior and intelligence, it would not be appropriate to substitute it completely for other variables—social, cultural, educational and psychological—which exert an influence on intellectual growth. Malnutrition never occurs alone. While the high frequency of mental retardation is not a consequence of malnutrition alone, a growing body of evidence suggests that this factor may be the most important.[8]

Miscellaneous Environmental Factors

Other maternal and neonatal environmental influences have been reported to cause significant effects on the growth, development, metabolism, behavior, congenital malformations, response to stress and aging of man and animal. Several such factors are briefly reviewed.

The growth of the fetus in utero is dependent not only on the supply of energy and of suitable amino acids from the maternal circulation, but also on fetal insulin, the excretion of which is stimulated by the insulinotrophic amino acids.[60] Any dietary deprivation which alters adversely the development of the endocrine system during pregnancy and early postnatally may have deleterious effects on the later life of progeny.[60]

The development of the central nervous system is severely retarded when there is a neonatal thyroid deficiency. The thyroid hormone may play a role for developing neurons to establish, during certain critical periods in early postnatal life, the complex interconnections that lay the foundation for the later behavioral pattern of animals.[55] Shapiro and Norman[55] have demonstrated that the administration of thyroxine to the newborn rat will accelerate the maturation of the pituitary-adrenal response to stress, accelerate the age at which the animal will respond behaviorally and neurophysiologically to acute environmental stimuli and increase the ability to learn a conditioned avoidance response.

Plasma concentrations of insulin, cortisol and growth hormone of a group of malnourished African village children was studied.[35] Plasma insulin concentrations were directly correlated with rate of growth in height and weight but cortisol concentrations showed reversed relations. Levine[34] reported on maternal and environmental influences on the adreno-cortical response to stress in weanling rats. He observed that handling of pups early after birth resulted in a reduction in plasma adreno-cortical steroids after exposure to novel stimuli. Bahlburg[3] reported on the significant effects of prenatal sex hormone administration on human male behavior.

The detrimental consequences of the fetal alcohol syndrome is well documented in the literature.[23,24,26,27,28,41] Effects of prenatal alcohol consumption has been studied in man and animal stressing learning and behavioral deficits.[10] Alcoholic women have a high risk of giving birth to children with growth retardation, congenital malformation and mental deficiency.[41]

Growth failure in fetal alcohol progeny occurs both prenatally and postnatally, with no catch-up in growth during infancy and early childhood. The persisting deficiency in growth appears not to be the consequence of the postnatal environment; affected babies raised in foster homes show no better growth or performance than those reared by alcoholic mothers.[23,26]

Meyers and Comstock[38] have demonstrated the adverse effects of cigarette smoking on perinatal mortality. Martin et al[37] observed that rat offspring whose mothers had received nicotine injections during gestation and lactation had shorter life spans than did the offspring of saline-injected controls. Offspring of mothers receiving methamphetamine during gestation remained significantly lower in weight than controls throughout the first 16 months of life. Developmental delay and significantly greater activity were also observed in this group of rats. Zemp and Middaugh[67] reported significant changes in developmental neurochemistry and behavior in progeny of animals administered D-amphetamine sulfate prenatally.

Caldwell et al[12] reported that a mild chronic zinc deficiency in the maternal animal will be expressed in the progeny (rat) as impaired behavior. Hambidge[21] reported that there is evidence that zinc deficiency may cause congenital malformations in humans. Sever and Emanuel[54] have observed a possible connection between maternal zinc deficiency and congenital malformation of the central nervous system in man. The need for studies focused on the behavioral effects to both mother and offspring resulting from chronic marginal deficiencies of other essential micro and macro nutrients is warranted.[12]

SELECTED HUMAN STUDIES

In view of the deleterious effects of early malnutrition in laboratory animals, it is most important to determine whether these alterations are relevant to humans. Retrospective studies indicate that changes in birth weight or infant mortality occurred with changes in nutrient intake of either general populations or segments which included pregnant women.[45]

World War II provided the circumstances for a study of two large groups of undernourished pregnant women.[59] One group consisted of women in Holland who were subjected to severe food restrictions for approximately eight months. The babies born during this time were shorter and lighter than those born before the period of restrictive food intake. This was reported to be the direct result of the mother's diet during the first half of pregnancy. However, there was not an increase in the rate of stillbirths, prematurity and malformation.[59] By contrast, babies born to another group of women in Russia during the siege of Leningrad were generally in poorer health, less resistant to infection, and did not nurse well.[1] Also, there was a higher rate of prematurity and stillbirths. The difference in the effect of the serious food shortages in the two countries is thought to relate to the nutritional status of the women prior to the hunger period. Women in Holland had generally been well nourished while the women in Leningrad had experienced chronic malnutrition. Thus, the importance of the nutritional status of the mother prior to conception is appreciated. Inadequate food, both in quality and quantity, represents only the most obvious and dramatic of the causes of low survival rates and poor growth performance in many deprived members of human populations in the world even during periods of peace.[14]

Graham and Adrianzen[19] reported that when severely malnourished Peruvian children were rehabilitated for a few months, by being placed in homes which provided better environment, regular meals and high standard of hygiene, they were able to make rapid linear growth. However, there were no significant improvements in IQ to parallel or match those observed in height changes. The human data are in agreement with the findings observed in rats.[24]

When a standard diet was fed to nine-year-old Taiwan children of undernourished mothers, they demonstrated a need for up to 30% more food to maintain body weight when compared to children whose mothers were well fed.[9] Other differences between the two groups include the following observations. The progeny of undernourished mothers weighed less at similar ages, excreted more nitrogen after a protein meal, wasted protein indicating poor food utilization and showed abnormal psychological patterns including slower learning and higher levels of emotion.[9]

In studies of young children from mothers receiving animal protein either approximately twice a month or at least twice a week (inadequate versus adequate protein), congenitally undernourished children have impaired food utilization.[14] Inadequate food, both in quantity and quality, represents only the most obvious and dramatic of the cause of low survival rates and poor growth performance in many deprived members of the human population of the world.[14]

The health of the newborn starts with the health and nutrition of the mother including when she was in utero, nutrition during her infancy, childhood, adolescence, adulthood and pregnancy. The health and nutrition of fetuses, of infants, of children, of adolescents and of adults in early, middle or of late maturity influence their health and nutrition during aging. Much evidence suggests that only during the first two-fifths of life can nutrition make substantial impact on the prevention of diseases which inflict the elderly.[61]

CONCLUSION

From diverse sources including clinical studies, field investigations and laboratory animal experiments come growing recognition that the timing and duration of nutritional and other stresses are the pertinent factors governing inhibited growth and development. In view of the deleterious effects of maternal and neonatal malnutrition, it is of great importance to determine whether findings in test animals are relevant to humans. Retrospective studies of man indicate that maternal malnutrition and other stresses do exhibit many similar metabolic and behavioral derangements observed in the test animal. The delivery of proper nutrients to pregnant women and to infants is essential not just for survival but for adequate early growth and development and as a preparation for health in later life.[14]

The views expressed in this paper are not necessarily those of the Veterans Administration.

REFERENCES

(1) Antonov, A.M. 1947. Children born during siege of Leningrad in 1942. *J. Pediat.* 30:250-256.
(2) Atallah, M.T., Barbeau, I.S., and Pellet, P.L. 1977. Metabolic and development changes in growing rats born to dams restricted in protein and or energy intake. *J. Nutr.* 107:650-655.
(3) Bahlburg, H.F., 1977. Prenatal effects of sex hormones on human male behavior: medroxyprogesterone acetate. *Psychoneuroendocrinology* 2:383-390.
(4) Barnes, R.H., Cunnold, S.R., Zimmermann, R.R., Simmons, H., MacLeob, R.B., and Krock, L. 1966. Influence of nutritional deprivation in early life on learning behavior of rats as measured by performance in a water maze. *J. Nutr.* 89:399-410.
(5) Barnes, R.H., Neely, C.S., Kwong, E., Iabaden, B.A., and Frankova, S. 1968. Postnatal nutritional deprivations as determinants of adult behavior toward food, its consumption and utilization. *J. Nutr.* 96:467-476.
(6) Barnes, R.H., Kwong, E., Morrissey, L., Vilhjalmsdottir, L., and Levitsky, D.A. 1973. Maternal protein deprivation during pregnancy or lactation in rats and the efficiency of food and nitrogen utilization. *J. Nutr.* 103:273:284.
(7) Barrows, C.H. 1968. Ecology of aging and the aging process—biological parameters. *Gerontologist* 8:84-91.
(8) Birch, H.G. 1972. Malnutrition, learning and intelligence. *Am J Publ Health* 62:773-784.
(9) Blackwell, R.Q., Chow, B.F., Chin, F.S.K., Blackwell, B.N., and Hsu, S.C. 1973. Prospective maternal nutrition study in Taiwan: rationale, study design, feasibility, and preliminary findings. *Nutr. Rep. Intl.* 7:517-532.
(10) Bond, N.W. 1977. Effects of prenatal alcohol consumption on shock avoidance learning in rats. *Psychol. Rep.* 41:1269-1270.
(11) Brozek, J. 1978. Nutrition, malnutrition and behavior. *J. Ann. Rev. Psychol.* 29:157-177.
(12) Caldwell, D.F., Oberleas, D., and Prasad, A.S. 1973. Reproductive performance of chronic mildly zinc deficient rats and the effects on behavior of their offspring. *Nutr. Rep. Intl.* 7:309-318.
(13) Chase, H.P., Welch, N.N., Dabiere, C.S., Vasan, N.S., and Butterfield, L.J. 1972. Alterations in human brain biochemistry following intrauterine growth retardation. *Pediatrics* 50:403-410.
(14) Chow, B.F., Blackwell, R.Q., Blackwell, B.N., Sherwin, R.W., Hsueh, M., and Lee, C.J. 1966. Problems of world nutrition: studies on the progeny of underfed mothers. In *Proc. 7th Intl. Cong. Nutr.*, Hamburg, pp 1-7.
(15) Chow, B.F., Blackwell, R.Q., Blackwell, B.N., Hou, T.Y., Aniline, J.K., and Sherwin, R.W. 1968. Maternal nutrition and metabolism of the offspring: Studies in rats and man. *Am. J. Public Health* 58:668-673.
(16) Dahlmann, N. 1977. Effect of environmental conditions during infancy on final body stature. *Pediatr. Res.* 11(5) 695-700.
(17) Davis, R.L., Hargen, S.M., and Chow, B.F. 1972. The effect of maternal diet on the growth and metabolic patterns of progeny (mice). *Nutr. Rep. Intl.* 6.1-7.

(18) Frisch, R.E. 1970. Present status of the supposition that malnutrition causes permanent mental retardation. *Am.J. Clin. Nutr.* 23:189-195.
(19) Graham, G.C., and Adrianzen, T.B. 1972. Late catch up growth in severe infantile malnutrition. *Johns Hopkins Med. J.* 131:204-209.
(20) Guthrie, H.A., and Brown, M.L. 1968. Effect of severe malnutrition in early life on growth, brain size, and composition in adult rats. *J. Nutr.* 54:419-426.
(21) Hambidge, K.M., Nelder, K.H., and Walravens, P.A. 1975. Zinc, acrodermatitis enteropathica, and congenital malformation. *Lancet* 1:577-582.
(22) Hanson, H.M., and Simonson, M. 1971. Effect of fetal undernourishment on experimental anxiety. *Nutr. Rep. Intl.* 4:307-314.
(23) Hanson, W.H., Jones, K.L., and Smith, D.W. 1976. Fetal alcohol syndrome: Experience with 41 patients. *JAMA* 235:1458-1462.
(24) Hsueh, A.M., Blackwell, R.Q., and Chow, B.F. 1970. Effect of maternal diet in rats on feed consumption of the offspring. *J. Nutr.* 100:1157-1164.
(25) Hurley, L.S., Cosens, G., and Therianet, J. 1976. Teratogenic effects of magnesium deficiency in rats. *J. Nutr.* 106:1254-1259.
(26) Jones, K.L., Smith, D., Ulleland, C.N., and Streissguth, A.P. 1973. Patterns of malformation in offspring of chronic alcoholic mothers. *Lancet* 1:1267-1272.
(27) Jones, K.L., Smith, D.W., Streissguth, A.P., and Myriarthopoulos, M.C. 1974. Outcome in offspring of chronic alcoholic women. *Lancet* 1:1074-1076.
(28) Jones, K.L., Smith, D.W., Ulleland, C.N., and Streissguth, A.P. 1976. A pattern of malformation in offspring of chronic alcoholic mothers. *Lancet* 1:1267-1271.
(29) Kahn, A.W. 1972. Development, aging and life duration: Effects of nutrient restriction. *Am. J. Clin. Nutr.* 25:822-828.
(30) Krueger, R.H. 1968. Some long-term effects of severe malnutrition in early life. *Lancet* 2:514-517.
(31) Kwong, E., and Barnes, R.H. 1977. Comparative contributions of dietary protein quality and quantity to growth during gestation, lactation and postweaning in the rat. *J. Nutr.* 107:420-425.
(32) Lee, C.J. 1970. Biosynthesis and characteristics of brain protein and ribonucleic acid in mice subjected to neonatal infection or undernutrition. *J. Biol. Chem.* 245:1998-2004.
(33) Lee, C.J. 1973. Neonatal infection, maternal malnutrition, and fetal brain metabolism. *Nutr. Rep. Intl.* 7:333-338.
(34) Levine, S. 1967. Maternal and environmental influences on the adreno cortical response to stress in weanling rats. *Science* 156:258-260.
(35) Lun, P.G. Whitehead, R.G., Cole, T.J., and Austin, S. 1979. The relation between hormonal balance and growth in malnourished children and rats. *Br. J. Nutr.* 41:73-84.
(36) McCay, C.M., Crowell, M.F., and Maynard, L.A. 1935. The effect of retarded growth upon the length of the life span and upon the ultimate body size. *J. Nutr.* 10:63-79.
(37) Martin, J.C., Martin, D.C., Radow, B., and Sigman, G. 1976. Growth, development and activity in rat offspring following maternal drug exposure. *Exper. Aging Res.* 2:235-251.
(38) Meyers, M.B., and Comstock, G.W. 1972. Maternal cigarette smoking and prenatal mortality. *Am. J. Epidemiology* 96:1-10.

(39) Monckeberg, F. 1973. Maternal diet, mental development and behavior: Summary. *Nutr. Rep. Intl.* 7:349-351.
(40) Morgan, B.L.G., and Winick, M. 1977. The effect of malnutrition on some aspects of RNA metabolism in the maternal liver and fetal tissues at different stages of pregnancy in the rat. *J. Nutr.* 170:1694-1701.
(41) Mulvihill, J.J., and Yeager, A.M. 1976. Fetal alcohol syndrome. *Teratology* 13:345-351.
(42) Naege, R.L. 1965. Prenatal nutrition. *J. Pediat.* 67:447-452.
(43) O'Connor, N. 1956. The evidence for the permanently disturbing effects of mother-children separation. *Acta. Psychol.* 12:174-191.
(44) Rider, A.A., and Simonson, M. 1973. Effect on rat offspring of maternal diet deficient in calories but not in protein. *Nutr. Rept. Intl.* 7:361-370.
(45) Roeder, L.M., and Chow, B.F. 1972. Maternal undernutrition and its long-term effect on offspring. *Am. J. Clin. Nutr.* 25:812-821.
(46) Roeder, L.M. 1973. Long term effects of maternal and infant feeding. *Am. J. Clin. Nutr.* 26:1120-1123.
(47) Ross, M.H. 1972. Length of life and caloric intake. *Am. J. Clin. Nutr.* 25:834-838.
(48) Ross, M.H., and Bras, G. 1975. Food preferences and length of life. *Science* 190:165-167.
(49) Ross, M.H., Lustbader, H.F., and Bras, G. 1976. Dietary practices and growth responses as predictors of longevity. *Nature*, London 262:548-553.
(50) Ross, M.H. 1976. Nutrition and longevity in experimental animals. *Curr. Concepts Nutr.* 4:43-57.
(51) Rosso, P. 1977. Fetal exchange during protein malnutrition in the rat: Placental transfer of α-amino isobutyric acid. *J. Nutr.* 107:2002-2005.
(52) Rosso, P. 1977. Maternal nutrition, nutrient exchanges and fetal growth, in: M. Winick (ed.) *Nutrition Disorders of American Women*, pp 3-25. John Wiley, New York.
(53) Sara, V.R. 1976. The influence of early nutrition and environmental rearing on brain growth and behavior. *Experientia* 32:1538-1540.
(54) Sever, L.E., and Emanuel, I. 1973. Is there a connection between maternal zinc deficiency and congenital malformation of the central nervous system in man. *Teratology* 7:117-122.
(55) Shapiro, S., and Norman, J.R. 1967. Thyroxine: Effects of neonatal administration on maturation, development and behavior. *Science* 155:1279-1281.
(56) Simonson, M., Sherwin, R.W., Anilane, J.K., Yu, W.Y., and Chow, B.F. 1969. Neuromotor development in progeny of underfed mother rats. *J. Nutr.* 98:18-24.
(57) Simonson, M., and Chow, B.F. 1970. Maze studies on progeny of underfed mother rats. *J. Nutr.* 100:685-690.
(58) Srivastara, U., Vu, M.L., and Goswami, T. 1974. Maternal dietary deficiency and cellular development of progeny in the rat. *J. Nutr.* 104:512-520.
(59) Stearns, G. 1958. Nutritional state of the mother prior to conception. *JAMA* 168:1655-1670.
(60) Turner, M.R. 1973. Protein deficiency, reproduction, and hormone factors in growth. *Nutr. Rept. Intl.* 7:289-296.

(61) Watkin, D.M. 1978. Logical base for action in nutrition and aging. *J. Am. Geriat. Soc.* 26:193-202.
(62) Watten, R.H. 1973. Implication to humans: Introduction: The effect of maternal nutrition on the development of the offspring. *Nutr. Rept. Intl.* 7:511-515.
(63) Whatson, T.S. 1972. Undernutrition in early life, lasting effects on activity and social behavior of male and female rats. *Dev. Psychol.* 9(6) 529-538.
(64) Winick, M. 1977. Early malnutrition: brain structure and function. *Prev. Med.* 6 (2) 358:360.
(65) Zamenhof, S., vanMarthens, E., and Grauel, L. 1971. DNA (cell numbers) in neonatal brain: alteration by maternal dietary caloric restriction. *Nutr. Rept. Intl.* 4:269-274.
(66) Zamenhof, S., and vanMarthens, E. 1978. The effects of chronic undernutrition over generations on rat development. *J. Nutr.* 108:1719-1723.
(67) Zemp, J.W., and Middaugh, L.D. 1975. Some effects of prenatal exposure to D-amphetamine sulfate and phenobarbital on developmental neurochemistry and behavior. *Addictive Dis.* 2:307-331.

–3–
Psychological Aspects of Nutrition in Aging

Sue V. Saxon and Mary Jean Etten

INTRODUCTION

Psychological variables have a decided impact on food ingestion and ultimately on health and well-being in older age. Pumpian-Mindlin[71] wrote "The way to a man's stomach is through his heart. By this I mean his feelings, his emotions, his attitudes, his prejudices, his food habits." Although published research reflects relatively little attention to psychological aspects of food ingestion in the elderly, we attempt to identify various factors playing strategic roles in nutritional patterns and behavior.

Topics to be considered range from a developmental perspective of food ingestion patterns and cultural-social meanings of food and food rituals to the specific impact of physical and psychological aging on nutrition. This chapter assesses the current status of the topics as well as clarifies the need for additional research on psychological implications of nutrition in older age.

PSYCHOSOCIAL DEVELOPMENT

Infancy and Childhood

According to Erik Erikson's[24] psychosocial stages of development, an early task for infants to learn is to develop a sense of basic trust versus a sense of mistrust. Since an infant's life experiences are necessarily restricted, essential caretaking activities such as feeding, changing, and bathing serve as prime data for infants in the development of either a sense of basic trust or mistrust toward humans. Similarly, Brammer and Shostrom[8] identify dependency as the first developmental stage in their life cycle view of behavior, emphasizing the complete dependency of the infant on others in order to have basic needs met. The manner in which needs are met—lovingly, roughly, hurriedly, etc.—provides early, but crucial, information to the infant about human relationships and may predispose him/her to react to ensuing relationship encounters with certain expectations based on these earliest experiences.

Savitsky[77] believes the feeding relationship between parent and child represents one of the first experiences with love for here the child can learn warmth, closeness, positive emotional responses, and a sense of security from the caretaker's behaviors surrounding feeding, or can be adversely affected in emotional development if the feeding relationship is not a pleasant, gentle, or comfortable one. As the child reacts to the way food is presented, so he/she will react to the individual presenting it and to foods themselves. Not only are lasting attitudes thus formed about specific foods, but feelings about human relationships formed in the context of feeding can be influenced as well.[2,58] The eating process represents a time of early training and learning for the child as food is used for reward, punishment, to express hostility or rejection, and a host of other subtle attitudes of caretaker to child and vice versa.[102] Such early experiences have substantial impact on later attitudes and behaviors toward foods and eating as well as on attitudes and behaviors toward interpersonal relations. Ginsburg[32] suggests both early eating and excretory experiences are very likely related to later personality traits. Personality traits and food habits of adulthood largely depend on whether in childhood food was viewed and offered by the caretaker figure as a mere biological necessity, a pleasure involving social relationships and relaxation, an outlet for aggression, a solace for frustration and rejection, or as a stage for the acting out of a number of other psychologically charged feelings and emotions. Such symbolic meanings may be difficult for others to interpret as they are idiosyncratic to the family unit, but they are necessary to an understanding of emotional responses to nutrition.

Freud's psychosexual stage theory stresses the importance of oral drives during the first year of life as exemplified by biting, chewing, and sucking behaviors. Theoretically, if the child does not have an opportunity to satisfy oral drives during the first year of life, he/she will be fixated in the oral stage, and later life development, including eating behaviors and attitudes towards food, are expected to be affected adversely.

In ongoing development as children strive to achieve autonomy and independence, food is used by both parents and children in an attempt to manipulate each other or to express covert attitudes. "You can't go out to play until you eat your carrots" promotes bargaining by the parent to obtain a desired behavior through encouraging the child to first complete a less desirable activity. On the other hand, a child may refuse to eat certain foods in an attempt to defy authority or to establish individual autonomy. Lurie[54] studied children without organic disease and concluded eating disturbances seem to serve three general functions:

(1) By compelling mother to administer to his/her needs the child is able to protect and prolong his/her dependent status.

(2) Eating or noneating can be used as revenge upon mother for past or present deprivations.

(3) Self-denial of food eases guilt arising from aggressive feelings toward mother who is not willing or able to cater to every whim of her child.

Regardless of theoretical developmental perspective, in our culture infants are indeed dependent and relatively helpless during the first year of life. Food and eating activities assume a role of no small significance in infancy and much is learned very early in life about the process of eating, appropriate foods to be eaten, food habits of the prevailing culture, and significance of food ingestion rituals.

Adolescence

Adolescents commonly use food and eating patterns as a means of demonstrating their independence and freedom of choice, whether it be in the form of specific food fads, unusual combinations of foods, sheer quantity, or undernutrition. Food ingestion assumes even greater significance as it becomes psychologically related to desired body shape and weight, attributes extremely important during the adolescent period.

Adulthood

At the other end of the life spectrum, attempts to manipulate others through the use of food and certain eating behaviors is not uncommon in some older persons who have lost a degree of mastery and control over their lives. They attempt to reinstate control and mastery by refusing to eat or by showing little attention to proper diet. This is by no means the only reason for disruptions in eating patterns observed in elderly persons, but it certainly is one familiar in long-term care institutions and in elderly living alone. Food has such obvious significance in most societies that it is used in a myriad of ways as a method for behavior control.

MEANING OF FOOD

Food and its diverse meanings exert a pervasive force in the lives of many people. According to Bogert, Briggs and Calloway[5], the importance of food is illustrated by the prevalence and variety of nonnutritive uses which it assumes in many cultures. For example, food has long been used by many pretechnological societies as an integral part of religious and magical rites; in our modern society we throw rice at weddings, use bread and water as punishment, withhold sweets to discipline children, and go on hunger strikes to make a point. Cameron[16] states

> Our language is full of ambiguous allusions to social acceptance and rejection, to verbal assaults, to gastric need for food, and to the spiritual need for sustenance. Thus we eat our words, and swallow our wrath, the Lord spews us forth, we sink our teeth into a problem, drink in a message, find an explanation indigestible, and reject it with biting comments.

Essentially, people learn to eat what their ancestors ate, especially foods readily available in the immediate environment. Food patterns and preferences are influenced by culture as well as family, and once formed they are often resistant to change in later life. To a large extent, culture or society influences attitudes and preferences about flavors, textures, combinations of food considered appropriate, the temperature of foods when eaten, what food is proper at various times of day, how often to eat, and the way in which various foods are eaten.[52,88] Cultural taboos and conditioned dislikes surrounding certain foods must also be appreciated as they tend to generate strong emotional responses from individuals. For example, we do not eat dog meat or grasshoppers in our particular culture and the thought of such a diet would be revolting to many who have been so conditioned. To others, such foods are quite acceptable and appetizing.

Regional and local food preferences must be taken into consideration as well when assessing psychological implications of eating and enjoyment of food.[27] Boykin[7] suggests soul foods, for instance, satisfy much more than hunger in that they hold deep emotional significance for many people. Satisfaction is derived not just from the food itself, but also from the way it is prepared and served. Similarly, regional food preferences such as grits in the South and baked beans in the Northeast are extremely influential in dictating enjoyment of food and should be taken into account to a greater extent by those concerned with nutrition.

Eating represents one of our earliest social activities. Troll[98] stated, "All through life, enjoyment of life is wrapped up with enjoyment of food." Learning to enjoy food, eating, and mealtime rituals traditionally occur early in life and usually within the context of the family. Not only are food habits and preferences profoundly influenced by individual family feeding patterns, but family meals undoubtedly serve psychologically to enhance morale and feelings of unity and security.[26,104]

Family traditions dictate number of meals a day, time of meals, significance of each (dinner is usually a more significant meal than breakfast or lunch), what foods are considered appropriate for specific meals, and the order in which foods are served and eaten. In many families, there are specific "family foods," those served when the family is alone, but not usually served to guests.[27] Obviously, all of these variables contribute to long-lasting psychological attitudes about nutrition.

Symbolism of Food

The psychological symbolism of food has been a subject of extensive discussion and analysis in the literature. Interpretations of symbolic meanings range through the following.

Congeniality and Sociability: Hospitality, friendliness, congeniality and sociability are all conveyed through food occasions. Degree of formality or informality may be determined by the type of food served and the hour; the dinner party or banquet is a more formal occasion than steaks on the back-yard grill. Tea, dinner, brunch, breakfast, cocktail party, coffee and dessert all have different connotations regarding the formality and type of social interaction expected.[53,56]

Security: Certain foods, especially those associated with pleasant experiences of childhood, symbolize security and safety. Milk, for example, is oftentimes associated with security and comfort although at other times related psychologically to dependency and helplessness (baby's food).[53,71,73,102,104] In times of stress, milk and milk product usage increases. During illness one may wish to revert to foods previously associated with comfort and security, as for example, cottage cheese and pears or milk toast.

Reward and Punishment: Reward foods tend to be sweets or delicacies used by parents to influence children's behaviors; however, reward foods are also used by adults to reward each other as well as to reward oneself.[71,102,104] For many persons, the frustrations and tensions of examinations, social isolation and loneliness, interpersonal conflicts, grief, and many other life stresses are eased by eating more highly desirable foods.[73] Moore[60] states that under stress most individuals prefer familiar foods with a past history of pleasant emotional connotations. Institutions attempting to meet the spectrum of nutritional needs of residents often seem unable to cater to such highly individualized dietary preferences. According to Stare,[92]

> Nursing homes can brighten the lives of their patients if the staff recognizes the role of good food as a link with the past. Mealtime may be the only pleasurable activity for some of the patients. Good food is a real morale builder.

In other situations food may serve as punishment for real or perceived wrongdoing. The "bread and water" cliche signifies a less than desirable meal in retribution for less than desirable behavior. Overly conscientious or guilt-ridden individuals commonly deny themselves enjoyable foods for diverse and complex psychological reasons related to feelings of worthlessness or to the lack of self-esteem.[53,60,61] Fleck[27] suggests feelings such as anger, resentment, indifference, or sadness can be clearly transmitted by the way one eats and by what one eats. Inadequate diet, refusal to eat, and in some instances overeating, can all be weapons to display hostility, rebellion, and to gain attention or concern from others.[53,73] Knowledge of such dynamics is of paramount importance in assessing nutritional problems in the elderly.

Age and Sex Symbolism: In our culture, meat is associated with masculinity, energy, activity, and even aggression, while vegetables and salads are more associated with femininity, gentleness, and refinement. Fruits generally symbolize love and affection (an apple for the teacher, gift basket of fruit, etc.). Some foods have a decided connotation of age or maturity in our society; for instance, coffee, tea, beer, olives, and gourmet dishes are usually considered adult, while milk, cereal, and peanut butter are more often associated with children.[1,27,56,60,102,104]

Idiosyncratic reactions to food symbolism must be considered in addition to societal and familial influences. Some foods have especially pleasurable associations in reminding us of people or happy situations of the past, while other foods have unpleasant associations based on individual past experiences.[102] All of these complex variables need to be recognized and understood as integral to the psychodynamics of nutrition, especially in elderly who have a long and diverse history of food experiences.

APPETITE AND HUNGER

The terms appetite and hunger are defined and used inconsistently in both the psychological and physiological literature. One distinction frequently drawn is to refer to hunger as physiologically based weakness, tension or pressure in the epigastric region, the familiar hunger pangs. Appetite refers to psychological phenomena associated with past life experiences affecting food ingestion. Food preferences, memories associated with food and eating, and other related psychological events of earlier life, logical or not, continue to influence eating behaviors throughout adulthood. Appetite often accompanies sensations of hunger, but may exist without it.[10,70]

Although not directly related to aging, numerous theories of hunger have been proposed throughout the years. Primarily derived from animal studies, no single theoretical formulation is yet accepted as comprehensive enough to satisfactorily explain eating behavior, hunger, and satiety in humans.[29,35,37,59] For an excellent review of the literature, see Grossman.[36]

Theories of Hunger

The most discussed theoretical views at present are:

(1) Thermostatic theory, in which temperature regulation mechanisms are considered important in the regulation of food ingestion. Animals eat to stay warm and cease eating to prevent overheating. Body temperatures do affect eating to some extent, but currently this view seems somewhat restrictive to adequately explain human food consumption.

(2) Lipostatic theory is based on the observation that animals maintain a relatively constant body weight over long time periods. It postulates some regulatory substance—perhaps a hormone—in the body which relays information to the brain on the amount of fat in body tissues and thereby regulates food intake.[37,50] There is still very little specific research evidence available to evaluate this theory.

(3) Glucostatic theory involves specialized cells in the brain (ventromedial hypothalamus) which respond to glucose and regulate food consumption. Decreases in blood sugar signal hunger and increased food intake results; satiety occurs when blood sugar levels increase to normal. Feeding behavior is then related to the rate of utilization of glucose rather than to absolute levels available.[40,59]

(4) Friedman and Stricker[29] believe each of these single-factor theories concerning the physiological basis of hunger is ignoring the most significant core phenomenon, fuel supply in the body necessary to maintain energy. They contend the physiological stimulus for hunger arises when the liver interacts with the intestines, adipose tissue, and brain to control food intake and body metabolism. According to Friedman and Stricker, hunger occurs when fuel supplies from the intestines are low. Since the liver is the organ most responsive to needs for metabolic fuels, whether from exogenous or endogenous sources, it presumably responds to levels of body fuel supplies and sends appropriate messages to the brain regulating food intake.

(5) The impact of conditioned responses on hunger and food ingestion has long been of interest. A historical perspective of this issue was prepared by Mursell,[63] while recently both Stunkard[96] and Booth[6] argue satiety and appetite are substantially conditioned reactions. This relationship should be considered more seriously in attempting to understand eating behavior patterns. For the elderly, years of past experiences with foods and the multitude of variables surrounding meals suggest a rich array of data worthy of greater attention than received thus far.

Appetite

In addition to physiological parameters influencing food ingestion, a number of psychological variables impact on eating behavior. Young[107] suggests the most relevant variables are (a) sensory discrimination between foods based on taste and smell, (b) likes and dislikes of specific foods associated as pleasant or unpleasant, (c) motivation by food incentives, (d) conditioned reactions based on various sensory cues, and (e) social and aesthetic factors surrounding food and feeding rituals. Appetite involves sensory perception, emotional state, food appearance, and previous eating habits.[10] All of these need to be evaluated in understanding nutritional patterns of the elderly.

OVERNUTRITION AND UNDERNUTRITION

According to Schlenker, Feurig, Stone, and Mickelson,[84] cross-sectional research on females in the U.S. shows average body weight increases until the 60s and then declines somewhat in older age. Body fat tends to increase in both sexes with age, but is usually greater in females compared to males. Experimental studies of overnutrition (obesity) and undernutrition in the elderly are sparse although substantial information has been accumulated about obesity in animals and in young humans. Definitions of obesity have not been consistent, further complicating inferences and generalizations based on empirical data. Body mass index, or weight in kilograms per height in meters squared, is one useful method for determining overweight status.[55] A body mass index of 27 or more indicates 30% or greater increase above ideal weight, or obesity. This is sufficient excess weight to adversely influence health status. Major health problems aggravated by obesity include, among others, cardiovascular disease, diabetes, hypertension, arthritis, gallstones, liver function, ovarian dysfunction, and endometrial cancer.[55,84] In fact, obesity has a detrimental impact on virtually every organ system in the body.[69]

Young's earlier studies[108] conclude food intake in animals depends not only on physiological hunger, but also on variables such as palatability, emotional reactions to foods, eating habits, and attitudes. Food ingestion in humans is a multifactored phenomenon involving highly complex interactions between physiologic, psychologic, cultural, and aesthetic variables. In environments where food is readily available, people eat not only to satisfy physiological hunger, but in response to personal-social needs and concerns as well.[31,41,57]

Schachter,[81] in a classic article, compared eating behavior in rats made obese by lesions in the ventromedial hypothalamus with obese humans. Similarities were striking in that both ate more food than normals, ate faster, were not motivated to work for food, but overate when food was present. Both humans and rats were more emotional in all experimental tests of emotional situations, but less emotional otherwise. Schachter theorizes eating behavior of the obese is triggered to a greater extent by external cues (such as sight, smell, taste) rather than being determined by internal physiologic states. Nisbett[64] reviewed the literature in this area and agreed with Schachter that obese humans are hyperresponsive to external cues relating to food (especially visual and gustatory cues) and relatively unresponsive to internal states signalling hunger and satiety.

In a later review of existing literature, Leon and Roth[49] emphasize the inconsistency of available data and conclude obesity cannot be considered a unitary syndrome. Further research is clearly needed to quantify the impact of highly specific variables. In the area of eating and emotional behavior, clinical data have been more consistent than results derived from laboratory studies. Many clinical studies describe an association between anxiety (and other emotional states) and food intake, although laboratory data do not support the generalization that food intake actually reduces anxiety. A number of clinical studies indicate obese people are more sensitive to external cues, especially taste, while other data do not lend strong support to this hypothesis. Some consistency exists in relation to effort expended in that obese subjects are likely to expend less effort in obtaining food than normal weight individuals, but eat more if food is present and easily available. Overall, Leon and Roth[49] question methodologies employed, definitions of obesity, and choices of highly selected subjects as particularly significant areas of variances between studies.

Cautioning against too simplistic, dichotomy-oriented views, Rodin[75] reviews existing research on eating behavior and obesity concluding both internal and external cues probably provide bi-directional feedback for each other. However, obese individuals have a tendency to respond more to external stimuli and cues than to internal cues in regulating eating behavior. Future research must assess complex physiologic, psychologic, and social interactions for a more comprehensive explanation of food ingestion in humans.

From a psychodynamic point of view, Freed[28] discusses the possibility of a variety of psychological factors influencing obesity, some of them unconscious motivations which are very strong and resistant to change. He identifies eight conditions associated with overeating:

(1) Environment. Growing up in a family in which meals and food are central activities to the family unit may build in habits and attitudes difficult to change in later years.

(2) Economics. In deprived homes where sufficient food is difficult to obtain, children may grow up with a need to eat all that is possible or not leave or waste any leftovers following a meal.

(3) Monotony. Monotonous daily routines, lack of interest or diversion all contribute to overeating as a way to escape.

(4) Occupation. Those in food-related occupations are constantly exposed to food and may be influenced to overeat.

(5) Organic disease. Those with disease or accident resulting in a sedentary life style sometimes eat as much as when active and thus gain weight. Monotony in the environment becomes a factor here also.

(6) Nervousness. Situations generating worry or anxiety lead to "psychic" hunger and to overeating.

(7) Glandular imbalance. Obesity is rarely of endocrine origin although certain endocrine functions increase tension and lead to overeating. Premenstrual tension and the climacteric are but two examples.

(8) Subconscious factors. Psychological factors such as frustration, overcompensation, lack of emotional satisfaction, escape from social competition, and regression to the oral stage all might lead to overeating. Freed suggests overeating in adulthood could be a latent trait from childhood showing itself in adulthood whenever undue stress is encountered.

In a later article, Hamburger[39] points to the role of emotions as particularly important factors relating to obesity in humans. From clinical studies of obese patients, he suggests four categories of events triggering overeating. First, nonspecific emotional tensions such as mild depression, nervousness, anger, anxiety, and boredom easily result in overeating. Secondly, life situations viewed as intolerable or frustrating call for overeating as a means of substitute gratification. Thirdly, overeating could be a symptom of an underlying emotional illness. Fourth, overeating may result from a compulsive addiction to foods. Some obese patients studied had a constant craving for food, almost uncontrollable, usually for sweets or ice cream. These patients were assumed to be substituting food for love, affection, and security in their lives. Stunkard[95] alludes to a milder form of addictive-like behavior as the "night-eating" syndrome when people eat little during the major part of the day, but begin eating heavily by the evening meal and continue until late at night. They report finding themselves "nibbling" and can't stop. Stunkard suggests such behavior is a reaction to periods of life stress and diminishes when the stressful situation is relieved. Another similar pattern is "binge-eating" in which huge amounts of food are consumed in a relatively short period of time. Also related to periods of high stress, eating behavior is compulsive and undoubtedly signifies highly personalized unconscious meaning to the person.

Weiss[102] states obesity occurs at any age when basic psychological needs for security, affection, and attention are so unfulfilled the immature personality can no longer cope and anxiety appears. If the individual personality pattern is that of oral gratification through eating, obesity will probably result. Sherwood[87] states according to available data obesity is a more pervasive problem in older age than is undernutrition.

In comparison to the research generated by obesity, undernutrition has received little attention, especially as it affects older persons. Siegel[90,91] identifies monotony in food as one important factor in lowered food acceptance. Older persons unable to shop adequately or unable to prepare a variety of foods may be subject to undereating based on monotony as a precipitating cause. Classic psychological responses to semistarvation are fatigue, loss of energy and initiative, irritability, depression, and intense preoccupation with food.[10,45,50,82] Such behaviors in elderly must be carefully evaluated for accurate assessment of etiology.

The psychosocial consequences of either overnutrition or undernutrition revolve primarily around feelings of self-esteem and social acceptance. In our culture, overnutrition or obesity probably has more serious psychological consequences than undernutrition because of our extensive concern for the trim, slim figure, no matter what age. According to Tobias and Gordon,[97] obese individuals are stigmatized and dealt with by insensitivity, cruelty, prejudice, and discrimination. Although some personalities are more resilient than others, the stigma attached to obesity probably makes a lasting impact on all involved.

PHYSICAL CHANGES

The need to ingest food is an invariant requirement for the maintenance of biological life and psychological well-being. With the progression of the aging process, human nutrition reflects increasing complexity as opposed to simplicity. Individual genetic differences, specific and unique age-related changes, personal and environmental factors, psychological insults, physical disease and injury, concomitant with varied idiosyncratic nutritional patterns contribute to highly diverse nutritional needs in this segment of our population.

Certain physical age-related decrements and chronic diseases occur more readily in the latter part of life, directly or indirectly affecting psychological aspects of food ingestion. Digestive system changes are a prime example of particular significance. Decreasing amounts of saliva in the mouth coupled with loss of natural teeth or ill-fitting dentures inhibits chewing some nutritious or preferred foods. Elders than choose moister or softer foods in an effort to compensate for age-related masticatory changes.[38] Such forced changes in eating patterns can appreciably diminish gratification and emotional support obtained from meaningful or personally significant foods. Furthermore, the inability to masticate foods properly is socially embarrassing and may decrease motivation to dine in public or to eat with others. Straus[94] reports a reduction in digestive enzymes and intestinal motility as a function of age and relates these changes to overall decreasing efficiency of the gastrointestinal system. Psychologically, aging manifestations in the gastrointestinal system often trigger overconcern and undue anxiety, especially about bowel functions and particularly constipation, to the extent that even essential foods and nutrients are excluded from the daily diet.

Other important age-related changes or diseases are those common to the skeletal and circulatory system leading to structural problems, pain, weakness, paralysis, and immobility. Limitations of this severity, in addition to sensory changes, present virtually insurmountable obstacles to shopping, opening food containers, and cooking.[48,74,86] Still another physical disorder, urinary incontinence, can be so psychologically embarrassing as to not only inhibit shopping activities but also dining in public, thereby effectively eliminating much needed emotional support and social interactions.[34]

Other specific health problems including diabetes, cardiac, gastrointestinal, or urinary disease frequently require therapeutic diets conceivably interfering with personally significant nutritional habit patterns or food preferences of a lifetime. Dramatic changes in diet deprive older adults of one of the few remaining consistent sources of satisfaction and psychological comfort. Cognizant of these behavioral dynamics, Berger[3] emphasizes the effects of major dietary modification may be more psychologically damaging to the elderly than the disease process itself; therefore, moderation in dietary change is advisable.

Disturbances in appetite are thought to be one of the first indicators of stress or physical illness, according to Howard and Schnell[42] so accurate assessment of appetite disruptions is essential for appropriate therapeutic intervention. Howard and Schnell also suggest the use of food as a form of psychological support and as helpful in instilling feelings of security during times of illness.

SENSORY SYSTEMS

> All knowledge of the world in which we live comes to us through our sensory systems. To survive, we must constantly be aware of the environment and changes taking place within it. We must also be able to interpret incoming information, to integrate it with knowledge about our body state at the moment, and to act upon it in an adaptive manner.[80]

Our ability to discriminate between tastes, to select foods, to eat appropriate amounts, and to enjoy the eating process is largely determined by the sensory constituents of food. Despite Engen's[22] belief that existing literature on taste and smell is insufficient to make a practical generalization about influences of these sensory modalities on diet in aging, we present relevant information currently available and suggest possible psychological implications pertaining to nutrition and older age.

Taste

Colavita[19] considers it important to distinguish between two usages of the word *taste*. In everyday vocabulary, taste implies temperature, smell, texture, color, perhaps pain, physiological gustatory sensations, and psychological expectations. For instance, lumpy gravy does not taste as flavorful as smooth gravy, and for most of us colored Easter eggs taste different than regular boiled eggs. In the lay person's definition of taste, a variety of sensory components, subjective feelings, and even environmental surroundings contribute to the psychology of eating and to taste per se.[19,73]

A more scientific and precise meaning of taste refers to sensations of sweet, salt, sour, and bitter originating from the four primary types of gustatory receptors that are found in taste buds on the tongue.[10,19] Physiologically, taste depends on stimulation of specific gustatory receptors sensitive to ions and molecules in solution, which, when stimulated, generate nerve impulses to the brain resulting in four primary sensations of salt, sour, bitter, or sweet. In combination, primary gustatory receptors produce myriad other taste qualities.[47,67]

Acuity and perception in the gustatory modality are highly individualized phenomena. Some persons possess excellent acuity and are able to perceive slight and subtle differences in taste qualities; others are unable to make highly specific differentiations in taste sensations.[73] Furthermore, individual taste thresholds and acuity fluctuate as adaption or sensory fatigue occurs readily in gustation after repeated presentations of the same food stimuli.[23] As long as gustatory acuity remains reasonably sensitive, pleasures of eating are enhanced appreciably through mastication which stimulates receptors and constantly produces new flavors, tastes, and different profiles of flavor in foods as they are chewed.[50,51]

Age-Related Gustatory Changes

Not only do taste buds vary in number from one person to another, but the number of active receptors decreases with age. Most older adults have as much as a 50% decrease in functioning taste buds compared to earlier years;[15,19] while in the later 70's it is estimated only about one-sixth of the number of taste buds characteristic of a 20 year-old remain functional.[89] Busse[13] reports in older age salt and sweet taste sensations decline earlier than bitter and sour sensitivities. Although research on the behavioral significance of gustatory changes and age is sparse, Schiffman[83] did find older adults substantially poorer in identifying blended foods when compared to younger subjects. Other studies indicate elders usually prefer and gain greater sensory satisfaction from more highly seasoned foods; these and other data reflect higher gustatory thresholds with advancing age. Related information describes elders' preferences for tart tastes over sweet in contrast with younger individuals,[19] while Eppright[23] states older women prefer tart tastes more than older men who tend to choose sweet tasting substances.

Alterations in taste sensations can be induced by numerous other factors possibly associated with older age. Smoking, for example, along with certain diseases, decreases gustatory sensitivity significantly.[13] Specific diseases of older adults including diabetes, cancer, and renal failure affect taste as do accidents or trauma to the head and mouth.[18] Likewise, various therapeutic interventions modify taste sensitivities in many individuals. Radiotherapy to the nasopharynx, a variety of medications frequently prescribed for older persons, such as insulin, lithium carbonate, 5-fluorouracil, as well as certain anesthetics all change taste sensations to varying degrees. Nutritionists need to be especially knowledgeable about disease, injury, and chemotherapeutic effects on taste to be able to alter diet accordingly and maintain elders' enjoyment and pleasures associated with eating.

Temperature, texture, and pain sensitivities are also mediated through touch receptors on the tongue and in the mouth cavity itself. Each of these produce pleasant or unpleasant sensations, enhancing or diminishing psychological pleasures of eating. We respond to the smoothness of Jello, the coarseness of grain, the toughness of meat, the greasiness of margarine, and the soft "velvet" taste of whipped cream. Similarly, food served at the appropriate temperature enhances taste while too hot or too cold foods cause physical pain. Serving differing textures and temperatures of various foods is helpful as one means of compensating for age-related sensory deficits in modalities crucial to taste and the psychological enjoyment of food.

Olfaction

Little empirical data exist on the relationship between olfaction and psychological implications of nutrition in older age. In spite of a scarcity of well-controlled experimental evidence, it is recognized that olfaction and gustation are closely allied sensory modalities; according to Shore[89] two-thirds of the sensation of taste depends on ability to smell. Schiffman[83] particularly stresses that any interference in olfactory perception drastically affects taste and enjoyment of food. The cold of watermelon enhances both flavor and odor, while hot dressing exudes a pungent odor not present in the same food served cold. The odor of spices suggests a feeling of warmth while the smell of mints suggests coolness.[27] As is true in other sensory modalities, olfactory acuity decreases with age, and higher olfactory thresholds require greater stimulation to activate remaining functional receptors.

Vision

Documented age-related visual changes include presbyopia, or decreased accommodation, slowing of dark adaptation, smaller pupil size, lessened visual acuity, and distortion of color vision.[20,80,89]

With age, less light enters the eye because of smaller pupil size and increasing opacity of the lens; therefore, additional illumination is needed for satisfactory visual perception. Color vision is modified so that elders perceive reds, oranges, and yellows more distinctly than blues, purples, and greens. Foods inappropriately colored (red cake) or drinks served in certain colored containers (tomato juice served in a blue glass) evidently have decreased psychological appeal to older persons.[27] Ideally, to enhance psychological appreciation, food should be served in adequately illuminated rooms with enough color, both in foods and surroundings, to stimulate older sensory modalities for maximal pleasure at mealtime.

Audition

Although rarely mentioned in the literature, the psychological appeal and enjoyment of some foods is enhanced by the sound produced while eating. Chewing carrots or potato chips produces a pleasant and expected crackling and crunching sound resulting in an added sense of satisfaction, according to Fleck.[27] With age, not only is auditory sensitivity altered by presbycusis (the reduced ability to hear higher frequency tones), but conduction losses and/or central hearing loss may develop and affect one's ability to perceive various kinds of relevant auditory information.[30,76] Decker[21] reports 79% of the patients over age 50 screened at a large metropolitan hospital had hearing losses severe enough to impair meaningful personal-social interaction with those around them. Of all the age-related perceptual changes, auditory impairment is considered to be one of the most devastating as it leads to social isolation, depression, paranoia, suspiciousness, embarrassment, or combinations of these problems. Hearing-impaired elderly tend to be reticent about eating in public where they are unable to participate in the usual social interactions. The hearing-impaired individual may be avoided to some extent by others, leading to social withdrawal and loss of social relationships. Mealtimes then become occasions of loneliness, isolation, and possibly eventually to emotional breakdown and loss of interest in living.[85,106]

Decreasing acuity in the special senses accompanying the aging process curtails not only the amount of information available to the individual from the environment, but also diminishes the accuracy of such information. Both amount and accuracy are necessary for effective coping and adaptation to a constantly changing world. As sensory data become less available to the aging individual, sensory deprivation and varying degrees of personal-social isolation often takes place. Sensory deprivation and isolation not infrequently produce profound alterations in behavior and in the ability to maintain reality contact with the world.[25,44]

Ernst et al[25] stress the majority of elderly suffer isolation of several kinds: emotional isolation, primarily from losses of significant people and activities; social isolation, from death of peers and retirement; and physiological isolation, from physical and sensory changes accompanying the aging process. Isolation tends to promote psychoneurotic reactions, especially depression, which in turn affects nutritional intake and eating patterns.

PSYCHODYNAMICS OF AGING AND NUTRITION

Loneliness

Feelings of loneliness may result from drastic change, separation, or loss. Aging involves gradual or sometimes dramatic changes in life style, work patterns, living arrangements, finances, and parent-child relationships. The presence of chronic disease along with age-related physical changes and loss of significant others through relocation or death further tax coping abilities in older age.

Loneliness, the loss of desired and necessary intimacy with significant others, has potentially serious physical and psychosocial consequences for the elderly.[17] Wilner and Sporty[105] document increasing psychological morbidity among older persons due in part to relocation of grown children, death of meaningful persons, and eventual dissolution of the immediate family constellation with no significant replacements. As physical, social, and psychological changes progress, mobility decreases, leaving elders housebound. Without appropriate community resources they are apt to gradually deteriorate physically and psychologically.

Other behavioral ramifications of loneliness appear more specifically in nutritional patterns of older persons, especially in the isolated lonely elderly. Poor appetite, disinterest in food, food selection and preparation frequently stem from prolonged feelings of loneliness with actual nutritional deficiencies a likely long-term outcome.[46,72] Pelcovits[66] reports loneliness not only decreases interest in preparing and eating nutritionally balanced meals, but leads to overall behavioral apathy and listlessness drastically affecting quality of life in the later years. Moreover, Weinberg[100] implies for some persons food and eating serves a psychological means of coping with loneliness, for in our society food and eating facilitates communication with others and has even been used symbolically as a substitution for love. Some people find the psychosocial meaning of food and eating rituals more personally significant than the physiological satisfaction derived.

Depression

Two basic types of depression are exogenous and endogenous. Exogenous, or reactive depression, occurs in response to external events such as loss of family or significant others, forced retirement, decreased income, loss of social status, and social isolation. Endogenous depression stems from intrapsychic dynamics or inner personality of the individual.[93]

There is substantial literature citing depression as a significant problem of the later years.[9,68,79] Jarvik,[43] among others, considers depression to be one of the most serious mental health problems of older age. Cumulating losses in physical, psychological and social spheres of life occur at a time in the life cycle when energy and adaptive capabilities are appreciably reduced. Brosin[10] suggests some depressions in the later years are reactions to an insatiable need for love.

Depression induced behaviors often include symptoms of anorexia.[14,99] Feelings of hopelessness, loss of security, and lack of energy affect shopping, food preparation, as well as food ingestion and are, therefore, contributory factors in poorer nutritional status in the later years.[98,101] From a psychodynamic perspective, loss of appetite or not eating can signify anger, protest, negativism, disappointment, rage, or an attempt to manipulate others.[33,77,78]

Severe depression manifests itself not only in reduced food intake but also in weight loss, constipation, and other gastrointestinal complaints.[77,103] In prolonged severe depression, an older adult may not talk, eat, or drink, but instead retire to bed. Such extreme behavioral manifestations threaten life itself as malnutrition, dehydration, fluid and electrolyte imbalance all occur more readily in older age.[4]

Death and Grieving

One of the most difficult losses to sustain throughout life is loss through death—of spouse, family and significant others. Responses to death range from preparatory grief emotions to myriad grief reactions after death has occurred. Grief symptomology includes gastrointestinal disruption of anorexia, dysphagia, indigestion, constipation, and diarrhea.[62] Such symptoms, coupled with the burdensome tasks of managing the home, shopping, food preparation, and other daily responsibilities difficult to accomplish in the grieving state, lead to nutritional deficiencies, especially when grief is prolonged for a year or more. In extreme grief reactions, individuals sometimes choose to isolate themselves from society and friends, leading to long-term inadequate dietary practices and consequent changes in health status.[11,12]

Conversely, overconsumption of food is another manifestation of grief in some people. In these instances, overeating symbolizes regression to childhood behavior when food and eating was probably used to reduce tensions and promote feelings of security.[27] If a person is psychologically soothed and comforted by food ingestion, the amount and frequency of eating will be expected to increase as substantial losses, especially deaths, occur with greater frequency in older age.[98]

Anxiety and Neurosis

Anxiety, a basic component of neurosis, is likely to develop as a response to perceived overwhelming cumulative loss associated with older age.[14,68,103] Oberleder[65] stresses the increase in anxiety-generating situations in older age and the decrease in tension-reducing opportunities, a relationship predisposing older individuals to emo-emotional breakdown.

Food habits present a sensitive indicator of anxiety or other emotional stresses. Reduced appetite, decided food preferences, food aversions, or food prejudices serve as possible defense mechanisms to protect individuals from overpowering anxiety. As indicated previously, certain foods or food rituals signify security and comfort to the individual and must be understood in an idiosyncratic functional context.[78]

The act of food ingestion itself, however, leads to feelings of anxiety if the individual is psychologically convinced the pleasures of eating should be denied self.[2] Other anxiety-ridden persons develop obsessive concerns about the physiological effects of various foods, specifically, which are gas forming, constipating, or irritating to the gastrointestinal system. Explanations concerning the usual physical effects of specific foods, psychological reassurance and support aids in reduction of anxiety and helps to alleviate stress-related gastrointestinal reactions generated by undue anxiety and excessive concerns about foods and food effects.[101] On the other hand, overeating as a means of reducing anxiety and tension, or even attracting attention from others, leads to obesity, a serious health problem at any age.[42,102] Because relatively few sources of pleasure may be available in the later years, especially for those institutionalized, foods and food ingestion assumes greater significance in daily living.

Symptoms associated with neurotic states, for example, hypochondrias, phobias, and obsessive-compulsive behaviors tend to interfere with the individual's ability to purchase, prepare, and eat nutritious diets. In a more extreme form, Weiner[101] explains the neurotic's refusal to eat as indicative of feelings of hopelessness or an unconscious wish to die.

Psychosis

Obviously, psychotic disorders disrupt normal patterns of food intake; therefore, nutritional status needs to be continuously assessed. In particular, hallucinations, extreme mood swings, and paranoid reactions warrant special attention to achieve proper nutritional intake during these episodes. Additional research is sorely needed in this area of behavioral and nutritional concern.

Organic Brain Syndromes

The presence of organic brain syndromes, either chronic or acute, drastically impairs ability to engage in food preparation and food ingestion, but relatively little has been written in this area. Reality orientation is proposed by Weiner,[101] with special emphasis on orienting confused elderly to meal schedules and to types of food served. Special attention is necessary to confused, disoriented, and psychotic individuals, and innovative procedures and programs for nutritional care are greatly needed.

Nutrition Programs

Various types of nutrition programs such as congregate dining, school lunch programs for seniors, and meals on wheels help offset some losses incurred by the aging process in psychological, social, and physical realms of life. Wilner and Sporty[105] maintain positive psychological effects of a community nutrition program for older adults are:

(1) The group becomes an active support system for members in personal and psychosocial emergencies.

(2) Programs offer service opportunities for older persons, especially widows who are energetic, outgoing, and eager to contribute to the well-being of others.

(3) Participants have a chance to develop individual potentials through meaningful associations with others.

(4) Many elders view the nutrition programs as a substitute for the family unit.

This chapter consists of a compilation of pertinent information concerning the psychological ramifications of nutrition as it relates to the aging process in humans. Although literature reviews show substantial data on the physiological aspects of nutrition, there is a dearth of empirical data on psychological implications of diet and food ingestion habits, especially concerning older adults. From the perspective of research currently available, it is obvious nutritional patterns do indeed impact to a substantial degree on the psychology of the elderly. We urge increased interest and research effort be devoted to this relatively neglected topic if we wish to fully understand behavior in older ages.

REFERENCES

(1) Anderson, L., Dibble, M.V., Mitchell, H., and Rynbergen, H.J. 1972. *Nutrition in Aging*. J.B. Lippincott, Philadelphia.
(2) Babcock, C.G. 1948. Food and its emotional significance. *J. Am. Diet. Assoc.* 24:390-393.
(3) Berger, R. 1976. Nutritional needs of the aged. In I.M. Burnside (ed.) *Nursing and the Aged*. McGraw-Hill, New York.

(4) Blumenthal, M.D. 1980. Depressive illness in old age: getting behind the mask. *Geriatrics* 35:34-48.
(5) Bogert, L.J., Briggs, G.M., and Calloway, D.H. 1973. *Nutrition and Physical Fitness*, 9th ed. W.B. Saunders, Philadelphia.
(6) Booth, D.A. 1977. Satiety and appetite are conditioned reactions. *Psychosom Med.* 39:76-80.
(7) Boykin, L.S. 1975. Soul foods for some older Americans. *J. Am. Geriatr. Soc.* 23:380-382.
(8) Brammer, L.M. and Shostrom, E.L. 1977. *Therapeutic Psychology*, 3rd ed. Prentice-Hall, Englewood Cliffs, NJ.
(9) Brink, T.L. 1979. *Geriatric Psychotherapy.* Human Sciences Press, New York.
(10) Brosin, H.W. 1968. The psychology of appetite. In M.G. Wohl and R.S. Goodhart (eds.) *Modern Nutrition in Health and Disease*, 4th ed. Lea and Febiger, Philadelphia.
(11) Brozek, J. 1970. Research on diet and behavior. *J. Am. Diet. Assoc.* 57:321-325.
(12) Brozek, J. 1978. Nutrition, malnutrition and behavior. *Ann. Rev. Psychol.* 29:157-177.
(13) Busse, E.W. 1978. How mind, body, and environment influence nutrition in the elderly. *Postgrad. Med.* 63:118-122; 125.
(14) Butler, R., and Lewis, M. 1977. *Aging and Mental Health*, 2nd ed. C.V. Mosby, St. Louis.
(15) Byrd, E., and Gertman, S. 1959. Taste sensitivity in the aging person. *Geriatrics* 14:381-384.
(16) Cameron, N. 1947. *Psychology of Behavior Disorders—A Biosocial Interpretation*, p. 110. Houghton-Mifflin, Boston.
(17) Carnevali, D. 1979. Loneliness, In D. Carnevali and M. Patrick (eds.) *Nursing Management for the Elderly*. J.B. Lippincott, Philadelphia.
(18) Carson, J., and Gormican, A. 1976. Disease-medication relationships in altered taste sensitivity. *J. Am. Diet. Assoc.* 68:550-552.
(19) Colavita, F. 1978. *Sensory Changes in the Elderly.* Charles C. Thomas, Springfield.
(20) Crouch, C.L. 1970. Lighting needs for older eyes. *J. Am. Geriatr. Soc.* 10:685-688.
(21) Decker, T.N. 1974. A survey of hearing loss in an older age hospital population. *Gerontologist* 14:402-403.
(22) Engen, T. 1977. Taste and smell. In J.E. Birren and K.W. Schale (eds.) *Handbook of the Psychology of Aging.* Van Nostrand Reinhold, NY.
(23) Eppright, E. 1947. Factors affecting food acceptance. *J. Am. Diet. Assoc.* 23:579-587.
(24) Erikson, E. 1950. *Childhood and Society.* Norton, New York.
(25) Ernst, P., Beran, B., Safford, F., and Kleinhauz, M. 1978. Isolation and the symptoms of chronic brain syndrome. *Gerontologist* 18:468-474.
(26) Fathauer, G.H. 1960. Food habits—an anthropologist's view. *J. Am. Diet. Assoc.* 37:335-338.
(27) Fleck, H. 1976. *Introduction to Nutrition*, 3rd ed. MacMillan, New York.
(28) Freed, S.C. 1947. Psychic factors in the development and treatment of obesity. *JAMA* 133:369-373.
(29) Friedman, M., and Stricker, E. 1976. The physiological psychology of hunger: a physiological perspective. *Psychol. Rev.* 83:409-431.
(30) Gaitz, C., and Warshaw, H. 1964. Obstacles encountered in correcting hearing loss in the elderly. *Geriatrics* 19:83-86.

(31) Galdston, I. 1952. Nutrition from the psychiatric viewpoint. *J. Am. Diet. Assoc.* 28:405-409.
(32) Ginsburg, S.W. 1952. The psychological aspects of eating. *J. Home Econ.* 44:325-328.
(33) Ginzberg, R. 1948. Nutrition in geriatrics—psychological and somatic aspects. *Am. J. Dig. Dis.* 15:339-346.
(34) Goldman, R. 1977. Aging of the excretory system: kidney and bladder. In C. Finch and L. Hayflick (eds.) *Handbook of the Biology of Aging.* Van Nostrand Reinhold, New York.
(35) Grossman, M.I. 1955. Integration of current views on the regulation of hunger and appetite. *Ann. NY Acad. Sci.* 63:76-89.
(36) Grossman, S.P. 1979. The biology of motivation. *Annu. Rev. Psychol.* 30:209-242.
(37) Groves, P. and Schlesinger, K. 1979. *Biological Psychology.* William C. Brown, Dubuque.
(38) Guthrie, H.A. 1979. *Introductory Nutrition*, 4th ed. C.V. Mosby, St. Louis.
(39) Hamburger, W.W. 1951. Emotional aspects of obesity. *Med. Clin. No. Amer.* 483-499.
(40) Hamilton, C.L. 1973. Physiologic control of food intake. *J. Am. Diet. Assoc.* 62:483-499.
(41) Hashim, S. 1977. Hunger and satiety in man. In M. Winick (ed.) *Nutritional Disorders in American Women.* John Wiley, New York.
(42) Howard, R., and Schnell, R. 1978. The psychology of diet and behavior modification. In R.B. Howard and N.H. Herbold (eds.) *Nutrition in Clinical Care.* McGraw-Hill, New York.
(43) Jarvik, L.F. 1975. The aging central nervous system: clinical aspects. In H. Brody, D. Harman and J.M. Ordy (eds.) *Aging*, Vol. 1. Raven Press, New York.
(44) Kammerman, M. (ed.) 1977. *Sensory Isolation and Personality Changes.* Charles C. Thomas, Springfield.
(45) Keys, A., Brozek, J., Henschel, A., Mickelsen, O., and Taylor, H.L. 1950. *The Biology of Human Starvation*, Vol. 11. Univ. Minn. Press, Minneapolis.
(46) Krehl, W.A. 1974. The influence of nutritional environment on aging. *Geriatrics* 29:65-76.
(47) Lamb, M.W. 1969. Food acceptance, a challenge to nutrition education— a review. *J. Nutr. Educ.* 1:20-22.
(48) Latchford, W. 1974. Nutritional problems of the elderly. *Community Health* 6:145-149.
(49) Leon, G., and Roth, L. 1977. Obesity: psychological causes, correlations, and speculations. *Psychol. Bull.* 84:117-139.
(50) Lepkovsky, S 1973. Newer concepts in the regulation of food intake. *Am. J. Clin. Nutr.* 26:271-284.
(51) Lepkovsky, S. 1975. Regulation of food intake. In C.C. Chinchester (ed.). *Advances in Food Research*, Vol. 21. Academic Press, NY.
(52) Lowenberg, M. 1974. The development of food patterns. *J. Am. Diet. Assoc.* 65:263-268.
(53) Lowenberg, M.E., Todhunter, E.N., Wilson, E.D., Savage, J.R., and Lubawski, J.L. 1974. *Food and Man*, 2nd ed. John Wiley, NY.
(54) Lurie, O.R.. 1941. Psychological factors associated with eating difficulties in children. *Am. J. Orthopsychiatry* 11:452-454.

(55) Mahan, L.K. 1979. A sensible approach to the obese patient. *Nurs. Clin. N. Am.* 14:229-245.
(56) Manning, M.D. 1965. The psychodynamics of dietectic. *Nurs. Outlook* 13:57-59.
(57) McBride, G. 1976. Human apetite, eating behavior complexities tantalize scientists. *JAMA* 236:1433-1445.
(58) Menzies, I.E.P. 1970. Psychosocial aspects of eating. *J. Psychosom. Res.* 14:223-227.
(59) Mogenson, G J. 1976. Neural mechanisms of hunger: current status and future prospects. In D. Novin, W. Wyrwicka and G. Bray (eds.) *Hunger: Basic Mechanisms and Clinical Implications.* Raven Press, New York.
(60) Moore, H.B. 1952. Psychologic facts and dietary fancies. *J. Am. Diet. Assoc.* 28:789-794.
(61) Moore, H.B. 1957. The meaning of food. *Am. J. Clin. Nutr.* 5:77-82.
(62) Murray, R., Huelshoettern, M.M., and O'Driscoll, D. 1980. *The Nursing Process in Later Life.* Prentice-Hall, Englewood Cliffs, NJ.
(63) Mursell, S.L. 1925. Contributions to the psychology of nutrition: hunger and appetite. *Psychol. Rev.* 32:317-333.
(64) Nisbett, R.E. 1972. Eating behavior and obesity in men and animals. In F. Reichsman (ed.) *Hunger and Satiety in Health and Disease. Adv. Psychosom. Med.* 7:173-193.
(65) Oberleder, M. 1969. Emotional breakdowns in elderly people. *Hosp. Commun. Psychiatry* 20:191-196.
(66) Pelcovits, J. 1972. Nutrition to meet the human needs of older Americans. *J. Am. Diet. Assoc.* 60:297-300.
(67) Pfaffman, C. 1964. Taste, its sensory and motivating properties. *Amer. Sci.* 52:187-206.
(68) Pfeiffer, E. 1977. Psychopathology and social pathology. In J.E. Birren and K.W. Schaie (eds.) *Handbook of the Psychology of Aging.* Van Nostrand Reinhold, New York.
(69) Price, J.H., and Pritts, C. 1980. Overweight and obesity in the elderly. *J. Geront. Nurs.* 6:341-347.
(70) Prugh, D.E. 1961. Some psychologic considerations concerned with the problem of overnutrition. *Am. J. Clin. Nutr.* 9:538-547.
(71) Pumpian-Mindlin, E. 1954. The meanings of food. *J. Am. Diet. Assoc.* 30:576-580.
(72) Rao, D.B 1973. Problems of nutrition of the aged. *J. Am. Geriatr. Soc.* 21:362-366.
(73) Robinson, C.H., and Lawler, M. 1977. *Normal and Therapeutic Nutrition*, 15th ed. MacMillan, New York.
(74) Rockstein, M. 1975. The biology of aging in humans—an overview. In R. Goldman and M. Rockstein (eds.) *The Physiology and Pathology of Human Aging.* Academic Press, New York.
(75) Rodin, J. 1977. Research on eating behavior and obesity: where does it fit in personality and social psychology? *Per. Soc. Psychol. Bull.* 3:333-355.
(76) Rupp, R. 1970. Understanding the problems of presbycusis. *Geriatrics* 25:100-110.
(77) Savitsky, E. 1953. Psychological factors in nutrition of the aged. *Social Casework* 34:435-440.
(78) Savitsky, E., and Zetterstrom, M. 1959. Group feeding for the elderly. *J. Am. Diet. Assoc.* 35:938-942.
(79) Savitz, H A. 1976. Mental hygiene for the aged. *NY State J. Med.* 76:1850-1853.

(80) Saxon, S.V., and Etten, M.J. 1978. *Physical Change and Aging—A Guide for the Helping Professions.* Tiresias, New York.
(81) Schachter, S. 1971. Some extraordinary facts about obese humans and rats. *Am. Psychologist* 26:129-144.
(82) Schiele, B.C., and Brozek, J. 1948. "Experimental neurosis" resulting from semistarvation in man. *Psychosom. Med.* 10:31-40.
(83) Schiffman, S. 1977. Food recognition by the elderly. *J. Gerontol.* 32: 586-592.
(84) Schlenker, E.D., Feurig, J.S., Stone, L.H., and Mickelson, O. 1973. Nutrition and the health of older people. *Am. J. Clin. Nutr.* 26: 1111-1119.
(85) Senturia, B., Goldstein, R., and Hersperger, W. 1976. Otorhinolaryngologic aspects. In F. Steinberg (ed.) *Cowdry's The Care of the Geriatric Patient,* 5th ed. C.V. Mosby, St. Louis.
(86) Shannon, B., and Smiciklas-Wright, H. 1979. Nutrition education in relation to the needs of the elderly. *J. Nutr. Educ.* 11:85-89.
(87) Sherwood, S. 1970. Gerontology and the sociology of food and eating. *Aging Human Develop.* 1:61-85.
(88) Shifflett, A. 1976. Folklore and food habits. *J. Am. Diet. Assoc.* 68: 347-349.
(89) Shore, H. 1976. Designing a training program for understanding sensory losses in aging. *Gerontologist* 16:156-165.
(90) Siegel, P.S. 1957. Repetitive element in the diet. *Am. J. Clin. Nutr.* 5: 162-164.
(91) Siegel, P.S., and Pilgrim, F.J. 1958. The effect of monotony on acceptance of food. *Am. J. Psychol.* 71:756-759.
(92) Stare F.J. 1977. Three score and ten plus more. *J. Am. Geriatr. Soc.* 25:533.
(93) Storandt, M. 1976. Psychologic aspects. In F. Steinberg (ed.) *Cowdy's the Care of the Geriatric Patient.* C.V. Mosby, St. Louis.
(94) Straus, B. 1979. Disorders of the digestive system. In I. Rossman (ed.) *Clinical Geriatrics,* 2nd ed. J.B. Lippincott, Philadelphia.
(95) Stunkard, A.J. 1961. Hunger and satiety. *Am. J. Psychiatry* 118:212-217.
(96) Stunkard, A.J. 1975. Satiety is a conditioned reflex. *Psychosom. Med.* 37:383-387.
(97) Tobias, A.L., and Gordon, J.B. 1980. Social consequences of obesity. *J. Am. Diet. Assoc.* 76:338-342.
(98) Troll, L.E. 1971. Eating and aging. *J. Am. Diet. Assoc.* 59:456-459.
(99) Watkin, D.M. 1978. Logical bases for action in nutrition and aging. *J. Am. Geriatr. Soc.* 26:193-201.
(100) Weinberg, J. 1972. Psychologic implications of the nutritional needs of the elderly. *J. Am. Diet. Assoc.* 60:293-296.
(101) Weiner, M.F. 1969. A practical approach in encouraging geriatric patients to eat. *J. Am. Diet. Assoc.* 55:384-386.
(102) Weiss, E. 1953. Psychosomatic aspects of dieting. *J. Clin. Nutr.* 1:140-149.
(103) Whitehead, T. 1979. *Psychiatric Disorders in Old Age.* Springer, NY.
(104) Williams, S.R. 1978. *Essentials of Nutrition and Diet Therapy,* 2nd ed. C.V. Mosby, St. Louis.
(105) Wilner, M., and Sporty, L. 1978. A nutrition program as surrogate for family life. *Psychiat. Qtr.* 50:59-62.

(106) Young, C.M. 1974. Nutritional counseling for better health. *Geriatrics* 29:83-91.
(107) Young, P.T. 1941. The experimental analysis of appetite. *Psychol. Bull.* 38:129-164.
(108) Young, P.T. 1957. Psychologic factors regulating the feeding process. *Am. J. Clin. Nutr.* 5:154-161.

–4–
Sociological Aspects of Nutrition and Aging

Albert J.E. Wilson III

INTRODUCTION

The sociological study of nutrition and aging has been described as pertaining to "... the ways in which food and eating patterns play a part in, affect, or result from human relationships and societal conditions."[33] Thus, the current and historical nutrition practices and nutritional status of older persons may be viewed as both independent and dependent variables relative to society and its subsystems. Societal conditions both influence and are influenced by nutrition related variables. Examples of these complicated interrelationships may be found in essentially all aspects of group life including, but not limited to, economic, political, religious, family, education, and health care systems.[2,3,8,10,33] Availability of food and culturally related habits also affect population size, growth or decline, and characteristics.[3,11,17,25]

It is apparent from the above comments that the area of nutrition is a logical and potentially productive one for sociological study. Yet much of the research which has yielded valuable sociological data has been carried out by nonsociologists. Sherwood comments that "... although sociologists have devoted relatively little effort to the sociology of food and eating, from the extensive research undertaken in the field of nutrition since the turn of the century, a great deal has been learned about the social aspects of food and eating."[33]

Regardless of the disciplinary affiliation of investigators, Sherwood is correct in her statement concerning the existence of a substantial body of material on the sociological aspects of food and nutrition. It appears that sociologists are developing a belated interest in this material and in their potential contributions in subsequent analyses. This chapter presents an overview of some of the major sociological factors as they relate to food, nutrition, aging and the aged. Hopefully, it may identify and clarify some areas for future sociological study.

FUNCTIONS OF FOOD

The basic function of food is to satisfy hunger and to nourish the body. But, food and eating serve a number of other functions for individuals and society which are basically social in nature. Bass et al and Kalish have identified some of these nonnutritional functions of food.[3,14] Meals are often used to initiate and maintain personal relationships and business associations. Food may be used as a vehicle for bringing people together for a specific purpose such as a political rally, birthday party, or church event. Some people use food to emphasize their individuality or to attract attention. Special meals are often prepared to express love or concern in times of bereavement or family disruption. Food may be used as a reward or punishment, as a status symbol, or as a basis for group identity. In all of these examples, the social function of "breaking bread together" is primary and the nutritional function is of relatively little importance.

RELATIONSHIP OF STUDIES OF EARLY LIFE NUTRITION TO AGING

Throughout the literature on nutrition and aging runs a thread of criticism that past investigations which have related sociological and nutritional variables have been directed toward pregnant women, infants, and children with little attention given to the aged.[6,11,19,33] For a number of reasons, it seems that these studies do, in fact, have a very direct relationship to nutrition in later life. There is a great deal of evidence that the food preferences and eating patterns developed in childhood are retained in later life.[1,3,22,33] There is also evidence that maternal and early childhood nutrition are important variables in (a) whether people survive to old age, and (b) their overall health status in old age.[6,19,22] For example, the National Advisory Council on Aging reports that "sketchy evidence thus far indicates that impaired childhood growth rates resulting from malnutrition lead to premature aging, while excess calorie consumption in childhood and early adulthood may lead to earlier death from age-associated diseases . . ."[18] Improvements in prenatal and childhood nutrition have resulted from the application of sociological concepts in the field of public health and in the food industry. These advances have affected the demographic structure of the "developed nations," increasing the quantity and the quality of the population as well as changing its age structure through changes in the "survivorship curve."

This recognition of the impact of advances in maternal and childhood nutrition upon aging and the aged is not intended to suggest that an additional focus on nutrition in later life is not needed. The sociological study of aging and nutrition has not received the attention which it merits. Such study can make substantial contributions to increasing understanding of the complicated interactions of social, psychosocial, physiological, and nutritional factors.

NUTRITIONAL STATUS OF OLDER AMERICANS

Assessments of the nutritional status and diet adequacy of older Americans have yielded variable and inconclusive findings on the prevalence of poor nutritional status among the aged. Estimates of the extent of nutritional deficiency range from 10 to 90 percent of the older population. Methodological differences, particularly in the areas of sampling and definitions, have contributed to this variability.[3,6,8,26] The lack of knowledge and lack of standards of nutritional adequacy for older persons have been major sources of inconsistency and uncertainty in assessment studies. Some investigators maintain that standards of adequate nutrition for younger populations may be inappropriate for the aged.[3,6,8] In spite of inconsistencies in methods and in findings, essentially all investigators agree that nutrition problems constitute a major area of concern.

Butler reports that admission data for nursing homes indicate nutritional deficiencies as high as 10 percent and that various studies of the Department of Agriculture and the National Center for Health Statistics ". . . confirm that the diets of older Americans are often below standard in quantity and quality."[6] Clark and Wakefield reported that over half of the 197 elderly persons whom they studied had inadequate diets with similar rates of inadequacy among the institutionalized and noninstitutionalized.[9] Clancy reported that about one third of her sample of 47 senior center participants had diets deficient in "at least three major nutrients."[8] Schlenker reported that a Michigan study of 201 females aged 40 to 80 found that 95 percent had intakes that were below 80 percent of the recommended dietary allowances of at least one nutrient.[31] A 1962 study in St. Petersburg, Florida revealed that about 60 percent of the households containing persons age 60 and over had food habit problems judged severe enough to require intervention. In this study of 1,800 households, ratings of food habits were based upon food served, presence of "finicky eaters," and presence of weight problems.[26] The Administration on Aging reports that at least a third of Americans age 65 and above lack basic essential nutrients in their diets.[1]

Numerous other assessments of dietary habits and nutritional status have been reported in the literature. The above examples seem to be representative and provide support for several propositions.

The first is that malnourishment and poor dietary practices are widespread among older Americans. The second is that methodological differences in prevalence studies have contributed to wide variations in reported rates. The third is that we lack knowledge of the actual nutritional needs of aged persons. The final and most general proposition to be noted is that we do not have an accurate picture of either the exact nature of nor the extent of dietary and nutritional deficiencies among the aged. As the National Advisory Council on Aging stated in 1978, "Some information on the nutritional adequacy of the diet of elderly individuals is beginning to accumulate, but our lack of experience in this area is vast. . . . There can be no rational basis for nutritional recommendations for the elderly until more information is obtained on the changing nutrient metabolism with age."[19]

SOCIOLOGICAL FACTORS AS INDEPENDENT VARIABLES

The most basic sociological variable affecting diet involves societal or subgroup definitions of food. What people eat depends upon what they have been socialized to define as edible. Societal definitions of food in general, and of preferred foods may have little relationship to the nutritional value of the substance defined as "good" food. Social prescriptions rather than physiological needs generally determine what people eat, when they eat, and where they eat. Meals are usually consumed at socially defined times which may or may not coincide with the timing of physiological needs for food. An illustration of the strength of internalized social definitions is that of the Englishman who found himself living with members of a primitive hunting and gathering tribe. Although totally out of step with his current social environment, he continued to observe "tea time" and maintained the tradition of "dressing" for dinner.[5,22]

The United States is a pluralistic society with wide variations in food behavior related to geographic, ethnic, racial, and religious subcultures. These variations have been observed in numerous studies of younger populations but few studies have been done on ethnic food behavior of the aging other than those participating in congregate meals programs. Studies of the meals program participants have shown definite patterns of ethnic preferences and have resulted in modification of rules for Administration on Aging programs to provide for these preferences.[23] It may be expected that the ethnic preferences reported in studies of younger persons also hold for the aged. Bass et al state that "the significance of a lifetime of food preferences and food behavior should not be overlooked by food and nutritional professionals developing food programs for the elderly."[3] A report of the Nutrition Foundation, Inc. states that ". . . the kinds of food designated as edible, the ways to prepare them, and the manner of consuming them are deeply imbedded in the behavioral systems of each culture."[22]

Religion also has a strong influence on dietary practices. Certain religions dictate what foods can or cannot be eaten and how certain foods must be prepared. Some prescribe special diets symbolic of the religion itself. Religious ceremonies may involve the giving of food, abstaining from food, or consumption of special foods as symbolic ways of showing respect and devotion to a supreme being.[17]

Some examples of religious influences on diet which are common in the United States are as follows:

Prior to 1966, Roman Catholics in the United States were required to observe a number of fast days and to abstain from eating meat on Fridays. This law was changed by the United States Catholic Conference in 1966 so that abstinence from meat was required only on Fridays during Lent. Many older Catholics still adhere to the traditional rules and continue to observe fast days and meatless Fridays.

The Greek Orthodox Church prescribes numerous fast days during which no meat or animal products (including milk, butter, and cheese) may be eaten. Fish, with the exception of shellfish, is also forbidden on these days.

Judaism prohibits the eating of meat from "unclean" animals including the pig. Only fish with both fins and scales are allowed, thus eliminating all shellfish and eels. Judaism also provides for strict supervision to assure compliance with prescriptions for the slaughter of animals and the preparation of food.[17]

The author is reminded of a recent incident which illustrates the impact of religious dietary rules. An elderly orthodox Jew from the American midwest was returning from a trip to Israel when he experienced a series of missed plane connections. Special Kosher meals had been ordered for his connecting flights but were not available on the alternate flights. Because of long delays and non-availability of Kosher food at airline terminals and on substitute flights, he went for nearly thirty hours without a meal.

Dietary habits based upon cultural definitions and values such as those described above affect all ages and all economic levels. They may have greater significance for older persons because of the tendency for the aged to adhere more rigidly to traditional customs, because of the cumulative effect of long term behavior patterns, and because of the higher prevalence of chronic disease among the elderly. Some of the socially defined food habits practiced over a lifetime may contribute to the development of nutrition related disorders in later life. The rigid adherence to traditional dietary prescriptions may complicate the treatment of such disorders and prevent compliance with therapeutic diets.

Up to this point, this discussion has emphasized sociocultural influences which reflect lifetime patterns and their consequences. Other sociological variables relate to conditions often associated with aging and to changes in status and role of older persons in American society. The existing body of research provides support for the idea that most older persons follow traditional food patterns (which may result in either positive or negative dietary habits) until medical and/or social conditions force change.[3] As in most areas of research concerning the aging, the literature is inconsistent on the extent of and the reaction of older persons to these changed conditions. Two reasons for these inconsistencies stand out. One is the great diversity of the older population. The other is the differences in methodology (particularly sampling and definitions) used by various investigators.

There seems to be general agreement that the following factors do contribute to dietary problems for large numbers of older persons:

(1) low income with restricted ability to purchase adequate amounts of needed foods and to provide for proper storage and preparation,

(2) loneliness, unhappiness, and bereavement contributing to loss of appetite,

(3) reduced incentive for eating related to loss of meaningful roles, reduced social interaction, and inactivity,

(4) deterioration of mental status because of social isolation and alienation,

(5) low resistance to sales pitches for fad foods and susceptibility to food fallacies,

(6) changes in physical environment (housing or neighborhood) which reduce ability to obtain and prepare food.[1,10,12,23,25,27]

The Older Americans Act nutrition program recognizes the impact of these variables in the statement of findings and purpose for Title VII (changed to Title III-C by the 1978 Amendments) as follows: "Many elderly persons do not eat adequately because (1) they cannot afford to do so; (2) they lack the skills to select and prepare nourishing and well-balanced meals; (3) they have limited mobility which may impair their capacity to shop and cook for themselves; and (4) they have feelings of rejection and loneliness which obliterate the incentive necessary to prepare and eat a meal alone."[38]

Changes in living arrangements and family roles are widely recognized as affecting food habits and resulting nutritional status. In some instances there is potential for positive nutritional changes. For example, an elderly widower may move from a single person household to a congregate facility. Much greater attention has been directed toward the negative changes associated with changes in family composition due to disability or death of a spouse; to moving from a house or an apartment to a single room out of financial necessity; or to institutionalization.[3,11,17,25,28,29]

When a man whose wife has always purchased and prepared his meals has to take over these duties because of the wife's disability, their diet is likely to suffer. Many men lack the knowledge to shop for the right foods and the skill to prepare them. A common adaptation is to switch to foods requiring little judgment in selection and a minimum of preparation. An additional barrier to sound nutrition may be introduced since illness and disability are usually associated with added expenses. Thus the couple whose income previously was barely adequate may find that they are no longer able to afford a nutritionally sound diet.[14,17,27]

In the case of widowhood, the surviving spouse may have little incentive to prepare and consume meals in solitude. Widowhood often results in reduced income which affects one's ability to establish and maintain social relationships as well as limits expenditures for food. Any one or a combination of disability of a spouse, death of a spouse, and reduction in income may necessitate a change of residence. The change is likely to be to a smaller, less expensive unit which may lack adequate facilities for the storage and preparation of food.[12,24,27] Figure 1 as presented in Rao[27] illustrates the "vicious cycle of low income, poor housing, lack of nutritional education, poor nutrition, and chronic disease."

Figure 1: Interrelation of Health, Economics, Social Conditions and Disease in the Aged.*

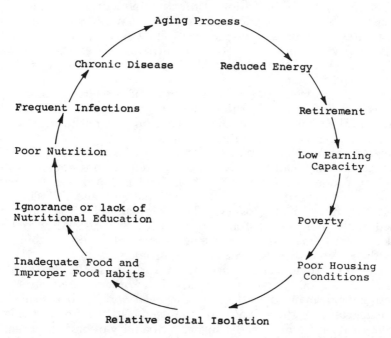

*Source: Rao, D.B., 1973, Problems of nutrition in the aged. *Am. J. of Geriatrics* 21:364.

It is assumed by many that relocation to a nursing home assures the older person of a sound diet. Nursing homes employ dieticians to plan menus and to assure that prescribed therapeutic diets are served. A number of studies have shown that the expectation of sound nutrition in institutions is not always achieved.[3,9,12] The serving of nutritionally adequate food in no way assures that the patient/resident will eat what is served. Few institutions take into account the food preferences and food behavior which patients have developed over their lifetimes. Nor do they consider ethnic, cultural, and social factors which influence the preferred environment and timing of meals.[3,9,12] Clark and Wakefield reported that more than half of the 102 nursing home residents that they studied ate diets which were nutritionally inadequate. They also reported that there was little difference in the adequacy of the diets of nursing home residents and persons having similar characteristics residing in the community.[9]

The media may be viewed as both an independent and a dependent variable relative to diet and aging. In this section the independent variable aspects are examined while the dependent variable aspects are to be discussed later. Studies of activities of retirees and of older persons show that they spend a greater proportion of their time watching television and reading than is true of other age groups.[1,12] The media have contributed substantially to public information and education on sound nutrition. Unfortunately, they have also provided the vehicle for highly influential advertising, much of which contributes to poor dietary practices. Junk foods (those with little nutritional value, but high levels of salt, sugar, chemical additives, and calories) and unneeded food supplements are vigorously promoted through the media.[3]

Ronald Deutsch quotes from a report of the White House Conference on Food and Nutrition that "no other area of the national health probably is as abused by deception and misinformation as nutrition. . . . The poor, in particular the old, the ill, and the least educated are cruelly victimized."[10] The National Retired Teachers Association/American Association of Retired Persons comment that "Making proper use of even the best food is a problem for many oldsters, who matured in virtual nutritional ignorance."[21] The relatively low nutritional knowledge level of many older persons combined with their hopes for easy solutions for difficult problems makes them prime targets for the advertising of products which frequently add to rather than reduce their problems.

The final sociological independent variable to be discussed in this section, rurality, cuts across all of the others. There are a few positives and many negatives in the relationships between rural residence and nutrition status of the aged. On the positive side is access to fresh produce and a higher probability of having more intimate relationships with neighbors. On the negative side is a list of characteristics of the rural aged and their environment which reduces their relative chances for adequate nutrition. Studies of the rural aged have described them as having depressed economic conditions, shortages of health care providers, substandard housing, inadequate transportation, social and geographic isolation, and high rates of physical and mental impairment.[39,42] Essentially all of the factors identified as contributing to poor nutrition for the general population are found with greater frequency among the rural aged.

NUTRITIONAL STATUS OF THE AGING AS AN INDEPENDENT VARIABLE

The presence of large numbers of older persons with relatively high rates of diet related concerns has affected almost every subsystem of society. Sociologists view the basic societal subsystems or social institutions as including the economic, political, family, educational, and religious subsystems. In addition to these basic subsystems, a number of other important subsystems exist in most societies. In the United States, the media subsystem is one of these which has particular significance for the aging.

Societal subsystems relate to the needs of subpopulations such as the aged in ways which may be either beneficial or detrimental to their well being. It is not the intent of this presentation to provide detailed analysis of these relationships, which could constitute a book in itself. However some selected examples are presented. Figure 2 illustrates the ways in which selected subsystems are influenced by the presence of older persons who have high rates of diet linked needs. This influence is mediated by pressures from special interest groups which promote various policies or courses of action and by individual reactions (intervening variables).

The impact on policies, programs, services, and products of the nutritional status of the aged may be positive or negative. The development of a national nutrition program under the Older Americans Act is generally viewed as a positive reaction by the political subsystem. The development and promotion of junk foods and unneeded food supplements is generally viewed as a negative reaction involving both the economic and the media sectors.[1,3,10,12] Special interest groups often bring pressure for government regulations which relate to production, transportation, storage, marketing, purity, quality, and price of food. The policies and programs which evolve reflect the interaction of varied consumer interests, producer interests, market interests, and individual responses.[3]

The Older Americans Act nutrition programs have stimulated the development of new technology and new specialties in business and industry. These are concerned with the preparation, packaging, storage, and delivery of foods which meet the Act's requirements for both congregate and home delivered meals.[15,37]

Figure 2: Dietary Needs of Older Persons as an Independent Variable Affecting Selected Societal Subsystems

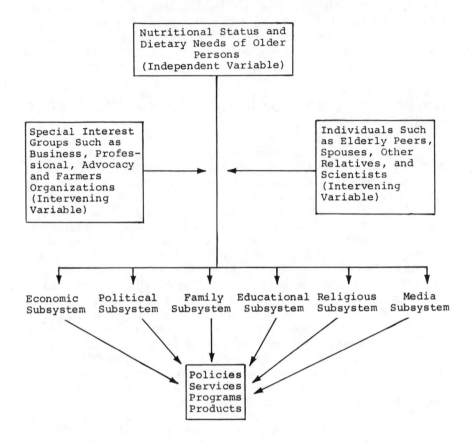

The educational subsystem has been influenced in several ways. One is the development of education and training for professionals and paraprofessionals specializing in nutrition and aging. Another is the initiation of educational programs on diet and nutrition directed toward the older population. A third example is the integration of adult education classes into congregate meal programs using food as the focus to involve older persons in educational experiences.

The religious subsystem has responded to nutritional needs of older persons through social ministries. Religious organizations sponsor housing developments which help to overcome some of the barriers to diet adequacy mentioned earlier. Church groups frequently sponsor and carry out home delivered meal programs and provide facilities for congregate meals. The religious segment has also responded by assisting in identification of specific problems in community nutritional needs assessments.

SUMMARY

Dietary practices evolve out of a complex interrelationship among social, sociopsychological, and environmental factors. Within the United States striking differences in food behavior are found among various subgroups of all ages. These patterns which are established in early life usually persist into advanced old age unless health or economic conditions force changes.

Maternal and early childhood dietary behaviors are important variables in determining whether individuals survive to old age and the health status of those who do survive. Thus improvements in prenatal and early childhood nutrition resulting in part from the application of sociological principles have affected the demographic structure of the "developed" nations. Most notable among these effects are increases in population size, improved general health, and population aging with a steadily increasing number and proportion of older persons.

Nutritional status and dietary practices are usually treated as dependent variables relative to sociological factors. That is, emphasis has been placed upon the way in which ethnicity, race, religion, income, family composition, residence, and other social characteristics affect food practices. Social prescriptions dictate what people eat, how it is prepared, when it is eaten, and where it is eaten. The dietary practices of older persons reflect a lifetime of food related behavior which has been influenced strongly by sociological variables.

Food and eating have numerous functions other than the provision of nutrients for the body. Meals may be used to initiate and maintain relationships, to bring groups together for some specific purpose, to express love or concern, to reward or punish, or to maintain group identity. Food has important symbolic meaning in certain religious practices.

The extent of diet related nutritional deficiency among older Americans is not known, but there is agreement that the rate is high. Because of variations in definitions and in research methodology, estimates of the proportion of the aged who have nutritional deficiencies and/or inadequate diets range from 10 to 90 percent. The presence in the United States of large numbers of older persons with actual or potential diet deficiencies has affected most of the basic subsystems of society including those concerned with politics, economics, the family, religion, education, and the media. Thus the nutritional status and the dietary habits of older persons may be viewed as independent variables affecting society as well as dependent variables affected by society.

Only in recent years has there been widespread recognition that malnutrition is as much a sociological and political problem as a medical problem. Sociologists have been slow to become involved in research on nutrition and diet although sociological concepts and orientations have been used by members of other disciplines such as dietetics, nutrition, and public health. Greater understanding of the role of sociocultural variables in nutrition and the application of sociological concepts can contribute significantly to efforts to improve the dietary practices and nutritional status of the aged.

REFERENCES

(1) Administration on Aging, 1980. *Older Americans Month Information Package.* U.S. DHEW, Washington, DC.
(2) American Dietetic Association, 1972. Position paper on nutrition and aging. *J. of the American Dietetic Association* 61:623.
(3) Bass, M.A., Wakefield, L. and Kolasa, K., 1979. *Community Nutrition and Individual Food Behavior.* Burgess, Minneapolis.
(4) Bernard, A., 1977. Sociological factors in nutrition for the elderly. In R. Kalish, *The Later Years.* Brooks/Cole, Monterey.
(5) Bredemeier, H.C. and Stephenson, R.M., 1964. *The Analysis of Social Systems.* Holt, Rinehart and Winston, New York.
(6) Butler, R.N., 1977. *Nutrition and Aging.* DHEW Pub. No. (NIH) 79-325. U.S. GPO, Washington, DC.
(7) Chinn, Austin B., 1971. (ed.)., *Working With Older People, Vol. IV, Clinical Aspects of Aging.* PHS Publication No. 1459, U.S. GPO, Washington, DC.
(8) Clancy, K.L., 1975. Preliminary observations on media use and food habits of the elderly. *The Gerontologist* 15:529-536.
(9) Clark, M. and Wakefield, L.M., 1975. Food choices of elderly persons. *J. of American Dietetic Assoc.* 66:600.
(10) Deutsch, R., 1977. *The New Nuts Among the Berries.* Bull, Palo Alto.
(11) Frankle, R.T. and Owen, A.Y., 1978. *Nutrition in the Community.* Mosby, St. Louis.
(12) Hess, B.B. and Markson, E.W., 1980. *Aging and Old Age.* McMillan, New York.
(13) Hickey,T., 1980. *Health and Aging.* Brooks/Cole, Monterey.
(14) Kalish, R.A., 1975. *Late Adulthood: Perspectives on Human Development.* Brooks/Cole, Monterey.
(15) Kirschner Associates and Opinion Research Corp., 1979. *Longitudinal Evaluation of the National Nutrition Program for the Elderly* (Executive Summary). DHEW Pub. No. (OHDS) 80-20250. U.S. GPO, Washington, DC.
(16) Lasky, M.S., 1977. *The Complete Junk Food Book.* McGraw Hill, New York.
(17) Lowenberg, M.E., 1968. *Food and Man.* Wiley, New York.
(18) Mann, G.V., 1973. Relationship of age to nutrient requirements. *The American Journal of Clinical Nutrition* 26:1096-1097.
(19) National Advisory Council on Aging, 1978. *Our Future Selves.* DHEW Publication No. 78-1443. U.S. GPO, Washington, DC.
(20) National Institute on Aging, 1979. *Nutrition and Aging.* U.S. GPO, Washington, DC.
(21) National Retired Teachers Association and American Association of Retired Persons, 1971. *Proposals for a National Policy on Aging.* NRTA/AARP, Washington, DC.

(22) Nutrition Foundation, Inc., 1969. *Food, Science, and Society.* The Nutrition Foundation, Inc., New York.
(23) Office of Human Development Services, Department of HEW, 1979. Proposed rules, grants for state and community programs on aging. *Federal Register* 44:45032-45063.
(24) Pelcovits, J., 1973. Nutrition education in group meals programs for the aged. DHEW Publication No. (OHD) 74-20238. Reprinted from *J. of Nutrition Education* V. U.S. GPO, Washington, DC.
(25) Percey, C.H., 1974. *Growing Old in the Country of the Young.* McGraw-Hill, New York.
(26) Pinellas County Health Department and Florida State Board of Health, 1964. *Final Report of a Study of Extra-Hospital Nursing Needs in a Retirement Area.* Pinellas County Health Department, St. Petersburg.
(27) Rao, D.B., 1973. Problems of nutrition in the aged. *J. of the American Geriatrics Society* 21:362-367.
(28) Rokstein, M. and Sussman, M. (eds.), 1976. *Nutrition, Longevity, and Aging.* Academic Press, New York.
(29) Rountree, J.L. and Tinklin, G.L., 1975. Food beliefs and practices of selected senior citizens. *The Gerontologist* 40:537-540.
(30) Sandoz Pharmaceutical Co., 1977. Nutrition and the elderly. *Care* 2:3-10.
(31) Schlenker, E.D., et al, 1973. Nutrition and health of older people. *The American Journal of Clinical Nutrition* 26:1111-1119.
(32) Sherwood, S., 1978. Malnutrition: a social problem. In M. Seltzer, S. Corbett and R. Atchley (eds.), *Social Problems of the Aging,* pp 212-225. Wadsworth, Belmont.
(33) Sherwood, S., 1973. Sociology of food and eating: implications for action for the elderly. *The American J. of Clinical Nutrition* 26:1108-1110.
(34) Shifflett, P.A., McIntosh, W.A. and Nyberg, K.C., 1978. The current status of nutritional research: a call for nutritional sociology. Paper presented at annual meeting of the Rural Sociological Society.
(35) Tenenbaum, Frances, 1979. *Over 55 Is Not Illegal.* Houghton Mifflin, Boston.
(36) U.S. Departments of Agriculture and HEW. Undated. *Food Is More Than Just Something to Eat.* Washington, DC.
(37) U.S. Department of HEW, 1972. Nutrition program for the elderly: rules and regulations. *Federal Register* 37:16844-16850.
(38) U.S. Department of HEW, 1976. *Older Americans Act of 1965, As Amended and Related Acts.* U.S. GPO, Washington, DC.
(39) U.S. House of Representatives, Committee on Government Operations, 1973. *Special Report on the Rural Aged.* U.S. GPO, Washington, DC.
(40) Watkin, D.M. and Mann, G.V., 1973. *Symposium: Nutrition and Aging.* DHEW Publication No. (OHO) 75-20240, U.S. GPO, Washington, DC.
(41) White House Conference on Aging, 1971. *Delegate Work Book on Nutrition.* U.S. GPO, Washington, DC.
(42) Wilson, Albert J.E. III, 1978. General status of the rural aged. Unpublished paper.

–5–

Protein Nutrition and Aging

Kung-Ying Tang Kao
and
Florence Lazicki Lakshmanan

Proteins are the most abundant organic molecules within cells and are indispensable to all aspects of cell structure and function. On a dry weight basis most cells consist of 50% or more protein. All proteins consist of amino acids. The dynamic status of protein in the animal body constantly demands a continuing supply of amino acids. The primary source of these amino acids is food protein. Protein nutritional status of any organism at any stage of life can only be appraised by first understanding protein and amino acid metabolism at each level of organization of the particular organism.

Aging is simply defined as the process of maturing and growing old. This life-long process begins before birth, at conception, and ends with death. It is not only an intricate process, but also a highly individual process resulting in a unique end-product, the aged subject. The complex nature of the aged subject is well-recognized and a challenge to the scientist.

Apropos, this chapter is divided into the following three main areas of consideration: protein metabolism and aging, aging-related diseases and protein metabolism, and finally a discussion of protein nutrition of the elderly.

PROTEIN METABOLISM AND AGING

Biological Function of Protein

Proteins have great functional versatility. Enzymes comprise the largest and most diverse class of proteins, in terms of biological functions. Still other proteins having intense biological activity are certain hormones, such as insulin, growth hormone, etc., and antibodies. Proteins also function as essential elements in the contractile and motile system. These proteins include actin and myosin. Other proteins, such as albumin, serum lipoprotein and hemoglobin have transport functions. In higher animals, the fibrous protein, collagen, is the major extracellular protein in connective tissue and in bone. Other fibrous proteins are α-keratin, elastin and various membrane proteins.

Biological Aging

Maturation of collagen, with its slow turnover and long half-life, has been suggested as one index of biological aging. Forty percent of all protein in the body is collagen which is an important component of connective tissue. This tissue permeates all organs and is the chief supporting tissue of the animal body. Tissues having collagen as their major protein component are skin, tendon, bone, teeth, cartilage, blood vessel, cornea, lens capsule, vitreous humor and basement membrane. It has been demonstrated that biologically old collagen has an increased number of hydrogen and ester bonds which form cross-links between molecules within the fibers and thereby increase their tensile strength.[8,151,152] The theory has been proposed that the number of cross-links between molecules increases throughout life. The low solubility of old collagen during thermally induced contraction[33,54,100] supports this hypothesis. It has been further suggested that more ester bonds requiring hexose for their formation are produced during aging.[40] Milch and his coworkers[101,102,103] have identified aldehydes as stabilizers of collagen against collagenase or pH shifts. The most potent stabilizer is the intermediate carbohydrate metabolite, glyceraldehyde. Subsequently, a group of reducible cross-links has been identified[3,12,41] in the collagen of various tissues. These cross-links are formed from aldehydes derived from lysine or hydroxylysine, reacting with itself to form an aldol-condensation product or with another lysine or hydroxylysine to form a Schiff base. The type and abundance of these reducible collagen cross-links vary with the species of animal and with the tissue studied.[3,4,79,80,97,139] Their amount per unit weight of tissue rises rapidly during early development and then decreases, also rapidly, with increasing age.[5,78,130] The fibrils become increasingly resistant to solvents and retain their tensile strength. It was proposed that the reducible cross-links formed during early development represent an intermediate stage. They are converted during the maturation process into nonreducible, more stable cross-links.[98] Furthermore, hexosyllysine and hexosylhydroxylysine increased with age indicating increased cross-links between collagen and glycoprotein.[78,129] Verzar,[152] Piez,[122] and Eyre[35] speculated that these cross-links represent a continuing and perhaps genetically programmed maturation process.

The effect of food intake on collagen solubility was reported but the results were not in agreement. When the food intake of rats is limited to about one-half of ad libitum consumption, the tensile strength of tendon fiber is lower for the restricted rats than for the controls.[24,34] The values for the restricted animals are similar to those for the young rats. It was also reported that with food restriction the decreases of neutral, salt-soluble collagen in skin resemble those observed in older animals.[51]

Tissue Protein and Aging

Much of the present knowledge about aging of tissues has been gained from work on experimental animals of various classes, and very little from man. Early work of ours[66,67,70] and of other investigators[88,91,108,109,137,140] has demonstrated that the changes in body proteins with increasing age in experimental animals and humans involve a decrease in soluble protein, an increase in collagen, and modification of elastin. These changes were most apparent in tissues rich in collagen.[66,67,70] Table 1 shows the influence of age on the soluble protein content of several tissues in female rats. Soluble protein content of tendon, heart muscle and skin decreased significantly with increasing age. The decreases were approximately 60% in tendon, 20% in heart muscle, and 30% in skin by 2 years of age. Soluble protein content of uterus also decreased 20 to 30% with age, but the difference was not statistically significant. No change, however, was observed in soluble protein content of skeletal muscle, lung, kidney, liver or spleen with increasing age.

Table 1: Influence of Age on Soluble Protein Content of Various Tissues in Female Rat

Tissue	3 to 5 Weeks	8 Months	2 Years
	g of soluble protein/100 g of tissue		
Upper leg muscle	17.6±1.1 (4)	19.6±1.6 (7)	19.4±2.3 (3)
Lower leg muscle	20.7±2.4 (4)	19.9±1.4 (7)	17.2±1.6 (3)
Abdominal muscle	18.6±3.6 (3)	17.9±1.0 (7)	20.0±1.1 (2)
Kidney	18.6±2.3 (4)	15.8±2.2 (7)	15.7±0.5 (3)
Lung	17.3±1.8 (3)	15.5±1.7 (7)	16.1±0.7 (3)
Liver	17.7±3.4 (4)	20.7±1.0 (7)	18.5±0.5 (3)
Spleen	19.1±2.0 (3)	19.3±0.9 (7)	18.3±0.8 (3)
Tendon	18.5±1.2 (3)	16.7±5.8 (4)	<u>6.5±1.8 (4)</u>
Heart	19.0±1.1 (3)	18.7±0.9 (4)	<u>14.8±0.3 (3)</u>
Aorta*	7.7±1.0 (4)	7.3±1.1 (6)	7.2±0.8 (4)
Skin	13.5±1.2 (3)	13.4±2.0 (4)	<u>9.1±0.9 (4)</u>
Uterus	14.5±1.5 (3)	9.3±2.5 (3)	11.6±3.1 (4)

Note: Values represent means ± standard deviation. Figures in parentheses represent number of animals or corresponding determinations. Figures showing significant differences within the 5% level of confidence are underlined.

*Each determination consists of pooled sample of 5 to 6 aortas.

Source: Reference 66.

The influence of age on insoluble collagen content of female rats is shown in Table 2. This fraction increased significantly, 50 to 100%, in abdominal muscle, tendon, aorta, skin, and uterus but decreased in cardiac muscle by two years of age. However, the insoluble collagen content of leg muscle, kidney, lung, liver, and spleen did not change with increasing age.

Table 2: Influence of Age on Insoluble Collagen Content of Various Tissues in Female Rat

Tissue	3 to 5 Weeks	8 Months	2 Years
		Age	
		g of insoluble collagen/100 g of tissue	
Upper leg muscle	1.0±0.5 (4)	1.2±0.3 (3)	1.0±0.4 (3)
Lower leg muscle	1.0±0.8 (4)	1.4±0.3 (4)	1.8±0.8 (3)
Abdominal muscle	1.5±0.2 (3)	1.7±0.4 (4)	2.2±0.4 (3)
Kidney	0.2±0.1 (4)	0.5±0.1 (7)	0.4±0.1 (3)
Lung	0.9±0.2 (4)	1.4±0.2 (5)	1.0±0.1 (3)
Liver	0.2±0.0 (3)	0.2±0.1 (7)	0.1±0.0 (3)
Spleen	0.8±0.2 (4)	0.5±0.1 (7)	0.6±0.3 (3)
Tendon	4.5±0.5 (3)	10.3±1.5 (4)	15.7±2.0 (4)
Heart	0.9±0.2 (3)	0.6±0.1 (4)	0.7±0.2 (3)
Aorta*	3.0±0.4 (4)	6.6±0.7 (6)	7.4±0.1 (4)
Skin	7.3±0.8 (3)	15.3±1.8 (4)	20.8±2.8 (4)
Uterus	0.8±0.1 (2)	3.3±0.9 (3)	3.2±0.5 (4)

Note: Values represent means ± standard deviation. Figures in parentheses represent number of animals or corresponding determinations. Figures showing significant differences within the 5% level of confidence are underlined.

*Each determination consists of pooled sample of 5 to 6 aortas.

Source: Reference 66.

Soluble collagen in the soluble protein fractions was calculated from their hydroxyproline content. Influence of age on the percentage of soluble collagen in total collagen in female rats is shown in Table 3. By two years of age, the percentage of soluble collagen of tendon decreased from 85% to approximately 20%; in aorta it decreased from 37% to 9%; in skin from 70% to 14%; and in uterus from 65% to 26%. Increased cross-linking of collagen and its resultant decreased solubility are indicative of maturation and the aging process. These results concurred with similar data from experiments with male rats.[66]

The effect of age on bone and cartilage protein fractions[70] is shown in Table 4. With increasing age soluble protein (or nonscleroprotein) and soluble collagen decreased in bone and cartilage. Insoluble collagen content of skull, femur, and rib did not change with

Table 3: Influence of Age on Percentage of Soluble Collagen in Total Collagen in Female Rat

	Age				
Tissue	3 Weeks	4 Weeks	5 Weeks	8 Months	2 Years
	g of soluble collagen/100 g of total collagen				
Tendon	85.6	82.1	68.0	48.8	18.6
Aorta*	37.4	29.2	36.6	17.6	9.0
Skin	70.0	45.1	32.0	25.7	14.0
Uterus	65.8	42.8	26.8	28.1	26.0

Note: Figures showing significant differences within 5% level of confidence are underlined.

*Each determination consists of pooled sample of 5 to 6 aortas.

Source: Reference 66.

Table 4: Influence of Age on Bone and Cartilage Protein Content of Female Rats

	Nonscleroprotein			Soluble Collagen		
	Age			Age		
	5 Weeks	2.5 Years		5 Weeks	2.5 Years	
Tissue	g/100 g dried tissue		P	g/100 g dried tissue		P
Skull	2.1±0.3	0.9±0.4	<0.001	0.51±0.10	0.28±0.07	<0.01
Femur	5.0±0.4	2.0±0.5	<0.001	0.70±0.05	0.36±0.07	<0.001
Rib	4.5±0.6	1.4±0.3	<0.001	0.64±0.10	0.35±0.06	<0.001
Vertebrae	7.0±0.7	3.5±1.6	<0.001	0.90±0.09	0.56±0.11	<0.001
Vertebral cartilage	4.9±1.4	4.0±1.1	—	1.28±0.27	0.86±0.05	<0.01
Costal cartilage	10.5±0.4	4.6±0.6	<0.001	1.41±0.10	0.71±0.18	<0.01

	Insoluble Collagen			Glycoprotein		
	Age			Age		
	5 Weeks	2.5 Years		5 Weeks	2.5 Years	
Tissue	g/100 g dried tissue		P	g/100 g dried tissue		P
Skull	18.4±1.7	17.8±0.8	—	1.71±0.13	1.20±0.05	<0.001
Femur	12.4±0.7	12.0±0.6	—	1.90±0.25	1.48±0.07	<0.01
Rib	15.7±1.0	16.3±1.5	—	1.73±0.15	1.20±0.07	<0.001
Vertebrae	14.0±0.9	15.8±0.8	<0.01	2.21±0.28	1.35±0.09	<0.001
Vertebral cartilage	25.4±2.4	22.1±1.8	<0.05	2.13±0.42	1.95±0.36	—
Costal cartilage	22.3±2.5	14.7±3.5	<0.02	3.01±0.59	5.94±0.27	<0.001

Note: Figures are mean ± standard deviation, calculated from 4 determinations each of 5-week-old skull and ribs; 3 determinations of 5-week-old rib cartilage and 6 determinations of all other tissues studied.

Source: Reference 70.

age, but it increased in vertebra and decreased in cartilage. A form of glycoprotein, not present in the connective tissues described previously, decreased in bone and increased in costal cartilage with increasing age.

In summary, during aging soluble protein and soluble collagen decreased while insoluble collagen increased in organs that were rich in connective tissue. An exception was the cartilages in which insoluble collagen decreased and glycoprotein concomitantly increased with age.

Protein Synthesis and Turnover During Aging

Protein synthesis and turnover of selected tissues have been studied (aorta, skin, tendon and uterus) by Kao and coworkers.[68] Rats at different ages were injected with a single dose of ^{14}C-lysine. In each tissue, incorporation of the radioactive amino acid was greatest in the soluble protein and decreased in order for soluble collagen, insoluble collagen, and elastin. The synthesis of collagen was greater in uterus, then skin, aorta, and tendon followed in decreasing order. Collagen turnover also was greatest in uterus followed by skin, but was greater in tendon than aorta.

The effect of age on the synthesis and turnover of proteins in these tissues has been studied in female rats[69] and the data are presented in Tables 5, 6, 7, and 8. Turnover rate of aortic noncleroprotein and soluble collagen (Table 5) was slower in 8-month- and 2-year-old rats than in 5-week-old rats. The relative specific activity 40 days after administration of ^{14}C-lysine was about 20% in the older animals and only 10% in the 5-week-old animal. Maximal specific activities of insoluble collagen and elastin in aorta were higher from 5-week-old rats than from 8-month- and 2-year-old rats. Turnover rates of insoluble collagen were similar in rats of all age groups studied. The specific activity of elastin from the 5-week-old rats decreased 30% forty days after ^{14}C-lysine administration. Very little synthesis and very little degradation were observed in the elastin of 8-month- and 2-year-old rats.

Synthesis and turnover of uterine proteins are shown in Table 6. The metabolism of nonscleroprotein and soluble collagen did not change with age. These fractions apparently were synthesized and catabolized efficiently. However, maximum specific activity of the insoluble collagen in rats at 8 months and 2 years was only twothirds and one-third, respectively of that at 5 weeks of age. Turnover rates were equal for rats at 5 weeks and 8 months. Forty days after administration of ^{14}C-lysine, the specific activities decreased to 20% of the maximal value. Turnover was slower in 2-year-old than in younger rats as indicated by a higher relative specific activity (40%) at the end of the experiment.

Previous work of ours has shown that a highly active protocollagen hydroxylase was present in porcine uterus.[75] This enzyme was capable of converting the unhydroxylated precursor, protocollagen, to collagen. Protocollagen hydroxylase activity decreased in porcine uterine homogenate with increasing age of the donor.[77] Furthermore we have shown that estrogens also stimulated synthesis and breakdown of collagen in rat uterine tissue[71,72,76] and in sponge biopsy connective tissue.[73]

Table 5: Synthesis and Turnover of Aorta Proteins of Female Rat (relative specific activity)*

Proteins	Age	No.	1	3	10	20	30	40
Nonscleroprotein	5 wk	1-3**	100±39 (222)	73±10	43±5	—	20	10
	8 mo	4-5	100±18 (154)	89±4	53±5	30±5	29±1	22±3
	2 yr	4	100±49 (259)	56±8	40±7	27±5	22±4	18±3
Soluble collagen	5 wk	1-3**	100±15 (140)	61±25	40±11	—	17	9
	8 mo	4-5	91±9	100±17 (77)	24±6	31±11	21±1	24±3
	2 yr	4	100±13 (198)	43±3	27±3	24±6	13±4	16±1
Insoluble collagen	5 wk	1-3**	71±12	93±30	100±12 (41)	—	78	54
	8 mo	4-5	100±61 (18)	78±44	57±6	57±6	57±11	57±17
	2 yr	4	64±21	100±7 (14)	50±7	29±21	32±7	43±28
Elastin	5 wk	1-3**	57±12	70±27	67±9	—	100 (33)	70
	8 mo	4-5	75±10	25±15	100±50 (4)	100±25	50±35	100±75
	2 yr	4	25±0	38±50	30±50	25±50	50±50	100±50 (2)

Column header group: ....Days After Injection..........

Note: Figures in parentheses are actual specific activity (cpm/mg of protein) observed.

*Assuming maximal specific activity as 100, the specific activity of the remaining days is expressed as % of maximal specific activity. Figures are mean ± standard deviation.
**Combined sample from 2 to 3 rats.

Source: Reference 69.

Table 7 presents the data on the synthesis and turnover of tendon proteins. No difference in the turnover of tendon nonscleroprotein was observed among groups. Values for maximal specific activity of soluble collagen in tendon at 8 months and 2 years were only 20% and 15% respectively, of the value at 5 weeks of age. Turnover rates of this fraction in tendons were also lower in 8-month- and 2-year-old than in 5-week-old rats. Insoluble collagen synthesis in the tendons of 8-month- and 2-year-old rats was extremely low and turnover was almost unmeasurable.

Table 6: Synthesis and Turnover of Uterine Proteins of Female Rat (relative specific activity)*

Proteins	Age	No.	Days After Injection					
			1	3	10	20	30	40
Nonscleroprotein	5 wk	1-3**	100±30 (328)	82±30	31±2	—	6	4±1
	8 mo	4-5	100±6 (386)	86±30	27±6	10±2	7±2	5±1
	2 yr	4	100±9 (560)	66±1	25±0	10±1	7±1	5±1
Soluble collagen	5 wk	1-3**	100±45 (220)	99±34	34±3	—	11±5	4±2
	8 mo	4-5	100±8 (262)	94±20	24±9	12±3	8±0	6±0
	2 yr	4	100±18 (330)	64±13	26±5	15±9	9±1	8±2
Insoluble collagen	5 wk	1-3**	56±22	100±67 (101)	88±21	—	30	19±4
	8 mo	4-5	34±8	100±52 (77)	33±13	38±14	35±4	21±6
	2 yr	4	43±28	100±46 (35)	86±28	100±32	76±20	40±14

Note: Figures in parentheses are actual specific activity (cpm/mg of protein) observed.

*Assuming maximal specific activity as 100, the specific activity of the remaining days is expressed as % of maximal specific activity. Figures are mean ± standard deviation.
**Combined sample from 2 to 3 rats.

Source: Reference 69.

Table 7: Synthesis and Turnover of Tendon Proteins of Female Rat (relative specific activity)*

Proteins	Age	No.	Days After Injection					
			1	3	10	20	30	40
Nonscleroprotein	5 wk	3-5**	69±9	100±10 (163)	56±14	—	20±2	13±2
	8 mo	4-5	100±14 (135)	95±8	52±7	27±7	21±2	18±6
	2 yr	4	100±46 (347)	46±8	22±2	19±1	13±9	9±1
Soluble collagen	5 wk	3-5**	48±4	100±18 (77)	66±18	—	34±4	20±5
	8 mo	4-5	18±8	92±38	100±39 (13)	77±15	54±15	77±30
	2 yr	4	90±30	100±60 (10)	70±80	50±50	30±20	40±10
Insoluble collagen	5 wk	3-5**	17±6	58±14	75±17	—	100±6 (36)	44±6
	8 mo	4-5	0	40±40	40±40	100±40 (2.5)	40±40	100
	2 yr	4	0	0	0	0	0	0

Note: Figures in parentheses are actual specific activity (cpm/mg of protein) observed.

*Assuming maximal specific activity as 100, the specific activity of the remaining days is expressed as % of maximal specific activity. Figures are mean ± standard deviation.
**Combined sample from 2 to 3 rats.

Source: Reference 69.

The data on the synthesis and turnover of skin protein are shown in Table 8. Age had no effect on the syntheses and turnover of nonscleroprotein of that tissue. However, syntheses of soluble and insoluble collagen in skin were lower, and the turnover rates of both fractions slower at 8 months and 2 years than at 5 weeks of age.

Table 8: Synthesis and Turnover of Skin Proteins of Female Rat (relative specific activity)*

Proteins	Age	No.	Days After Injection					
			1	3	10	20	30	40
Nonscleroprotein	5 wk	3–6**	100±17 (221)	87±8	41±9	–	10±1	7±1
	8 mo	4–5	100±16 (220)	87±15	30±11	19±4	13±1	10±2
	2 yr	4	100±32 (334)	61±10	34±6	19±2	9±3	8±3
Soluble collagen	5 wk	3–6**	100±8 (160)	80±17	40±11	–	10±1	9±2
	8 mo	4–5	100±26 (73)	96±22	43±11	34±6	18±3	16±3
	2 yr	4	100±38 (102)	47±12	26±5	20±5	17±4	15±5
Insoluble collagen	5 wk	3–6**	51±10	100±26 (61)	84±16	–	41±11	41±3
	8 mo	4–5	42±17	58±25	67±33	100±25 (12)	67±8	83±21
	2 yr	4	25±25	63±37	43±25	95±25	100±75 (4)	50±50

Note: Figures in parentheses are actual specific activity (cpm/mg of protein) observed.

*Assuming maximal specific activity as 100, the specific activity of the remaining days is expressed as % of maximal specific activity. Figures are mean ± standard deviation.
**Combined sample from 2 to 3 rats.

Source: Reference 69.

Collagen metabolism of the femur, rib, vertebra and skull has also been reported for female rats.[74] Maximal specific activity of femur soluble collagen was not different for the three age groups (Table 9). Turnover rate was apparently lower in the 2-year-old than in younger rats. Maximal specific activities and turnover rates of the soluble collagen in rib, vertebra and skull were slightly lower in 2-year-old rats than in 5-week- and 6-month-old rats. Insoluble collagen synthesis was very active in the bones of 5-week-old rats regardless of bone type. The rate of synthesis decreased with age. Specific activity of insoluble collagen in 5-week-old rats reached its maximum in 3 to 10 days and then decreased slowly. This maximum occurred later in 6-month- and 2-year-old rats. Virtually no degradation was apparent in the 2-year-old rats. Insoluble collagen metabolism was relatively less active in skull than in the other tissues studied.

Table 9: Collagen Metabolism of Femur, Rib, Vertebra and Skull of Female Rat* (specific activity)

Tissue	Age	Days After Injection					
		1	3	7	11	20	31
		cpm/mg nitrogen					
Femur							
Soluble collagen	5 wk	956±94**	569±54	300±9	138±13	81±9	56±4
	6 mo	1277±133	774±34	306±34	208±10	137±3	85±3
	2 yr	975	426	196	132	106	72
Insoluble collagen	5 wk	481±62	751±21	590±43	685±39	471±52	360±60
	6 mo	92±10	75±17	65±24	68±3	85±10	58±5
	2 yr	11	14	13	16	27	35
Rib							
Soluble collagen	5 wk	775±71	556±39	325±18	138±19	113±11	75±18
	6 mo	1010±89	611±38	331±17	331±31	222±55	130±14
	2 yr	507	365	192	140	115	106
Insoluble collagen	5 wk	593±14	620±11	656±12	633±44	538±75	380±37
	6 mo	129±17	120±20	95±20	72±3	99±10	62±7
	2 yr	22	15	19	17	44	31
Vertebra							
Soluble collagen	5 wk	988±119	663±69	363±20	169±22	113±10	69±9
	6 mo	1379±222	597±55	396±14	144±14	191±10	126±7
	2 yr	613	367	170	136	102	81
Insoluble collagen	5 wk	510±33	684±10	611±62	734±60	587±65	417±66
	6 mo	61±10	41±7	34±10	112±10	89±10	72±7
	2 yr	14	18	14	27	53	43
Skull							
Soluble collagen	5 wk	781±194	506±56	388±31	219±19	75±19	63±6
	6 mo	1044±96	732±65	365±49	296±7	341±17	123±10
	2 yr	379	183	85	115	64	111
Insoluble collagen	5 wk	288±37	403±8	459±24	476±37	421±36	348±51
	6 mo	38±7	38±10	30±7	85±3	100±10	68±10
	2 yr	11	14	13	21	21	21

*Five week- and 6 month-old rats were sacrificed in groups of four: 2 year-olds, in groups of two.
**Mean ± standard error.

Source: Reference 74.

Mokrasch and Manner[107] have studied protein synthesis in brain slices of rats by incorporation of ^{14}C-valine. Their data (Table 10) indicated a high perinatal synthesis. After 10 days of age synthesis continuously decreased until it reached a plateau at 90 days of age. It has since been reported that protein metabolism of muscle[106] also declined with increasing age.

Whole-body protein synthesis of rat declined with increases in either adult body size[110,156] or age.[156,170]

Whole-body protein metabolism of humans has been reviewed by Waterlow.[157] In a concerted effort to quantify the protein needs of the elderly population, whole body protein metabolism has been studied in the aged subject.[161,165] Protein synthesis and breakdown have been determined by use of ^{15}N-glycine infusion. After a constant level of isotope enrichment has been achieved, urinary urea was analyzed for ^{15}N. Turnover rates were calculated according to Picou and Taylor-Roberts.[121] Body protein synthesis declined rapidly during the first year of life.[165] The rate in young adults was about one-sixth of that in premature infants. During later adult years protein synthesis continued to decline but more gradually; the value for elderly subjects was 63% of that for young adults. Rates of total body protein synthesis and breakdown were lower in women than in men and significantly lower in elderly women than in young women.[161] Golden and Waterlow[48] have compared methods and labels for assessing total protein synthesis in the elderly. Their results also suggested that the rate of protein turnover was lower (20 to 30%) in elderly than in middle-aged or young adult subjects.

Table 10: ^{14}C-Valine Incorporation as a Function of Rat Age

Age (days)	Body Weight (g)	No. of Experiments	Amino Acid Incorporation (cpm/mg protein)
Fetus	3.75	3	27,950±4,950
2	10	4	31,240±4,770
4	13	3	34,420±2,360
7	20	8	25,560±13,100
10	30.5	3	8,923±3,610
16	34	2	3,418±1,250
25	72.5	3	3,570±3,110
28	87.5	2	5,755±318
90	177	3	462±242
600	532	4	300±113

Source: Reference 107.

Those investigators also evaluated their data in relation to muscle mass, based on urinary creatinine excretion, and to body cell mass, based on whole body ^{40}K determination. They have concluded that the decline of whole body protein synthesis and breakdown with age in man was associated with body cell loss and that the visceral organs make a progressively greater contribution to whole body protein metabolism than the skeletal musculature.[161] Urinary 3-methylhistidine excretion, also used as an index of muscle protein breakdown, confirmed that the rate of muscle protein breakdown per unit body weight was lower in the elderly than in the young adult.[149]

AGING-RELATED DISEASES AND PROTEIN METABOLISM

Both structural and functional proteins play prominent roles in health. Diseases that are prevalent among the elderly and involve structural and functional protein alterations include atherosclerosis/arteriosclerosis, osteoporosis, senile dementia, diabetes and cancer. The following discussion will be confined to those diseases that might be influenced by protein nutrition.

Atherosclerosis and Arteriosclerosis

Atherosclerosis and arteriosclerosis are the most common diseases of the western population that are associated with the aging process. Atherogenesis is a progressive disorder beginning early in life. Moon and his coworkers[108,109] reported that the degree of fibrosis of the arterial intima increases with age. Narrowing of the arterial lumen results in higher blood pressure and greater work load on the heart. Recently it has been reported that collagen increased and elastin decreased with aging in the arterial intima of humans.[64] The authors proposed that this chemical change causes the arterial intima to become more brittle and may be one of the first events in atherogenesis. In contrast venous intima became more flexible with age and the collagen/elastin ratio correspondingly decreased. Environmental factors can induce systolic injury and damage to the arterial intima, and consequentially, the development of atherosclerotic lesions. Atherosclerosis induced by environmental factors is considered reversible and curable.[7,128] The disease also appears to be genetically controlled and influenced by the sex hormones. It is more prevalent in the western than in the oriental populations, and is more common in males than in females. These factors, including age, however, are regarded as irreversible.[7]

Early in this century diet was implicated as an important environmental factor in the etiology of atherosclerosis. It has been demonstrated in animals that the protein component of the diet can affect incidence and severity of the disease. The role of dietary protein in atherogenesis has been recently reviewed by Carroll[22] and by Kritchevsky.[85] According to Carroll[22] plasma cholesterol level was reduced and atherosclerosis development inhibited by replacing dietary casein with isolated soy protein or mixed protein.[58] The role of the amino acid composition of these proteins on the development of atheromata has been investigated by Kritchevsky.[85] The ratios of arginine/lysine in casein and soy protein are 2.00 and 0.90 respectively. The addition of lysine to soy protein and arginine to casein changed the degree of atheromata and the level of serum cholesterol. In studies of amino acid mixtures, glutamate as the source of nonessential nitrogen, lowered serum cholesterol.[25,117,118] Additional research, particularly with humans, is needed to clarify the influence of dietary protein *per se* not only on blood serum cholesterol but also on blood lipoproteins. Ultimately it is the interactions of protein with the other components of the diet that influence the levels of blood lipids and lipoproteins and their effects on the arterial wall.[85]

Osteoporosis

Osteoporosis, another complex disease involving structural protein, is a chronic disease of aging and results in 40 to 50% loss of bone tissues between the fifth and ninth decades.[42] Characteristics of the disease are loss of cortical bone substance, reduced trabecular bone collagen and demineralization. Bone mass is reduced to such degree that mild physical stress can result in skeletal fractures. Yearly, osteoporosis causes 190,000 hip fractures, 180,000 vertebral fractures and 90,000 broken arms.[93] Total skin collagen is also decreased in osteoporotic subjects.[9] Although the disease occurs in both sexes it appears earlier and progresses more rapidly in females. In women it presents soon after menopause. Currently, investigators are studying the effects of lack of estrogen on the development of osteoporosis.[94]

The causes of the disease have been controversial for some time. Formerly, the primary phenomenon was considered impaired formation of the proteinaceous collagen matrix; poor mineralization was considered a secondary phenomenon. It is now evident that the primary defect is increased bone resorption and not decreased bone formation; i.e. increased collagen degradation and not decreased collagen synthesis.[42,127] Bone catabolism is excessive and an imbalance occurs in the dynamic equilibrium between bone formation and resorption.

Collagenase activity correlates well with bone resorption.[132] Collagenase synthesis and release apparently are associated with the loss of collagen during bone resorption. Bone metabolism as well as serum calcium homeostasis are controlled by two peptide hormones, calcitonin and parathyroid hormone (parathormone). Parathyroid hormone stimulates production of 1,25-dihydroxy vitamin D. Those two, in turn, induce bone resorption, the release of calcium into the blood stream and the reabsorption of calcium by the renal tubules.[29] In addition, increased production of 1,25-dihydroxy vitamin D results in increased intestinal calcium absorption. As a result serum calcium levels are increased. Calcitonin, on the other hand, inhibits bone resorption by blocking osteoclast function and induces diuresis, which increases urinary calcium excretion. Calcitonin does not change serum calcium levels. Parathyroid hormone has also been shown to depress collagen synthesis in osteoblast cultures.[119] Calcitonin on the other hand, decreases collagen degradation as indicated by its effect on reduction of urinary hydroxyproline-containing peptide.[84] Urinary hydroxyproline excretion has also been correlated with bone resorption rate.[136] Recently Milhaud et al.[104] have demonstrated that calcitonin levels in blood of osteoporotic patients are lower than in blood of control subjects. No differences in parathyroid hormone levels, however, were observed between the two groups. They hypothesize that the etiology of this disease could be related to a calcitonin deficiency.

The role of nutrition in the development or control of osteoporosis has not been clearly defined. One of several nutritional factors implicated in this structural protein disease is dietary protein. Sulfate production and excretion in the urine have been implicated as factors contributing to hypercalciuria.[160] Rats fed different sources of high-protein diets varying in their sulfur amino acid contents exhibited varying degrees of calciuria suggesting a relationship between extent of calciuria and sulfur amino acid content of the diet. Studies with humans have shown that hypercalciuria and negative calcium balance occur in the young adult,[2,23,63,154] in the elderly,[134] and in the osteoporotic subject[89] fed high-protein diets. It has been demonstrated that renal function was altered when hypercalciuria was induced by feeding a high-protein diet.[1,57,82] Altered parathyroid hormone function was not considered as the cause of hypercalciuria.[82] These investigators also observed an increase of urinary hydroxyproline excretion and speculated that the high-protein diet increased bone resorption. These investigations suggested that dietary management might be helpful for osteoporotic patients.

Senile Dementia

Senile dementia is, of course, particularly associated with age. Of individuals over 64 years of age, about 15% exhibit some degree of senile dementia.[155] Typical symptoms include memory loss, judgment deterioration, low initiative, and the inability to calculate or to make plans. The disease affects the functional protein of the central nervous system of the body and is most frequently attributed to arteriosclerosis. Autopsies, however, have shown that parenchymatous damage due to infarcts or atheroma accounted for only a relatively small percentage (20%) of the cases.[142]

During the course of normal aging, brain morphology is modified. Slight atrophy of the brain occurs;[27] the ratio of gray to white matter changes;[26] cortical neurons decrease in number;[14] dendrites and dendritic spines are lost.[36] Comparisons of brains between demented subjects and age-matched controls, however, have revealed no significant differences in the number and size of neurons between the two groups on the whole.[142] Nevertheless, neurofibrillary tangles and neuritic plaques appeared to be correlated with the presence of senile dementia.[81,95,141]

Recently, Buell and Coleman[19] have suggested that growth and not death was the predominant feature of the normal aged brain cell population. They examined specifically the pyramidal neurons of layer-II from parahippocampal gyrus. Brains were obtained at autopsy from the following three groups of individuals: neurologically normal aged, 68 to 92 years old; neurologically normal middle-aged, 44 to 55 years old; and demented patients, 70 to 86 years old. Of the two populations of neurons existing in the tissue, the dominant population in normal aged subjects was the neuron with an increased number of dendritic trees and increased length of terminal segments, and not the dying neuron with shrinking dendritic trees. Actually, dendritic trees were more extensive in the brains of the normal elderly group than in the brains of the middle-aged group. In the dementic brain, dendritic tree growth apparently fails during the course of aging. Terminal segments were fewer and shorter than in brains of normal subjects.

Chemically, 90% of the organic matter of brain is protein or lipoprotein. All complex functions performed by the brain depend on these materials in the role of information processing[126] and on glycoprotein and complex carbohydrates for information storage.[10,16,17,31] The glycoproteins are present at all cell surfaces and in the synaptic region, acting as receptors for transmitter action. For current information and literature review on the chemistry of the nervous system, the readers are advised to consult the *Neurochemistry of Amino Sugars*.[18]

Flexner[39] demonstrated the direct effect of protein metabolism on brain and nerve function. Mice had been trained to turn either right or left to avoid electric shock. Injection of puromycin, an inhibitor of protein synthesis, into the cerebrum produced amnesia with respect to the previous training.

Brain neurons synthesize, at their terminals of the nerve, low-molecular weight, water-soluble molecules known as neurotransmitters. These molecules are stored in the terminus and released into the synaptic cleft when the neuron is depolarized. Included among these relatively simple-structured chemicals are serotonin, dopamine and norepinephrine, metabolites of the amino acids, tryptophan and tyrosine, respectively. Their biochemical functions as neurotransmitters are very different. The influence of aging on the catecholamine (dopamine and norepinephrine) system has been well documented.[92] Dopamine and norepinephrine concentrations of different regions in the brain are lower in aged humans[131,133] and animals[38,105,131,133] than in the young. Tyrosine hydroxylase and aromatic L-amino acid decarboxylase activities, involved in the production of these neurotransmitters, are reduced in brains of senescent humans.[28,90,96] Moreover, the activities of two major enzymes involved in catecholamine catabolism, monoamine oxidase and catechol-o-methyltransferase increase with age in rats[138] and in humans.[131,145] Increased concentrations of homovanillic acid, the principal o-methylated deaminated dopamine, have been found in the hindbrain and in the cerebrospinal fluid[49,131] of elderly subjects.

The indoleamine neurotransmission system is likewise age-dependent. Tryptophan hydroxylase, the enzyme required for the synthesis of the neurotransmitter serotonin, is much lower in the brains of aged rats than in the brains of young rats.[99] Correspondingly, serotonin level is reduced in aged brains. Decreased net synthesis of indoleamine transmitters associated with aging is further supported by the observation that the concentration of the deaminated metabolite of serotonin, 5-hydroxyindole acetic acid, is increased in the hindbrain[131,145] as well as in the cerebrospinal fluid of elderly humans.[49] The association of impaired catecholamine or of indoleamine neurotransmission with psychological and behavioral changes of the elderly remains to be evaluated.

In the last few years, research on amino acid nutrition as related to brain function has clearly indicated the vital dependence of the brain on an adequate amino acid supply[37] for proper neurotransmission. Sufficient amounts of tryptophan or of tyrosine cannot be synthesized by the brain and must depend on the circulating plasma which in turn is chiefly dependent on the food ingested. Deficits of neurotransmitters of amino acid origin have been implicated in insomnia, some cases of depression, migraine headaches, schizophrenia and Parkinson's disease. Several of these conditions occur in the elderly population. Fernstrom and Wurtman[37] reported that a high carbohydrate diet induced the synthesis of serotonin in the brain of rats. The same group of investigators have also tried to use dietary precursors of neurotransmitters to treat brain disease.[53] Information on protein nutrition and brain biochemistry is extremely limited; research should be expanded to include the aged individual.

PROTEIN REQUIREMENTS AND AGING

Protein/Amino Acid Requirements

In two reviews known information on human protein and amino acid requirements was coordinated and evaluated.[60,61] The major emphasis of each conspectus was discussion of requirements as related to different stages of life: infancy, preschool, school-age, adolescence, young adulthood, and elderly; middle-aged was not named as a category. Delineation of this group by years is difficult and arbitrary, but middle-age is an integral part in the continuum of aging. Physiological and psychological stresses occurring during this period of life can significantly affect body metabolism. A considerable amount of information on requirements is available for infants and young adults but not for adolescents. A very few data are available on individuals between 35 and 64 years of age (an arbitrary division for middle-age), but are buried in the literature among the young adults or the elderly. Conclusions conflict concerning the protein needs of the elderly compared to the needs of the young adults. According to Irwin and Hegsted[60] most elderly people have diverse abnormalities that may be more important than age in determining their nutritional needs. The heterogeneity of this group thus imposes limitations on the nutritional studies. Nevertheless, the size of the elderly population is increasing and the "diverse" dietary needs of this "group" require better understanding.

Significant public and federal interest in nutrition and aging were generated about 10 years ago[150,158,159] and resulted in the expansion of federally-funded research programs in both of these areas. Addi-

Table 11: Protein/Amino Acid Requirements of Elderly Subjects

Investigating Institution	Number of Subjects	Age Range (yrs)	Sex	Weight Range (kg)	Height Range (cm)	Criterion	Experimental Periods (days)	Dietary Protein/Amino Acid Pattern	Physical Ailments	Suggested Intake (g/kg body wt/day)
Protein Requirements M.I.T.[135]	11	67–91	F	47–85	148–168	Factorial	8–10	Protein-free	Mild symptoms of arthritis, hypertension, occasional constipation; thyroid therapy (due to surgery)	0.42
M.I.T.[147]	8	68–72	M	56–105	160–177	Factorial	10	Protein-free	Some subjects recruited from chronic disease hospital	0.55
M.I.T.[148]	7	70–84	F	62–78	153–164	N balance	10	Whole egg	—	0.83
M.I.T.[148]	7	68–74	M	56–105	160–178	N balance	10	Whole egg	Peptic ulcer (partial gastrectomy) diuretic drug treatment; isomized	0.70–0.85
Berkeley, U. Cal.[171]	6	63–77	M	77–91	168–180	Factorial	17	Protein-free	Obese; no medications	Baseline info only
Berkeley, U. Cal.[171]	6	63–77	M	77–91	168–180	N balance	15	Egg white	Obese; no medications	0.59
Tryptophan Requirement M.I.T.[143]	10	64–83	F	55–88	149–163	N balance Plasma tryptophan	4	Egg	—	2
M.I.T.[143]	4	71–78	M	67–72	166–173	N balance Plasma tryptophan	4	Egg	—	2
Threonine Requirement M.I.T.[144]	12	62–82	F	50–87	145–173	N balance Plasma threonine	5	Egg	—	8
M.I.T.[144]	1	70	M	70	170	N balance Plasma threonine	5	Egg	—	8

tional studies have since been made with elderly subjects in attempts to define more clearly the protein/amino acid needs of people over 65 years of age. Excellent coverages on protein metabolism and nutrition of the aged have been published.[111,146,166-168]

Pertinent information and the results of the most recent investigations on protein and amino acid requirements of the elderly are summarized in Table 11. The suggested intakes for protein requirements range from 0.42 to 0.85 g/kg body weight/day. Superimposed on the heterogeneity of this population group were the differences among these studies with respect to criteria, and the design and conditions of the experiments. Determinations of nitrogen balance and obligatory nitrogen losses were used for estimating protein requirements. The limitations of these methods are well-known and have been reviewed.[55,56,123,169] According to researchers at the Massachusetts Institute of Technology (MIT) the factorial method grossly underestimated the protein needs of the elderly[148] and of young adults.[20,44,164] The MIT investigators suggested, on the extremely limited data, a protein intake of 1 g/kg body weight/day as a wise level for the elderly.[146] This intake was the first international standard. A lower value 0.59 g/kg body weight/day was proposed by the group at Berkeley.[171] Because conditions and types of individuals studied differed between the two research institutions, acceptance of one value over the other would not be justified.

In the last ten years, alternative approaches to nitrogen balance have been explored for estimating amino acid requirements. They included plasma amino acid response,[86,120,144,162,163] urinary sulfate excretion,[86,87] and urinary amino acid excretion.[32,86,87] The plasma amino acid response method has been used to estimate tryptophan[143] and threonine[144] requirements of the elderly. Based on the response curves in plasma the authors have concluded that the tryptophan requirement is about 2 mg/kg body weight/day for the elderly and about 3 mg/kg body weight/day for young adults, but that the threonine requirement did not differ significantly between elderly women and young men. Several other biochemical and physiological parameters now are measured in studies of requirements to further evaluate the physical and metabolic condition of the subjects. These parameters include blood enzymes, blood proteins, blood hormones, lean body mass, and basal metabolic rate. Young et al.[167] hypothesized that essential amino acid requirements per unit body cell mass may be higher for the elderly than for the young adult, since lean body mass decreased and the proportion of body fat increased during aging.

Dietary Survey

Another approach in assessing human protein requirements has been the evaluation of food consumption data of healthy populations. Two of the methods used in surveys, 24-hour recall[47] and dietary history[115] have been reported to overestimate the amount of protein consumed. Standards applied in food consumption surveys also varied, but the Recommended Dietary Allowance (RDA) set by the National Research Council, U.S. was used in most studies. The RDA itself however, has been revised periodically. For the population group over 55 the suggested protein intake has been lowered from 1 g/kg body weight/day in 1964 to 0.9 in 1968, and 0.8 in 1974. In the latest edition (1980) of the RDA the suggested protein allowance remained at the 1974 value.

Recently O'Hanlon and Kohrs[115] reviewed the publications on dietary surveys involving the aged. They concluded that the mean protein intake was adequate according to each survey, except for the Ten-State Survey. Table 12 provides information on protein and calorie intakes of "healthy" elderly individuals primarily living in noninstitutionalized settings. All surveys expressed the data as total intake/day. For women the RDA is 46 g and for men, 56 g (1974). None of the studies made during the last few years of escalating inflation indicated that the total daily protein intake of the elderly population was inadequate. Only a few investigators reported the data on a body weight basis and inadequacies were observed in some instances. In the Colorado study[62] 29% of the women, and in the Maine study,[65] 40% of the women and 12% of the men, were considered to be eating an insufficient amount of protein, based on the current accepted RDA of 0.8 g protein/kg body weight/day. In both of these studies serum total protein and albumin were measured to determine dietary protein adequacy. According to the Colorado investigators 38% of the elderly women had total serum protein values below the acceptable level and only 4% had serum albumin values below normal. Hair was also examined by the Maine investigators and this tissue appeared to be a more sensitive indicator of protein insufficiency than serum albumin or total protein.

Contributing to the generation of data on the elderly are reports from other industrialized nations[30,112,114,153] that were based on dietary surveys. The reports from Canada[112] and Sweden[114] indicated that the protein intake is lower for the aged than for the young adults of their countries. Whether the intakes can be considered poor depends, however, on the standard used.[112] In the Swedish study[11] the protein intakes were considered sufficient because the amino acid analysis of the diets indicated a high chemical EAA/NEAA (essential amino acid/nonessential amino acid) score for the protein consumed and blood serum protein levels were normal. In Great Britain[30] protein consumption of the elderly was also considered more than adequate by a standard for minimum protein requirement set at 39 g for aged men and 35 g for aged women by the United Kingdom in 1969. Information from a more recent study made in Northern Ireland[153] indicated that a small percentage of the aged women, 11%, had protein intakes of less than ⅔ RDI (recommended daily intake) of 34 g, which would be considered poor if eaten habitually.

Protein Nutrition and Aging

Table 12: Selected Dietary Surveys of the Elderly*

Investigating Institution	Number of Subjects	Age Range (yrs)	Sex	Intake Protein (g/day)	Intake Energy (kcal/day)	...% of RDA... Protein (%)	...% of RDA... Energy (%)	% of Subjects Eating < RDA for Protein on a Body Wt Basis
Colorado State Univ.[62]	24	62-99	F	54.8	1,381	119	77	29
Purdue Univ.[50]	26	52-89	F	77	1,846	167	103	—
	18	52-89	M	86	1,722	153	72	—
Penn. State Univ.[52]	49	<62	F	56.8	1,363	123	76	—
	15	52-89	M	80.4	1,801	144	75	—
Univ. of Rhode Island[15]	14	70-96	F	60.3	1,633	131	91	—
	9	72-90	M	72.6	2,166	130	88	—
Univ. of Maine[65]	78	65-89	F	57.9	1,356	126	75	40
	17	65-82	M	93.8	2,017	165	84	12
Univ. of Tenn.**[6]	4,262	<60	F	56	1,373***	122	76	—
	3,432	<60	M	76	1,888	136	79	—

*Surveys selected from those in which the subjects were not institutionalized. RDA for those over 55 years of age: Protein (females, 55 g; males, 65 g)[115]; Energy (females, 1,450 kcal; males 2,000 kcal).[115]
**Combined data from Household Food Consumption Survey (1965); Ten-State Nutrition Survey (1968-1970); Health and Nutrition Examination Survey (1971-1974).
***Data for 4,286 subjects.

Protein-Energy Interaction

It has been well-established and accepted that protein utilization and energy intake are closely interrelated. One cannot be considered without the other in nutrition. Amino acids not needed for the formation of structural tissue proteins, enzymes, etc. are converted in the liver to ammonia and keto acids. The ketone bodies and glucose formed from the keto acids are consequently integrated into fat and carbohydrate metabolism. Accordingly protein can also act as a source of energy for the body. Protein, therefore, is more readily utilized with an excess energy intake. Both protein and amino acid requirements have always been estimated on the basis of an ample supply of calories to assure maintenance of body weight. Interest has been renewed in the last few years concerning the interrelated impact of dietary protein-energy intake on protein requirement values.[21,43,45,46,59,83,113,124,125] Those investigators studied young adult males only.

Inoue et al.[59] reported that values for nitrogen protein utilization (NPU) and protein requirement for young men fed egg protein, were 63% and 0.46 g/kg body weight/day, respectively, with excess energy and, were 44% and 0.65 g/kg body weight/day, respectively, with maintenance energy. According to the studies of Calloway[21] energy intake has a greater effect on nitrogen balance than protein intake when either protein or energy intake was marginal. This has also been shown by Nageswara Rao and coworkers[113] with Indian men fed two adequate levels of mixed protein source (rice, milk and dal) with varying levels of energy. Dietary periods of these studies were relatively short, less than two weeks. Long-term studies allowing adaptation to the diet had been advocated.

Longer dietary periods have been used by Garza et al.[43,45,46] who fed varying amounts of energy with egg protein provided at the 1973 FAO/WHO safe level of 0.57 g/kg body weight/day. Excess energy was necessary at this level of protein intake for persistent positive nitrogen balance.[43] Further studies indicated that this level of protein was not sufficient for most healthy young men fed dietary energy intakes that maintained body weight on a long-term basis.[45] Most subjects actually gained weight when energy was provided at levels necessary for maintaining nitrogen balance. Moreover, they demonstrated that additional dietary nitrogen supplied as nonessential amino acids can decrease the amount of energy needed to support nitrogen equilibrium.[46] The ratio of carbohydrate to fat in the diet apparently was also important in protein utilization and nitrogen balance.[124] With a carbohydrate to fat ratio of 2:1 NPU was higher and nitrogen balance improved in young men fed milk protein. Replacement of dietary sucrose with dextrimaltose did not alter protein metabolism, nitrogen balance, or protein utilization.[125] Therefore, efficiency of nitrogen utilization apparently depended on total calories and total nitrogen consumed, energy requirements of the individual and on the ratio of carbohydrate to fat, but not on the type of dietary carbohydrate.

According to the dietary survey information the energy intake of the elderly population particularly of aged women, was generally below the RDA. Very likely protein is being utilized less efficiently and a protein intake of 1 g/kg body weight/day, as recently suggested by the MIT investigators,[146] might not be unrealistic for the elderly. Energy requirements decline continuously after 20 years of age.[13] Studies on energy-protein interaction are needed and well justified not only in the elderly but also in the middle-aged who generally are less active physically than young adults and are increasing body fat stores.

CONCLUDING REMARKS AND SUMMARY

In the process of aging, significant changes occur in body protein. The most prominent is the increase in the amount and the content of collagen in tissues rich in connective tissue. Furthermore, the intermolecular cross-links of collagen increase with age resulting in decreased solubility and increased tensile strength. Generally, rates of protein synthesis and degradation decrease with increasing age. The rates of whole body protein synthesis and degradation also decrease with increasing age. These changes may be among the causes of normal aging or age-related diseases. Since all tissue proteins originate from dietary protein, dietary adjustment might help modify the aging process.

The state of knowledge concerning the protein needs of the elderly can be considered in its infancy. Protein nutrition experimentation started more than a century ago and much information has been generated over the years. But, application and comparison of this information to the elderly population are relatively recent and questionable. The protein requirements for different ages have been determined on a cross-sectional basis. Aging, however, is a longitudinal process. Longitudinal observations in man are impractical in feeding experiments but can be realized in survey studies. One such attempt on a longitudinal basis was made by Ohlson and Harper[116] and provides some data on the eating habits of women from young adulthood through middle-age. The protein intakes of these women increased between ages 18 and 36 years and then decreased after 42 years of age to the values at 18 years. Caloric intakes, however, decreased from 18 to 36 years, remained constant until 42 years, and then further decreased at 56 years of age. Although longitudinal survey studies have many problems, as indicated by this report, their contribution to our understanding of nutrition and aging, or of nutrition and longevity can be valuable, particularly when combined with biochemical measurements and health assessment of the subjects.

REFERENCES

(1) Allen, L.H., Bartlett, R.S. and Block, G.D. 1979. Reduction of renal calcium reabsorption in man by consumption of dietary protein. *J. Nutr.* 109:1345-1350.

(2) Anand, C.R. and Linkswiler, H.M. 1974. Effect of protein intake on calcium balance of young men given 500 mg calcium daily. *J. Nutr.* 104:695-700.

(3) Bailey, A.J. and Peach, C.M. 1968. Isolation and structural identification of a labile intermolecular crosslink in collagen. *Biochem. Biophys. Res. Commun.* 33:812-819.

(4) Bailey, A.J., Fowler, L.J. and Peach, C.M. 1969. Identification of two interchain crosslinks of bone and dentine collagen. *Biochem. Biophys. Res. Commun.* 35:663-667.

(5) Bailey, A.J., Peach, C.M. and Fowler, L.J. 1970. Chemistry of the collagen cross-links. Isolation and characterization of two intermediate intermolecular cross-links in collagen. *Biochem. J.* 117:819-831.

(6) Beauchene, R.E. and Davis, T.A. 1979. The nutritional status of the aged in the U.S.A. *Age* 2:23-28.

(7) Bierman, E.L. 1978. Atherosclerosis and aging. *Fed. Proc., Fed. Am. Soc. Exp. Biol.* 37:2832-2836.

(8) Bjorksten, J. 1958. A common molecular basis for the aging syndrome. *J. Am. Geriatr. Soc.* 6:740-748.

(9) Black, M.M., Shuster, S. and Bottoms, E. 1970. Osteoporosis, skin collagen, and androgen. *Br. Med. J.* 4:773-774.

(10) Bogoch, S. 1968. Toward a comprehensive biochemical theory of memory. In *The Biochemistry of Memory*, pp. 213-218. Oxford Univ. Press, New York.

(11) Borgstrom, B., Norden, A., Akesson, B., Abdulla, M. and Jagerstad, M., eds. 1979. Nutrition and old age. Chemical analyses of what old people eat and their states of health during 6 years of follow-up. *Scand. J. Gastroenterol.* 14, Suppl. 52:13-274.

(12) Bornstein, P., Kang, A.H. and Piez, K.A. 1966. The nature and location of intermolecular cross-links in collagen. *Proc. Nat. Acad. Sci. U.S.A.* 55:417-424.

(13) Bray, G.A. 1977. Energy requirements of the aged. In Symposium on nutrition of the aged. *Nutr. Soc. of Canada*, pp. 45-52.

(14) Brody, H. 1955. Organization of the cerebral cortex. III. A study of aging in the human cerebral cortex. *J. Comp. Neurol.* 102:511-556.

(15) Brown, P.T., Bergan, J.G., Parsons, E.P. and Krol, I. 1977. Dietary status of elderly people. *J. Am. Diet. Assoc.* 71:41-45.

(16) Brunngraber, E.G. 1969a. The possible role of glycoproteins in neural function. *Perspect. Biol. Med.* 12:467-470.

(17) Brunngraber, E.G. 1969b. Glycoproteins. In A. Lajtha (ed.), *Handbook of Neurochemistry*, Vol, 1, pp. 223-244. Plenum Press, New York.

(18) Brunngraber, E.G. 1979. *Neurochemistry of Amino Sugars*, Charles C. Thomas, Springfield.

(19) Buell, S.J. and Coleman, P.D. 1979. Dendritic growth in the aged human brain and failure of growth in senile dementia. *Science* 206:854-856.

(20) Calloway, D.H. and Margen, S. 1971. Variation in endogenous nitrogen excretion and dietary nitrogen utilization as determinants of human protein requirement. *J. Nutr.* 101:205-216.

(21) Calloway, D.H. 1975. Nitrogen balance of men with marginal intakes of protein and energy. *J. Nutr.* 105:914-923.
(22) Carroll, K.K. 1978. The role of dietary protein in hypercholesterolemia and atherosclerosis. *Lipids* 13:360-365.
(23) Chu, J.-Y., Margen, S. and Costa, F.M. 1975. Studies in calcium metabolism. II. Effects of low calcium and variable protein intake on human calcium metabolism. *Am. J. Clin. Nutr.* 26:1028-1035.
(24) Chvapil, M. and Hruza, Z. 1959. The influence of aging and undernutrition on chemical contractility and relaxation of collagen fibers in rats. *Gerontologia* 3:241-252.
(25) Coles, B.L. and Macdonald, I. 1972. The influence of dietary protein on dietary carbohydrate: lipid interrelationships. *Nutr. Metab.* 14:238-244.
(26) Corsellis, J.A.N. 1976. Some observations on the purkinje cell population and on brain volume in human aging. In R.D. Terry and S. Gershon (eds.), *Neurobiology of Aging, Aging* Vol. 3, pp. 205-209, Raven Press, New York.
(27) Corsellis, J.A.N. 1978. Posttraumatic dementia. In R. Katzman, R.D. Terry and K.L. Bick, (eds.), *Alzheimer's Disease-Senile Dementia and Related Disorders, Aging* Vol. 7, pp. 125-133, Raven Press, New York.
(28) Cote, L.J. and Kremzner, L.T. 1974. Changes in neurotransmitter systems with increasing age in human brain. *Trans. Am. Soc. Neurochem.* 5:83.
(29) DeLuca, H.F. 1979. Vitamin D metabolism and function. In F. Gross, A. Labhart, T. Mann and J. Zander, (eds.), *Monographs on Endocrinology*, Vol. 13, pp. 30-38, Springer-Verlag, New York.
(30) Dept. of Health and Soc. Security. 1972. A nutrition survey of the elderly. Reported by the panel on nutrition of the elderly. Reports on Health and Social Subjects No. 3, pp. 28-29, Her Majesty's Stationery Office.
(31) Dische, Z. 1966. In H. Peeters (ed.), *Peptides of the Biological Fluids*, Vol. 13, pp. 1-20. Elsevier Publishing Co., Amsterdam.
(32) Doherty, R.F.; Bodwell, C.E., Vaughan, D., Young, V.R., Perera, W.D.A. and Scrimshaw, N.S. 1976. Urinary excretion of branched chain amino acids in humans at graded intake levels of leucine. *Fed. Proc., Fed. Am. Soc. Exp. Biol.* 35:261.
(33) Everitt, A.V., Gal, A. and Steele, M.G. 1970. Age changes in the solubility of tail tendon collagen throughout the lifespan of the rat. *Gerontologia* 16:30-40.
(34) Everitt, A.V. 1971. Food intake, growth and the ageing of collagen in rat tail tendon. *Gerontologia* 17:98-104.
(35) Eyre, D.R. 1980. Collagen: molecular diversity in the body's protein scaffold. *Science* 207:1315-1322.
(36) Feldman, M.L. and Dowd, C. 1974. Aging in rat visual cortex: light microscopic observations on layer V pyramidal apical dendrites. *Anat. Rec.* 178:355.
(37) Fernstrom, J.D. and Wurtman, R.J. 1974. Nutrition and the brain. *Sci. Am.* 230:84-91.
(38) Finch, C.E. 1973. Catecholamine metabolism in the brains of ageing male mice. *Brain Res.* 52:261-276.
(39) Flexner, J.B., Flexner, L.B. and Stellar, E. 1963. Memory in mice as affected by intracerebral puromycin. *Science* 141:57-59.

(40) Gallop, P.M., Seifter, S. and Meilman, E. 1959. Occurrence of 'ester-like' linkages in collagen. *Nature* 183:1659-1661.
(41) Gallop, P.M., Blumenfeld, O.O. and Seifter, S. 1972. Structure and metabolism of connective tissue proteins. *Ann. Rev. Biochem.* 41: 617-672.
(42) Garn, S.M. 1975. Bone-loss and aging. In R. Goldman and M. Rockstein (eds.), *The Physiology and Pathology of Human Aging*, pp. 39-57. Academic Press, New York.
(43) Garza, C., Scrimshaw, N.S. and Young, V.R. 1976. Human protein requirements: the effect of variations in energy intake within the maintenance range. *Am. J. Clin. Nutr.* 29:280-287.
(44) Garza, C., Scrimshaw, N.S. and Young, V.R. 1977a. Human protein requirements: a long-term metabolic nitrogen balance study in young men to evaluate the 1973 FAO/WHO safe level of egg protein intake. *J. Nutr.* 107:335-352.
(45) Garza, C., Scrimshaw, N.S. and Young, V.R. 1977b. Human protein requirements: evaluation of the 1973 FAO/WHO safe level of protein intake for young men at high energy intakes. *Br. J. Nutr.* 37:403-420.
(46) Garza, C., Scrimshaw, N.S. and Young, V.R. 1978. Human protein requirements: interrelationships between energy intake and nitrogen balance in young men consuming the 1973 FAO/WHO safe level of egg protein, with added nonessential amino acids. *J. Nutr.* 108:90-96.
(47) Gersovitz, M., Madden, J.P. and Smiciklas-Wright, H. 1978. Validity of *J. Am. Diet. Assoc.* 73:48-55.
(48) Golden, M.H.N. and Waterlow, J.C. 1977. Total protein synthesis in elderly people: a comparison of results with [^{15}N] glycine and [^{14}C] leucine. *Clin. Sci. Mol. Med.* 53:277-288.
(49) Gottfries, C.G., Gottfries, I., Johansson, B., Olsson, R., Persson, T., Roos, B.-E. and Sjostrom, R. 1971. Acid monoamine metabolites in human cerebrospinal fluid and their relations to age and sex. *Neuropharmacology* 10:665-672.
(50) Greger, J.L. and Sciscoe, B.S. 1977. Zinc nutriture of elderly participants in an urban feeding program. *J. Am. Diet. Assoc.* 70:37-41.
(51) Gross, J. 1958. Studies on the formation of collagen. II. The influence of growth rate on neutral salt extracts of guinea pig dermis. *J. Exp. Med.* 107:265-277.
(52) Grotkowski, M.L. and Sims, L.S. 1978. Nutritional knowledge, attitudes, and dietary practices of the elderly. *J. Am. Diet. Assoc.* 72:499-506.
(53) Growdon, J.H., Cohen, E.L. and Wurtman, R.J. 1977. Treatment of brain disease with dietary precursors of neurotransmitters. *Ann. Intern. Med.* 86:337-339.
(54) Hamlin, C.R. and Kohn, R.R. 1972. Determination of human chronological age by study of a collagen sample. *Exp. Gerontol.* 7:377-379.
(55) Hegsted, D.M. 1976. Balance studies. *J. Nutr.* 106:307-311.
(56) Hegsted, D.M. 1978. Assessment of nitrogen requirements. *Am. J. Clin. Nutr.* 31:1669-1677.
(57) Hegsted, M. and Linkswiler, H.M. 1980. The long term effect of level of protein intake on calcium balance in young adult women. *Fed. Proc., Fed. Am. Soc. Exp. Biol.* 39:901.
(58) Hermus, R.J.J. 1975. Experimental atherosclerosis in rabbits on diets with milk fat and different proteins. *Agric. Res. Rep.* 838, Centre for Agricultural Publishing and Documentation, Wagenigen, The Netherlands.

(59) Inoue, G., Fujita, Y. and Niiyama, Y. 1973. Studies on protein requirements of young men fed egg protein and rice protein with excess and maintenance energy intakes. *J. Nutr.* 103:1673-1687.
(60) Irwin, M.I. and Hegsted, D.M. 1971a. Monograph: A conspectus of research on protein requirements of man. *J. Nutr.* 101:385-429.
(61) Irwin, M.I. and Hegsted, D.M. 1971b. Monograph: A conspectus of research on amino acid requirements of man. *J. Nutr.* 101:539-566.
(62) Jansen, C. and Harrill, I. 1977. Intakes and serum levels of protein and iron for 70 elderly women. *Am. J. Clin. Nutr.* 30:1414-1422.
(63) Johnson, N.E., Alcantara, E.N. and Linkswiler, H.M. 1970. Effect of level of protein intake on urinary and fecal calcium and calcium retention of young adult males. *J. Nutr.* 100:1425-1430.
(64) Johnson, W.T.M., Himelstein, A.L., Farmer, D.B. and Horwitz, O. 1980. Chemical changes on aging in aortic and venous intima in relationship to atherosclerosis. *Fed. Proc., Fed. Am. Soc. Exp. Biol.* 39:428.
(65) Jordan, V.E. 1976. Protein status of the elderly as measured by dietary intake, hair tissue, and serum albumin. *Am. J. Clin. Nutr.* 29:522-528.
(66) Kao, K.Y.T. and McGavack, T.H. 1959. Connective tissue I. Age and sex influence on protein composition of rat tissue. *Proc. Soc. Exp. Biol. Med.* 101:153-157.
(67) Kao, K.Y.T., Hilker, D.M. and McGavack, T.H. 1960. Connective tissue III. Collagen and hexosamine content of tissues of rats of different ages. *Proc. Soc. Exp. Biol. Med.* 104:359-361.
(68) Kao, K.Y.T., Hilker, D.M. and McGavack, T.H. 1961a. Connective tissue IV. Synthesis and turnover of proteins in tissues of rats. *Proc. Soc. Exp. Biol. Med.* 106:121-124.
(69) Kao, K.Y.T., Hilker, D.M. and McGavack, T.H. 1961b. Connective tissue V. Comparison of synthesis and turnover of collagen and elastin in tissues of rat at several ages. *Proc. Soc. Exp. Biol. Med.* 106:335-338.
(70) Kao, K.Y.T., Hitt, W.E., Dawson, R.L. and McGavack, T.H. 1962. Connective tissue VII. Changes in protein and hexosamine content of bone and cartilage of rat at different ages. *Proc. Soc. Exp. Biol. Med.* 110:538-543.
(71) Kao, K.Y.T., Hitt, W.E., Bush, A.T. and McGavack, T.H. 1964a. Connective tissue XI. Factors affecting collagen synthesis by rat uterine slices. *Proc. Soc. Exp. Biol. Med.* 115:422-424.
(72) Kao, K.Y.T., Hitt, W.E., Bush, A.T. and McGavack, T.H. 1964b. Connective tissue XII. Stimulating effects of estrogen on collagen synthesis in rat uterine slices. *Proc. Soc. Exp. Biol. Med.* 117:86-97.
(73) Kao, K.Y.T., Hitt, W.E. and McGavack, T.H. 1965a. Connective tissue XIII. Effect of estradiol benzoate upon collagen synthesis by sponge biopsy connective tissue. *Proc. Soc. Exp. Biol. Med.* 119:364-367.
(74) Kao, K.Y.T., Vernier, C.M. and McGavack, T.H. 1965b. Connective tissue IX. Metabolism of collagen in bone of rat. *Proc. Soc. Exp. Biol. Med.* 119:584-587.
(75) Kao, K.Y.T., Treadwell, C.R., Previll, J.M. and McGavack, T.H. 1968. Connective tissue XVII. Protocollagen hydroxylase of pig uterus. *Biochim. Biophy. Acta* 151:568-572.
(76) Kao, K.Y.T., Arnett, W.M. and McGavack, T.H. 1969a. Effect of endometrial phase, ovariectomy, estradiol and progesterone on uterine protocollagen hydroxylase. *Endocrinology* 85:1057-1061.

(77) Kao, K.Y.T. and McGavack, T.H. 1969b. Connective tissue XVIII. Age differences in protocollagen hydroxylase of porcine uterine homogenate. *Proc. Soc. Exp. Biol. Med.* 130:491-492.
(78) Kao, K.Y.T. and Hitt, W.E. 1974. The intermolecular cross-links in rat uterine collagen. *Biochim. Biophys. Acta* 371:501-510.
(79) Kao, K.Y.T., Hitt, W.E. and Leslie, J.G. 1976. The intermolecular cross-links in uterine collagens of guinea pig, pig, cow, and human beings. *Proc. Soc. Exp. Biol. Med.* 151:385-389.
(80) Kao, K.Y.T. and Leslie, J.G. 1979. Intermolecular cross-links in collagen of human placenta. *Biochim. Biophys. Acta* 580:366-371.
(81) Kidd, M. 1963. Paired helical filaments in electron microscopy in Alzheimer's disease. *Nature* 197:192-193.
(82) Kim, Y. and Linkswiler, H.M. 1979. Effect of level of protein intake on calcium metabolism and on parathyroid and renal function in the adult human male. *J. Nutr.* 109:1399-1404.
(83) Kishi, K., Miyatani, S. and Inoue, G. 1978. Requirement and utilization of egg protein by Japanese young men with marginal intakes of energy. *J. Nutr.* 108:658-669.
(84) Krane, S.M., Harris, E.D., Jr., Singer, F.R. and Potts, J.T., Jr. 1973. Acute effect of calcitonin on bone formation in man. *Metab., Clin. Exp.* 22:51-58.
(85) Kritchevsky, D. 1979. Vegetable protein and atherosclerosis. *J. Am. Oil Chem. Soc.* 56:135-140.
(86) Lakshmanan, F.L., Perera, W.D.A., Scrimshaw, N.S. and Young, V.R. 1976. Plasma and urinary amino acids and selected sulfur metabolites in young men fed a diet devoid of methionine and cystine. *Am. J. Clin. Nutr.* 29:1367-1371.
(87) Lakshmanan, F.L., Vaughan, D.A. and Barnes, R.E. 1978. Urinary inorganic sulfate excretion in human subjects fed purified amino acid mixtures with and without threonine. *Fed. Proc., Fed. Am. Soc. Exp. Biol.* 37:359.
(88) Lansing, A.I. 1955. Ageing of elastic tissue and the systemic effects of elastase. In G.E.W. Wolstenholme and M.G. Cameron (eds.), *Ciba Found. Colloq. on Ageing*, pp. 88-10 ʾ. Little, Brown and Co., Boston.
(89) Licata, A.A., Bou, E. and Bartter, F.C. 1976. Effects of dietary protein on calcium metabolism in normal subjects and in patients with osteoporosis. In fifty-seventh annual session of the American College of Physicians, pp. 5-8. Philadelphia.
(90) Lloyd, K.G. and Hornykiewicz, O. 1972. Occurrence and distribution of aromatic L-amino acid (L-dopa) decarboxylase in human brain. *J. Neurochem.* 19:1549-1559.
(91) Ma, C.K. and Cowdry, E.V. 1950. Aging of elastic tissue in human skin. *J. Gerontol.* 5:203-210.
(92) Makman, M.H., Ahn, H.S., Thal, L.J., Sharpless, N.S., Dvorkin, B., Horowitz, S.G. and Rosenfeld, M. 1979. Aging and monoamine receptors in brain. *Fed. Proc., Fed. Am. Soc. Exp. Biol.* 38:1922-1926.
(93) Marx, J.L. 1979. Hormones and their effects in the aging body. *Science* 206:805-806.
(94) Marx, J.L. 1980. Osteoporosis: new help for thinning bones. *Science* 207:628-630.

(95) Matsuyama, H., Namiki, H. and Watanabe, I. 1966. Senile changes in the brain in the Japanese. Incidence of Alzheimer's neurofibrillary change and senile plaques. In F. Luthy and A. Bischoff (eds.) *Proceedings of the Fifth International Congress of Neuropathology*, pp. 979-980. Excerpta Medica Series #100, Amsterdam.

(96) McGeer, E.G. and McGeer, P.L. 1975. Age changes in the human for some enzymes associated with metabolism of catecholamines, GABA, and acetylcholine. In J.M. Ordy and K.R. Brizzee (ed.), *Neurobiology of Aging, Advances in Behavioral Biology*, Vol. 16, pp. 287-305. Plenum Press, New York.

(97) Mechanic, G. and Tanzer, M.L. 1970. Biochemistry of collagen cross-linking. Isolation of a new crosslink, hydroxylysinohydroxynorleucine, and its reduced precursor, dihydroxynorleucine, from bovine tendon. *Biochem. Biophys. Res. Commun.* 41:1597-1604.

(98) Mechanic, G.L., Gallop, P.M. and Tanzer, M.L. 1971. The nature of crosslinking in collagens from mineralized tissues. *Biochem. Biophys. Res. Commun.* 45:644-653.

(99) Meek, J.L., Bertilsson, L., Cheney, D.L., Zsilla, G. and Costa, E. 1977. Aging-induced changes in acetylcholine and serotonin content of discrete brain nuclei. *J. Gerontol.* 32:129-131.

(100) Meyer, A. and Verzar, F. 1959. Altersveranderungen der hydroxyprolinabgabe bei der thermischen kontraktion von kollagenfasern. *Gerontologia* 3:184-203.

(101) Milch, R.A. and Murray, R.A. 1962. Studies of collagen tissue aging: thermal shrinkage of metabolite-treated collagenous tissues. *Proc. Soc. Exp. Biol. Med.* 111:551-554.

(102) Milch, R.A., Murray, R.A. and Kenmore, P.I. 1962. Studies of collagen tissue aging: degradation of glyceraldehyde-treated hide collagen. *Proc. Soc. Exp. Biol. Med.* 111:554-556.

(103) Milch, R.A. 1963. Studies of collagen tissue aging: interaction of certain intermediary metabolites with collagen. *Gerontologia* 7:129-152.

(104) Milhaud, G., Benezech-Lefevre, M. and Moukhtar, M.S. 1979. Deficiency of calcitonin in age related osteoporosis. *Biomed. Express (Paris)* 29:272-276.

(105) Miller, A.E., Shaar, C.J. and Riegle, G.D. 1976. Aging effects on hypothalamic dopamine and norepinephrine content in the male rat. *Exp. Aging Res.* 2:475-480.

(106) Millward, D.J., Garlick, P.J., Stewart, R.J., Nnanyelugo, D.O. and Waterlow, J.C. 1975. Skeletal-muscle growth and protein turnover. *Biochem. J.* 150:235-243.

(107) Mokrasch, L.C. and Manner, P. 1963. Incorporation of ^{14}C-amino acids and ^{14}C-palmitate into proteolipids of rat brains in vitro. *J. Neurochem.* 10:541-547.

(108) Moon, H.D. and Rinehart, J.B. 1952. Histogenesis of coronary arteriosclerosis. *Circulation* 6:481-488.

(109) Moon, H.D. 1957. Coronary arteries in fetuses, infants and juveniles. *Circulation* 16:263-267.

(110) Munro, H.N. 1969. *Mammalian Protein Metabolism*, vol. 3, 158-173. Academic Press, New York.

(111) Munro, H.N. and Young, V.R. 1978. Protein metabolism in the elderly, observations relating to dietary needs. *Postgrad. Med.* 63:143-152.

(112) Murray, T.K. and Nielsen, H. 1978. Nutritional status of aged Canadians. *Can. Home Econ. J.* 6-12.

(113) Nageswara Rao, C., Nadamuni Naidu, A. and Narasinga Rao, B.S. 1975. Influence of varying energy intake on nitrogen balance in men on two levels of protein intake. *Am. J. Clin. Nutr.* 28:1116-1121.
(114) Nair, B.M. and Andersson, I. 1978. Quantitative and qualitative evaluations of protein intake in a geriatric subpopulation from a southern Swedish community. *Am. J. Clin. Nutr.* 31:1280-1289.
(115) O'Hanlon, P. and Kohrs, M.B. 1978. Dietary studies of older Americans. *Am. J. Clin. Nutr.* 31:1257-1269.
(116) Ohlson, M.A. and Harper, L.J. 1976. Longitudinal studies of food intake and weight of women from ages 18 to 56 years. *J. Am. Diet. Assoc.* 69:626-631.
(117) Olson, R.E., Bazzano, G. and D'Elia, J.A. 1970a. The effects of large amounts of glutamic acid upon serum lipids and sterol metabolism in man. *Trans. Assoc. Am. Physicians* 83:196-210.
(118) Olson, R.E., Nichaman, M.Z., Nittka, J. and Eagles, J.A. 1970b. Effect of amino acid diets upon serum lipids in man. *Am. J. Clin. Nutr.* 23:1614-1625.
(119) Parfitt, A.M. 1976. The actions of parathyroid hormone on bone: relation to bone remodeling and turnover, calcium homeostasis, and metabolic bone disease. Part III of IV parts. PTH and osteoblasts, the relationship between bone turnover and bone loss, and the state of bones in primary hyperthyroidism. *Metab., Clin. Exp.* 25:1033-1069.
(120) Perera, W.D.A., Miller, M., Bilmazes, C., Young, V.R. and Scrimshaw, N.S. 1974. Plasma amino acid approach for estimation of the leucine requirements in young men. *Fed. Proc., Fed. Am. Soc. Exp. Biol.* 33:683.
(121) Picou, D. and Taylor-Roberts, T. 1969. The measurement of total protein synthesis and catabolism and nitrogen turnover in infants in different nutritional states and receiving different amounts of dietary protein. *Clin. Sci.* 36:283-296.
(122) Piez, K.A. 1968. Cross-linking of collagen and elastin. *Ann. Rev. Biochem.* 37:547-570.
(123) Rand, W.M., Scrimshaw, N.S. and Young, V.R. 1977. Determination of protein allowance in human adults from nitrogen balance data. *Am. J. Clin. Nutr.* 30:1129-1134.
(124) Richardson, D.P., Wayler, A.H., Scrimshaw, N.S. and Young, V.R. 1979. Quantitative effect of an isoenergetic exchange of fat for carbohydrate on dietary protein utilization in healthy young men. *Am. J. Clin. Nutr.* 32:2217-2226.
(125) Richardson, D.P., Scrimshaw, N.S. and Young, V.R. 1980. The effect of dietary sucrose on protein utilization in healthy young men. *Am. J. Clin. Nutr.* 33:264-272.
(126) Richter, D. 1970. In A. Lajtha (ed.). *Protein Metabolism Of The Nervous System.* pp. xiii-xv. Plenum Press, New York.
(127) Robbins, S.L. and Angell, M. 1971. *Basic Pathology*, pp. 535-536, W.B. Saunders Co., Philadelphia.
(128) Roberts, J.C., Jr. and Strauss, R. (eds.) 1965. *Comparative Atherosclerosis.* Harper and Row, New York.
(129) Robins, S.P. and Bailey, A.J. 1972. Age-related changes in collagen: the identification of reducible lysine-carbohydrate condensation products. *Biochem. Biophys. Res. Commun.* 48:76-84.

(130) Robins, S.P., Shimokomaki, M. and Bailey, A.J. 1973. The chemistry of the collagen cross-links. Age-related changes in the reducible components of intact bovine collagen fibers. *Biochem. J.* 131:771-780.
(131) Robinson, D.S., Nies, A., Davis, J.N., Bunney, W.E., Davis, J.M., Colburn, R.W., Bourne, H.R., Shaw, D.M. and Coppen, A.J. 1972. Ageing, monoamines, and monamine-oxidase levels. *Lancet* 1:290-291.
(132) Sakamoto, S., Sakamoto, M., Goldhaber, P. and Glimcher, M. 1975. Collagenase and the bone resorption: Isolation of collagenase from culture medium containing serum after stimulation of bone resorption by addition of parathyroid hormone extract. *Biochem. Biophys. Res. Comm.* 63:172-178.
(133) Samorajski, T. and Rolsten, C. 1973. Age and regional differences in the chemical composition of brains of mice, monkeys and humans. *Prog. Brain Res.* 40:262-265.
(134) Schuette, S.A., Zemel, M.B. and Linkswiler, H.M. 1980. Studies on the mechanism of protein-induced hypercalciuria in older men and women. *J. Nutr.* 110:305-315.
(135) Scrimshaw, N.S., Perera, W.D.A. and Young, V.R. 1976. Protein requirements of man: obligatory urinary and fecal nitrogen losses in elderly women. *J. Nutr.* 106:665-670.
(136) Sjoerdsma, A., Udenfriend, S., Keiser, H. and LeRoy, E.C. 1965. Hydroxyproline and collagen metabolism. Clinical implications. *Ann. Intern. Med.* 63:672-694.
(137) Sobel, H., Gabay, S., Wright, E.T., Lichenstein, I. and Nelson, N.H. 1958. The influence of age upon the hexosamine-collagen ratio of dermal biopsies from men. *J. Gerontol.* 13:128-131.
(138) Stramentinoli, G.M., Gualano, M., Catto, E. and Algeri, S. 1977. Tissue levels of S-adenosylmethionine in aging rats. *J. Gerontol.* 32:392-394.
(139) Tanzer, M.L. and Mechanic, G. 1970. Isolation of lysinonorleucine from collagen. *Biochem. Biophys. Res. Commun.* 39:183-189.
(140) Tattersall, R.N. and Seville, R. 1950. Senile purpura. *Q. J. Med.* 19:151-159.
(141) Terry, R.D. 1963. The fine structure of neurofibrillary tangles in Alzheimer's disease. *J. Neuropathol. Exp. Neurol.* 22:629-642.
(142) Terry, R.D. 1978. Senile dementia. *Fed. Proc., Fed. Am. Soc. Exp. Biol.* 37:2837-2840.
(143) Tontisirin, K., Young, V.R., Miller, M. and Scrimshaw, N.S. 1973. Plasma tryptophan response curve and tryptophan requirements of elderly people. *J. Nutr.* 103:1220-1228.
(144) Tontisirin, K., Young, V.R., Rand, W.M. and Scrimshaw, N.S. 1974. Plasma threonine response curve and threonine requirements of young men and elderly women. *J. Nutr.* 104:495-505.
(145) Tryding, N., Tufvesson, G. and Nilsson, S. 1972. Ageing, monoamines, and monoamine-oxidase levels. *Lancet* 1:489.
(146) Uauy, R., Scrimshaw, N.S. and Young, V.R. 1977. Human protein metabolism in relation to nutrient needs in the aged. In *Nutrition of the Aged. Proc. of a Symposium by the Nutrition Society of Canada*, pp. 53-71.
(147) Uauy, R., Scrimshaw, N.S. and Young, V.R. 1978a. Human protein requirements: nitrogen balance response to graded levels of egg protein in elderly men and women. *Am. J. Clin. Nutr.* 31:779-785.

(148) Uauy, R., Scrimshaw, N.S., Rand, W.M. and Young, V.R. 1978b. Human protein requirements: obligatory urinary and fecal nitrogen losses and the factorial estimation of protein needs in elderly males. *J. Nutr.* 108:97-103.
(149) Uauy, R., Winterer, J.C., Bilmazes, C., Haverberg, L.N., Scrimshaw, N.S., Munro, H.N. and Young, V.R. 1978c. The changing pattern of whole body protein metabolism in ageing humans. *J. Gerontol.* 33:663-671.
(150) U.S. Senate. Nutrition and Human Needs. Part 14. Nutrition and the aged. Hearings before the select committee on nutrition and human needs, 90th Congress, 2nd Session, and 91st Congress, 1st Session, September 9, 10, and 11, 1969. Washington, D.C.: Government Printing Office, 1969. pp. 5287-5292.
(151) Verzar, F. 1963. The aging of collagen. *Sci. Am.* 208:104-114.
(152) Verzar, F. 1968. Intrinsic and extrinsic factors of molecular aging. *Exp. Gerontol.* 3:69-75.
(153) Vir, S.C. and Love, A.H.G. 1979. Nutritional status of institutionalized and noninstitutionalized aged in Belfast, Northern Ireland. *Am. J. Clin. Nutr.* 32:1934-1947.
(154) Walker, R.M. and Linkswiler, H.M. 1972. Calcium retention in the adult human male as affected by protein intake. *J. Nutr.* 102:1297-1302.
(155) Wang, H.S. 1977. Dementia of old age. In W.L. Smith and M. Kinsbourne (eds), *Aging and Dementia*, pp. 1-24. Spectrum, New York.
(156) Waterlow, J.C. and Stephan, J.M.L. 1967. The measurement of total lysine turnover in the rat by intravenous infusion of L-(U-^{14}C) lysine. *Clin. Sci.* 33:489-506.
(157) Waterlow, J.C. 1966. In H.N. Munro (ed.), *Mammalian Protein Metabolism*, pp. 326-390. Academic Press, New York.
(158) White House Conference on Food, Nutrition and Health, Final Report. Report of Panel II-4: Aging. Washington, D.C.: Government Printing Office. 1970.
(159) White House Conference on Aging, Final Report, Washington, D.C.: Government Printing Office. 1972.
(160) Whiting, S.J. and Draper, H.H. 1980. The role of sulfate in the calciuria of high protein diets in adult rats. *J. Nutr.* 110:212-222.
(161) Winterer, J.C., Steffee, W.P., Perera, W.D.A., Uauy, R., Scrimshaw, N.S. and Young, V.R. 1976. Whole body protein turnover in aging man. *Exp. Gerontol.* 11:79-87.
(162) Young, V.R., Hussein, M.A., Murray, E. and Scrimshaw, N.S. 1971. Plasma tryptophan response curve and its relation to tryptophan requirements in young adult men. *J. Nutr.* 101:45-59.
(163) Young, V.R., Tontisirin, K., Ozalp, I., Lakshmanan, F. and Scrimshaw, N.S. 1972. Plasma amino acid response curve and amino acid requirements in young men: valine and lysine. *J. Nutr.* 102:1159-1169.
(164) Young, V.R., Taylor, Y.S.M., Rand, W.M. and Scrimshaw, N.S. 1973. Protein requirements of man: efficiency of egg protein utilization at maintenance and submaintenance levels in young men. *J. Nutr.* 103:1164-1174.
(165) Young, V.R., Steffee, W.P., Pencharz, P.B., Winterer, J.C. and Scrimshaw, N.S. 1975. Total human body protein metabolism in relation to protein requirements at various ages. *Nature* 253:192-194.
(166) Young, V.R. 1976. Protein metabolism and needs in elderly prople. In M. Rockstein and M.L. Sussman (eds.), *Nutrition, Longevity, and Aging*, pp. 67-102. Academic Press, New York.

(167) Young, V.R., Perera, W.D., Winterer, J.C. and Scrimshaw, N.S. 1976. Protein and amino acid requirements of the elderly. In M. Winick (ed.), *Nutrition and Aging*, pp. 77-118. John Wiley and Sons, New York.
(168) Young, V.R. 1978. Diet and nutrient needs in old age. In J.A. Behnke, C.E. Finch and G.B. Moment (eds.), *The Biology Of Aging*, pp. 151-172. Plenum Press. New York.
(169) Young, V.R. and Scrimshaw, N.S. 1978. Nutritional evaluation of proteins and protein requirements. In M. Milner, N.S. Scrimshaw and D.I.C. Wang (eds.), *Protein Resources and Technology*, pp. 136-173. AVI Publishing Co., Westport.
(170) Yousef, M.K. and Johnson, H.D. 1970. ^{75}Se-selenomethionine turnover rate during growth and aging in rats. *Proc. Soc. Biol. Med.* 133:1351-1353.
(171) Zanni, E., Calloway, D.H. and Zezulka, A.Y. 1979. Protein requirements of elderly men. *J. Nutr.* 109:513-524.

—6—

Cholesterol Metabolism and Atherosclerosis in Aging

Mushtaq Ahmad Khan and Alif M. Manejwala

INTRODUCTION

Atherosclerosis is the most ubiquitous and disabling disease afflicting mankind especially among the affluent societies of modern ages. Although clinical manifestations of this disease become evident in old age, the seeds for the development of atherosclerosis are sown very early in life. It has been reported that at least among American males, the disease is already established in the second decade of life; by the age of 20 years, advanced lesions can be found in the arteries.[77] In the United States, more than 80% of the cases of atherosclerotic disease are found in individuals over the age of 65 years.[35] The disease progresses through ages 30 to 40 and becomes established in the late 40s and 50s when its clinical manifestations (coronary occlusion, myocardial infarction, cerebral infarction, gangrene, stroke, etc.) start to appear (Figure 1). The developmental horizons for atherogenesis are operative over the life span of an individual and the progression of the disease is dependent upon multifactorial insults that one is continuously exposed to in life. It has been shown by many investigators that both severity of atherosclerosis[5,76,103] and risk of coronary heart disease (CHD)[37] are closely associated with age. It is in this regard that some very important questions regarding the process of aging and the development of atherosclerosis have been raised.[16] What is the relationship between aging and atherosclerosis? Stated more specifically, does the fact that atherosclerosis is age-related indicate that it is the end result of an aging process? Or more importantly, is it associated with age-related alterations in metabolism or both? In this chapter pertinent literature will be reviewed in an attempt to explore these questions.

Figure 1: Evolution of Atherosclerosis with Age

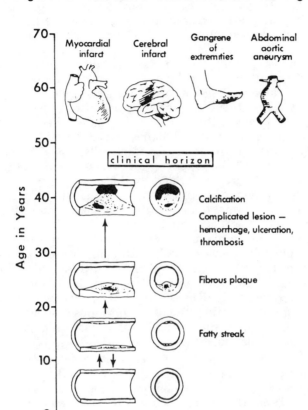

After McGill et al.[77] Reproduced with permission from *Atherosclerosis and Its Origin*, Academic Press.

AGE-RELATED CHANGES IN THE ARTERIES

Before considering age-related changes, it is important to briefly review the structural features of the three reasonably well-defined layers of the normal arterial wall; the intima, the media and the adventitia. The intima is the innermost layer consisting of endothelial cells that line the lumen. The endothelial cells are covered on the outer aspect by a perforated tube of elastic tissue called the internal elastic lamina. The endothelium normally modulates the passage of water and other substances from blood plasma into the tissues by an active transport process. It is the intima that is affected by atherosclerosis.

The middle layer (media) is the artery's main supporting layer and is composed of either a single layer (muscular arteries) or multiple lamellae (elastic arteries) of only one cell type, the smooth-muscle cells. They are arranged in a diagonal spiral fashion and are surrounded by small amounts of collagen and elastic fibers.

The outermost layer (adventitia) consists of a loose admixture of thick bundles of collagen and elastic fibers of various sizes, and a mixture of smooth-muscle cells and fibroblasts. Blood vessels (vasa vasorum) and nerve endings are present in this layer.[10]

Like any other tissue, the arteries are also subject to wear and tear and other changes that occur with aging. There is a slow and symmetrical increase in the thickness of the intima caused by a gradual accumulation of smooth-muscle cells. The fibrous proteins (elastin, collagen and the mucopolysaccharides) which surround the smooth-muscle cells are all secreted by this cell type.[114]

The amount and composition of various lipids in the arteries vary with age. In normal human aorta, there is a progressive accumulation of phospholipid (sphingomyelin), which is the result of a gradual decrease in sphingomyelinase activity in the aging arterial wall.[34,92] This rise in phospholipid synthesis with aging is perhaps in response to the need for enhanced membrane synthesis for intracellular organelles like plasma membranes, vesicles, lysosomes, etc. While most of the phospholipid in the normal arterial wall seems to be synthesized *in situ* by the medial smooth-muscle cells, the cholesterol esters that enter the normal artery with aging appear to be derived from the plasma.[18] There is evidence to suggest that accumulation of cholesterol in the normal aged artery is a reflection of the entry of lipoprotein moiety into the aortic endothelium.[99] Using specific fluorescent antibody techniques, selected areas of the arterial intima have been shown to accumulate large quantities of beta-lipoproteins from plasma in both intra- and extracellular locations. From the findings of this study, it has been suggested that failure of the arterial wall to metabolize lipids carried by plasma beta-lipoprotein is a basic cause of atherosclerosis.[109]

Intermediary metabolism of human arterial tissue and the changes that take place with age and the process of atherogenesis have been extensively reviewed.[65] Adequate interpretation of the variations in enzyme activities in connection with aging has not been provided. In the case of some enzymes, the changes in tissue composition of the arterial wall with age may affect the levels of enzyme activity. For example, the decrease in the activities of glycogen phosphorylase and creatine phosphokinase of the aorta, with age, is probably associated with atrophic changes in the muscular component of the vessel wall.

Age-related changes in the ground substance, elastin, collagen fibers, lipids and calcium have been reviewed by Bertelsen.[13] Age-associated changes in the aortic wall denote changes of a uniform nature and appearance. Such changes take place both in normal and in atherosclerotic vessels thus making it difficult to distinguish between changes due to aging and the atherosclerotic changes. Age-associated changes in the media include proliferation of cells, accumulation of ground substance, increase in number of collagen fibrils, splitting of elastic membranes and the deposition of calcium salts. In the intima, the proliferation of fibroblasts, the accumulation of ground substance and diffuse lipodosis are among the age-related changes whereas the formation of lipoidal lumps and plaques with or without secondary fibrosis and the formation of primary fibrous plaques comprise specific atherosclerotic changes of the arterial wall. As mentioned previously, proliferation of cultured arterial cells can be influenced by age *in vivo* (donor) and *in vitro* (longevity in culture). There is a decline in growth response to serum in direct relation to the number of passages of cultured arterial smooth-muscle cells.[101] In addition, cells from arterial explants of older donors were difficult to grow and had a shorter lifespan in culture.[75]

It has also been reported[97] that with aging, degenerative changes involving peptide bond cleavage in elastins from nonatherosclerotic aorta may occur, thus complicating the situation.

EVOLUTION OF ATHEROSCLEROTIC LESION

Although there is a great deal of information on the pathogenesis of adult atherosclerosis in humans and animals, much remains to be learned about its stages of development at younger ages. Factors that initiate the process remain uncertain as do those influencing its progression. There is evidence to suggest that the process of atherogenesis begins early in childhood and that lesions progress through several stages before clinical manifestations become evident in middle and late adult life (Figure 1).[50,104] Data collected by the International Atherosclerosis Project[78] along with the results of the autopsy examinations done on U.S. casualties during the Korean[35] and Vietnam[79] conflicts leave little doubt as to the early onset of the disease.

The first change recognizable as atherosclerosis is the "fatty streak" characterized by deposition of lipid (predominantly cholesterol) in the musculoelastic intima. Lipid deposition begins in the aorta of virtually every child in all populations studied in the first year of life; coronary arteries are affected in the second decade and the intracranial arteries in the third decade.[77] In general, these lipid deposits grow in size and number covering approximately 10% of the aorta at age 10 and 30 to 50% by age 25.[6] Because the lipid component of fatty streaks is remarkably similar in composition to the lipids in blood, migration of lipids from blood into the vessel wall is the most likely possibility. The cells of the arterial wall do not seem to produce these lipids since plaques have been observed to form even on synthetic vascular prostheses.[58] The ultimate role of fatty streaks in the production of complicated lesions of athero-

sclerosis is still obscure. It is reasonable to consider the following three pathways which fatty streaks could follow:
(1) They may remain as harmless lipid deposits of the intima without further progression.
(2) They may progress to form fibrous plaques and complicated lesions especially under the influence of so-called atherogenic stimuli such as hypercholesterolemia, hypertension, anxiety, cigarette smoking, pollution, etc.
(3) They may regress completely and leave no signs of their ever being present.

It is known that fatty streaks are clinically harmless and potentially reversible. According to Aschoff[7] lipid deposits in the mitral valve and at the root of the aorta found in infants disappear leaving no scar. On the other hand, fatty streaks located on the posterior wall of thoracic aorta regress after reaching a peak during puberty, but there persists the thickened intima at these sites. Early experimental lesions produced by high fat-high cholesterol diet have also been shown to regress almost completely in rhesus monkeys fed the so-called "regression diet," which is low in fat and low in cholesterol.[110] In this study, however, regressed lesions showed endothelial changes suggestive of reparative processes. Although the progression from fatty streaks through fibrous plaques to complicated ulcerated lesions cannot be documented by longitudinal studies in human subjects, there are enough cross-sectional observations to permit the assumption that such a progression does occur.[58] This conclusion is strengthened by the following observations: In a histological study of sections from a standardized location in the cornary arteries, no distinction could be made histologically between the pure fatty streak and the typical fibrous plaque suggesting a gradual transition of fatty streak to a fibrous plaque.[43] In addition, the close topographic association between fatty streaks and fibrous plaques at different ages is consistent with the hypothesis that fatty streaks are precursors of complicated fibrous plaques.[104] The histologic examination of fatty streaks from young adult North American white males has shown that there is an inflammatory reaction around the abnormal lipid deposits. Inflammatory reaction—a response to injury, is accompanied by increased capillary permeability, increased transudation or exudation of protein-rich plasma, and migration of leukocytes into the injured area. Based on these observations, injury to the endothelium has been suggested as the first initiative step for the inception of the lesions.[77] As previously noted, the disease progresses with age under the continuous influence of a complex interplay of many factors including the physical properties of vessel and rheologic characteristics of the flow within them, as well as the lipid and other metabolic processes in the blood and the vessel wall. If the insults by the so-called risk factors (environmental and biological) continue or the individual possesses increased genetic propensity to the development of the disease, the lipid is prematurely surrounded by a capsule of connective tissue giving it the name "fibrous plaque." In subsequent years, blood vessels grow into the plaque from below, which in some cases rupture and bleed into the plaque. Necrosis followed by sloughing of the tissue brings about formation of a thrombus on the surface of the vessel. This reduces the size of the lumen, ultimately causing reduction in blood flow to a part of or the entire organ which brings about death of the tissue. Unfortunately, this point is reached in most cases without prior symptoms.[77]

RISK FACTORS ASSOCIATED WITH ATHEROGENESIS

Despite tremendous research efforts in this area, the etiology of atherosclerosis still remains unknown. From the available data, it seems that the disease develops through the interactions of multiple factors, referred to as the "risk-factors." Many studies conducted over the past 35 years have stressed associations between certain biochemical, physiological and environmental factors and the development of premature atherosclerotic diseases. Among the major risk factors are hyperlipidemia, hypertension and cigarette smoking. Other risk factors which may play a role in the pathogenesis of atherosclerosis include unknown genetic factors (family history of premature coronary heart disease), hyperglycemia (diabetes mellitus), obesity, sedentary habits and psychosocial stresses.[16,90]

The overall effects of the risk factors are roughly additive, so that a person exposed to multiple risk factors is at a higher risk of developing CHD. The morbidity and mortality from atherosclerotic diseases increase when all or most of these factors act in combination. For example, presence of one of the factors increased the probability of a fatal event by twofold, while presence of two of the factors increased the risk fourfold. In persons exposed to all three factors, the risk was eight times that of the nonsmoker with normal serum cholesterol and normal blood pressure.[98] Of all these risk factors, many in themselves are age-related such as hypercholesterolemia, hyperglycemia, hypertension, obesity, and sedentary living habits. In this section, age-related factors causing hypercholesterolemia will be discussed along with the changes associated with metabolism of cholesterol leading to atherosclerosis, a natural concomitant of aging. However, it is pertinent to briefly discuss the other potent age-associated risk factors (such as hypertension, diabetes mellitus, obesity and sedentary living) before going into a detailed discussion of the role of cholesterol.

Hypertension

Hypertension, an age-related risk factor, has been shown to

accelerate the progression of atherosclerosis in humans and experimental animals, particularly in the presence of hypercholesterolemia. Atheroma of the aorta, coronary and cerebral arteries were found to be more extensive and severe in hypertensive than in normotensive individuals in autopsy finding.[71,90] Daly and coworkers[29] reported a greater occurrence of hypercholesterolemia in experimentally induced hypertension in rats than their normotensive counterparts, which were used as controls. Hypertension is believed to enhance the process of atherogenesis by directly producing local injury at specific high-pressure sites by mechanically stretching the vessel wall. This may be the initiative first step for start of the disease according to the chronic injury hypothesis of atherogenesis. It is still not known if the sustained high pressure within the arterial lumen produces changes to enable smooth-muscle cells or stem cells to proliferate. However, hypertension may alter permeability of the endothelial lining thereby facilitating the transport of large amounts of lipoproteins through the intact endothelial lining. In addition, Wolinsky[113] has reported that hypertension greatly increases the lysosomal enzyme activity which may lead to increased cell degeneration by the action of highly destructive enzymes released into the arterial wall. It should be pointed out that although hypertension by itself has not been proven to cause atherosclerosis, the experimental and clinical evidence clearly indicates that the early diagnosis and treatment of hypertension could substantially lower the risk of CHD.[24] From these findings it appears that the risk of CHD increases with an increase in blood pressure.[71]

Diabetes Mellitus

Diabetes mellitus has been suggested as a serious age-related risk factor for atherosclerotic disease involving coronary, cerebrovascular and peripheral arteries for many years.[36,48] It is also reported that atherosclerotic diseases are more prevalent and severe in diabetic individuals than in nondiabetics, especially if the arterial endothelium and smooth-muscle cells are intrinsically defective. Although hyperglycemia is known to affect aortic wall metabolism, the role of glucose in atheroma formation, if any, is poorly understood. It has been reported that insulin in physiological concentrations can stimulate growth of cultured smooth cells up to the sixth passage and fails to stimulate proliferation of cells that had aged *in vitro*.[101] As previously noted, cells obtained from arterial explants from older donors were more difficult to grow and had shorter life spans in culture.[75] Also, hyperlipidemia is found with increased frequency in diabetics than in nondiabetics. This could be due to the impaired lipoprotein removal from the circulation as diabetes progresses leading to a higher concentration of lipids which consequently accelerates the process of atherogenesis.[36]

Vavrik et al[107] have pointed out a close association between hypercholesterolemia and two other risk factors of atherogenesis, namely, hypertension and diabetes mellitus. The study on 509 residents of homes for aged showed that the prevalence of serum cholesterol above 250 mg/dl was significantly higher (41.9%) in diabetic females than in nondiabetic females (20.6%). They also found that serum cholesterol concentration 250 mg/dl or above occurred more often in CHD patients than the normal group of both males and females. The fact that the physiologic hepatic catabolism of cholesterol is depressed in diabetes mellitus could explain the correlation between diabetes and hypercholesterolemia.[107] Such metabolic derangements of carbohydrate and lipid metabolism are accompanied by vascular abnormalities of the blood vessels of retina and the kidney (diabetic microangiopathy) along with atherosclerotic changes of larger blood vessels (diabetic macroangiopathy).[72]

Obesity

The reports on the relationship of obesity to atherosclerosis are conflicting. However, obesity is closely associated with other important age-related risk factors for atherosclerosis, which include diabetes mellitus, hypertension, hypertriglyceridemia, and hypercholesterolemia. Sterol balance and kinetic analysis studies have indicated that cholesterol production is markedly enhanced in obesity.[81,85] These workers claim that a gain in body weight of 1 kg of adipose tissue increases daily cholesterol production by 20 to 22 mg. While some of this increased production of cholesterol is balanced by enhanced elimination of neutral steroids, the expanded adipose tissue serves as an important site for deposition of dietary or endogenously produced cholesterol.[4] The degree of adiposity has also been linked to triglyceride concentrations in serum which, in turn, is associated with the process of atherogenesis.[17,22,44] Liver biopsies from obese subjects showed an increased conversion of acetate to mevalonate which showed a positive correlation with relative body weight of the subjects.[86] Higher circulating levels of insulin (basal and after stimulation) have been observed in obese patients[8] and also with increasing age[18] which is probably responsible for increased hepatic lipogenesis. An association between increased circulating insulin and atherogenesis has also been suggested[102] because of its direct effect on arterial wall metabolism.

Some epidemiological and clinical studies have failed to show a clear relationship between obesity and CHD.[21,95] Epstein[36] has suggested that obesity in populations of developed countries (where most studies have been done) is so common that the possible association between CHD and obesity may have been masked. It should also be pointed out that gross obesity in persons over the age of 70 years is a rare phenomenon and thus a relationship of this syndrome to CHD in this age group may not be found.[16]

Sedentary Life-Style

Epidemiological studies support the view that sedentary living may be a risk factor of CHD in the advanced countries like the United States.[83] It is known that physical exercise is associated with the increase in the level of high density lipoprotein (HDL)[115] which is claimed to protect against CHD. Although the lack of physical exercise does not have a major impact on CHD, it certainly adds to the prevalence of CHD when it exists in combination with other risk factors.

Hypercholesterolemia

Since one important focus of this chapter is cholesterol metabolism in the aged, it is pertinent to include a brief discussion of the physiological roles of cholesterol in the body. Cholesterol is present in practically every cell of the body.[23,80] Between 75 to 150 g of cholesterol are present in the body of a 70 kg male; of which less than 20% of cholesterol exists in a rapidly exchanging pool and the remaining is present in tissues with slower or no cholesterol turnover. Various regulatory mechanisms influence cholesterol metabolism to maintain cholesterol homeostasis. Although there is no known "safe" cholesterol level, it has been reported that, in general, men with serum cholesterol levels of over 260 mg/dl have three or four times the risk of suffering from CHD as compared to men having cholesterol values of 220 mg/dl or less.[60]

Functions of Cholesterol:

(1) Cholesterol is essential to maintain the normal structure and function of a cell.[23,25]

(2) As much as 50% of myelin which surrounds the nerve fiber is made up of cholesterol which is very important for the proper functioning of the nervous system.[23]

(3) About 40 to 50% of cholesterol is converted to bile acids and bile salts which play a vital role in digestion and absorption of fats in the body.[82]

(4) A small amount of cholesterol is used in the synthesis of steroid hormones, which are essential in the functioning of the various systems of the body.[23,82]

(5) Cholesterol is a precursor of vitamin D that plays a major role in intestinal absorption of calcium and thus bone formation.[23]

Before going into further discussion of cholesterol metabolism and its regulation, it is quite reasonable to briefly review the "cholesterol connection" to the process of atherogenesis. The following observations are summarized from various studies.

(1) Serum cholesterol levels are higher than normal in most of the patients with CHD.[64,68,107] In a recent longitudinal study of youthful cholesterolemia, it was observed that the medical students who were hypercholesterolemic as compared to their normocholesterolemic classmates were 30 times as susceptible to the episodes of acute myocardial infarction occurring 13 to 21 years after the higher cholesterol levels were discovered.[106]

(2) A definite correlation has been observed between hypercholesterolemia and aging up to the sixth decade of life.[28,111]

(3) When hypercholesterolemia was experimentally induced in laboratory animals, they developed atheromatous plaques in their arteries at an earlier age.[62,70,108]

(4) Although it has not yet been proven with certainty in man[68,90,110,115] early lesions of atherosclerosis are known to regress in experimental animals when the levels of plasma cholesterol are controlled at low levels.

Regulation of Cholesterol Metabolism: The amount of cholesterol present in the body depends upon the dietary intake, amount absorbed, endogenous cholesterol synthesis, and degradation to bile acids and excretion. Unlike other metabolites cholesterol cannot be oxidized to oxygen and water as mammalian tissues do not possess the enzymes which can destroy the steroid nucleus. Cholesterol synthesis takes place primarily in the liver and intestine and is regulated by a sensitive feedback mechanism by hydroxymethylglutaryl Coenzyme-A reductase (HMG-CoA reductase). Bile acid production is regulated through 7-alpha hydroxylase.[31] Intestinal absorption and hepatic synthesis are interrelated. In case of decreased bile salt production due to diminished intestinal absorption of cholesterol, the negative feedback system is released which triggers enhanced synthesis causing elevation of serum cholesterol and thus completing the loop by increased production of bile salts and thereby depressing cholesterol synthesis.

Virtually every mammalian cell except the erythrocyte has the capacity to synthesize cholesterol but at a different rate. More than 90% of cholesterol produced in the body is synthesized in the liver and the intestine (ileum)—liver being the major site of synthesis.[31] Acetyl Coenzyme-A which is derived from amino acids, fats and carbohydrates is the main substrate in the biosynthesis of cholesterol. Dietary cholesterol has an inhibitory feedback effect on hepatic biosynthesis of cholesterol by depleting HMG-CoA reductase, a rate-limiting enzyme in the biosynthetic pathway.[80] Nevertheless, unlike liver, the intestine which is not influenced by the negative feedback control of cholesterol may be a significant site of non-hepatic cholesterol synthesis.[112] With regard to hepatic cholesterol synthesis in humans, however, the picture is not very clear. Although virtually complete inhibition has been noticed after the cholesterol-rich diet in adults[14] as well as children[41], a partial inhibition of biosynthesis of cholesterol has also been reported.[83] There is, however, little doubt that the absorption of cholesterol influences the rate of its synthesis at least in the liver by a negative feedback system. On the other hand, when cholesterol absorption is suppressed with plant sterol beta-sitosterol, cholesterol synthesis rises strikingly.[82]

Cholesterol within the intestinal lumen consists of a mixture of free and esterified cholesterol from both exogenous (dietary) and endogenous sources. Cholesterol esters are hydrolyzed to free cholesterol by the action of pancreatic cholesterol esterase. Free cholesterol is then solubilized in mixed micelles in the presence of bile acids and other amphipathic substances. This mixture moves across the cell membrane (presumably by passive diffusion), of brush border of small intestine and mixes with the intracellular pool of unesterified cholesterol (synthesized de-novo from acetate by the epithelial cells). A large portion of this pool is next esterified with long-chain fatty acids, is then released into the intestinal lymph where it is incorporated in chylomicrons which end up either in slow or rapidly exchanging pools.[32] Depending upon the intake, between 200 to 500 mg of cholesterol is absorbed mainly in the small intestine of an average person.

Aging in man is associated with a variety of changes in lipid metabolism such as an increase in the level of serum cholesterol. It is known that body composition of man changes with increase in age. The amount of fat increases at the expense of muscle mass and bone mass, which is probably responsible for the slight decline in oxygen consumption. In other words, there is a gradual decline in the rate of basal energy metabolism after the age of about 25 years in both males and females; therefore, it is assumed that the overall rate of metabolism is slowed down with age which causes retardation of cholesterol metabolism.[96] As a result, cholesterol will accumulate in tissues and blood.

Research on age-related changes of various facets of cholesterol metabolism in humans has been virtually nonexistent. Studies using rats as experimental animals have produced conflicting results. Using isotopic tracer techniques, Yamamoto and Yamamura[116] have reported an age-related decrease in gastrointestinal absorption of cholesterol in Wistar rats from 2 to 18 months of age. These workers reported that hepatic cholesterogenesis, biliary and fecal excretion of cholesterol and its metabolites, and gastrointestinal absorption of cholesterol decreased in aged rats and thus the turnover of cholesterol was slowed down. However, in a recent study[49] with Sprague-Dawley rats varying in age from 1 to 42 months, cholesterol absorption was shown to be increased in a linear fashion with increase in age of the animals. When cholesterol absorption was expressed as a percentage of the amount infused, 14% was absorbed at the age of 1 month, as compared to 38% at the age of 42 months. These workers postulate that parallel age-related changes in cholesterol absorption may occur in humans. In another study, prolonged alimentary cholesterol loading in 2 to 5 year old dogs led to the development of degenerative and atrophic changes in the mucosa of small intestine.[67] Such atrophic changes in the villi have previously been shown to correspond with the inhibition of absorption of lipids observed in patients with atherosclerosis.[1]

Cholesterol by itself is insoluble in aqueous solution, therefore, both endogenous and exogenous cholesterol in the blood is carried on or by proteins along with triglycerides and phospholipids in the form of lipoproteins. Major classes of lipoproteins have been separated by means of ultracentrifugation and electrophoresis into four groups, namely, chylomicrons, low-density lipoproteins (LDL), very-low-density lipoproteins (VLDL) and high-density lipoproteins (HDL). Detailed discussion of lipoprotein metabolism is beyond the scope of this chapter and excellent reviews on the subject exist in the literature.[12,33]

Inverse Correlation Between HDL and LDL: Recently, evidence has been accumulating linking increased concentrations of serum HDL with a decreased risk of atherosclerosis, perhaps by facilitating excretion of cholesterol by HDL from the body.[12,59] An inverse relationship between serum LDL and HDL has been shown by many recent studies.[59,91,110,115] HDL was found to be significantly lower in CHD patients than in healthy subjects. On the other hand, the prevalence of CHD and myocardial infarction fell with increasing level of HDL even when such variables like cigarette smoking, hypertension, age and obesity were taken into account.[91] How HDL protects against heart attacks is still the subject of much debate. It is assumed that HDL prevents cholesterol deposition in cells by blocking directly the uptake of LDL and by facilitating cholesterol excretion.[27] However, this postulation is not very convincing. HDL is influenced by many factors such as running exercises,[115] cigarette smoking,[42] sex,[20] age,[12] etc. It is believed that HDL level is about the same in men and women until age 17 or 18, after which there is an abrupt fall of 10 to 20% in the male, which is sustained at lower levels for the next 3 or 4 decades;[20] whereas, in women it is stable or even rises during the reproductive years.

Generally, women have higher levels of HDL than men which may explain why they are less susceptible to atherosclerotic diseases than men. The average HDL concentration for men was 45 mg/dl as compared to that for the women, which was 55 mg/dl. These differences were attributed to the levels of male and female sex hormones. While mechanisms are still not clear, estrogens are known to increase the hepatic synthesis of many transport proteins.[87]

The relationship of habitual physical exercise to HDL concentration has been established.[115] Fasting plasma HDL levels of 41 active men (running about 15 miles per week for the previous year) between the ages of 35 to 59 years were significantly higher than the comparison group of similar ages (mean HDL cholesterol 64 versus 43 mg/dl).

Variations in HDL concentration have been noted with age. It is unlikely that these changes are invariable concomitants of aging; they may be partly determined by environmental and inherited biological factors. In general, among the adult population, the ratio of LDL:HDL-cholesterol appears to rise steadily with age.[45] In a recent study[66] however mean HDL-cholesterol concentration for the age group 50 to 59 years was significantly lower than for the corresponding controls. This was in contrast to the substantially higher mean HDL-cholesterol levels in older age groups. The reasons for such variations are not yet understood.

The concept that the plasma cholesterol level continues to rise with age is supported by the following observations from different geographical locations. Cholesterol levels of civil servants in India showed slow age-related increase with a plateau between the sixth and seventh decades of life (Table 1.)[30]

Table 1: Serum Cholesterol Levels in Indian Males

Age (yr)	Number	Average Weight (lb)	Serum Cholesterol (mg/dl ± SEM)	Ester (%)
15-19	5	97	114 ± 6	77
20-29	10	101	134 ± 7	75
30-39	10	123	166 ± 13	64
40-49	10	133	163 ± 12	70
50-59	10	121	149 ± 11	69
60-69	6	108	179 ± 21	65

After Das and Bhattacharya.[30]

Kipshidze[64] studied quantitative changes in various lipid parameters in sera of 260 persons in Adjaria, U.S.S.R., along with 51 long-lived persons (aged 90 years or older) and showed that the mean cholesterol level steadily increases with age, reaching its maximum between 50 and 59 years and subsequently falls to normal levels.

Werner et al[111] studied about 3,000 American males and females in which males showed steady rise in serum cholesterol levels until about the sixth decade of life, followed by a decline. Cholesterol levels in females rose slower than those of males and reached maximum by the sixth and seventh decade of life. Similar observations (Figure 2) have been made by other investigators.[69,93]

Figure 2: Five-Year Age-Group Means of Serum Cholesterol and Standard Errors of the Means

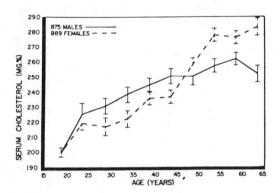

After Schilling et al.[93] Reproduced with permission from *Am. J. Clin. Nutr.*

One suggestion about the increasing concentration of cholesterol is its slower elimination from the body in advancing age.[117] As most of the cholesterol excreted in feces is excreted in bile acid fraction, this was measured in young and old rats.[55] Four micrograms of tritiated cholesterol per 100 grams body weight was injected in young and old rats and their feces were collected every day for 3 to 4 days. (Since a negligible amount of cholesterol is excreted through urine, the total radioactivity of feces was measured and the results were expressed as total excreted radioactivity as percentage of the injected dose.) Calculated in this way, the clearance index of tritiated cholesterol showed a decrease of about 70% in the older animals as compared to the younger ones. It is true that the older animals eat less food in proportion to their body weight and hence excretion should be less in these rats. Older animals excreted much less amounts of cholesterol than young rats, even when the food intake of young animals was restricted to correspond with that of the old rats.

It can be concluded from the above observation that the excretion of cholesterol in old animals is much less efficient than that of young animals. This could be due to the functional inefficiency of the cells involved in the excretion of cholesterol, or partly due to the decrease in reactivity to the hormones.

The turnover of cholesterol is apparently slowed down in the old animals.[51,54] When radioactive cholesterol was injected in Sprague-Dawley rats, the elimination of cholesterol (synthesis and degradation) was much faster in younger rats than the older rats. The half-life of cholesterol in the fast-exchanging pool was 0.7 day in young rats and 0.9 day in old rats. The half-life of the slow-exchanging pool was 22 days in young rats and 32 days in the old rats.[53] The slow turnover of cholesterol means slow disappearance of cholesterol from the site where it is deposited. This may be one of the reasons why older animals are more susceptible to atherogenesis.[53]

Endocrine glands regulate cholesterol metabolism through their hormonal secretions. The interrelationships between thyroxine, insulin and cholesterol metabolism were studied extensively in experimental animals[46,115] and in man.[94] It is assumed that with aging the activities of these endocrine glands decrease and so does the sensitivity of cholesterol metabolism to the hormonal factors. These changes could partly play a role in the pathogenesis of atherosclerosis. Influence of genetic factors on plasma cholesterol concentration has been extensively reviewed.[26,84] Most investigators have suggested additive effects of gene action and a much stronger influence of the environmental factors. Despite considerable variations, three levels of responses have been described in experimental animals (normoresponders, hyporesponders, and hyperresponders) after feeding of high cholesterol diets.[26] In addition, familial tendency of coronary heart disease has been recognized;[3] however, both age and environmental factors make it extremely difficult to establish the true heritability of plasma lipids and atherogenesis.

In short, it is very reasonable to assume that although the balancing mechanism between absorption and synthesis on one side and excretion and catabolism on the other do exist, the increasing levels of cholesterol with age could represent just a long-term accumulation from a very small excess of input over the output. This may be due to the progressive inability of the cell to metabolize cholesterol as it ages. It is also true that the 20th century man has become more reluctant to accept deterioration of coronary arteries and subsequent development of atherosclerosis as a necessary accompaniment of the aging process or his personal being or so-called constitution. In a search for associated factors from his mode of life, diet has become one of the focal points of enquiry.[39]

Excellent reviews have been published on the subject of dietary interrelationships of serum cholesterol and atherogenesis.[18,31,39] Such studies in man have produced much conflicting and often confusing data on the effects of dietary cholesterol on the concentrations of circulating cholesterol. While a great deal of individual variations exist in most biological responses, many investigators claim dietary cholesterol has an undeniable hypercholesterolemic effect.[20,25,38] The conflicting observations regarding the effects of dietary cholesterol on the levels of serum cholesterol were perhaps misinterpreted because cholesterol was not fed in a form suitable for absorption.

Dietary fat may enhance absorption of cholesterol, by stimulating flow of bile in addition to increasing the solubility of cholesterol in the intestinal lumen.[105] It has been shown that cholesterol absorption is proportional to cholesterol ingestion. Patients fed as much as 3 grams of cholesterol per day have been shown to absorb up to 1 gram of cholesterol.[89] Absorption of cholesterol and changes that take place with aging have been dealt with elsewhere in this chapter and thus will not be discussed here.

The relationship of aging process per se to quantitative or qualitative intake of dietary cholesterol is complicated in man by the ubiquitous nature of atherosclerosis among the adult population. However, carefully controlled clinical investigations have shown that a variety of dietary practices including the substitution of unsaturated for saturated fats,[40,47,63] use of increased amounts of fiber,[100] consumption of yogurt,[74] etc. will depress serum cholesterol levels. Based on recent observations, it has been claimed that pure crystalline cholesterol does not produce any deleterious effects upon the intimal surface. It has been shown that as little as 250 mg/kg of oxidation products, when fed to the rabbits by gastric gavage, caused over 7% of the aortic and arterial smooth-muscle cells to undergo necrosis in as short a time as 18 hours. When an equal amount of purified cholesterol was similarly fed to another group of rabbits, there was no arterial and aortic smooth-muscle cell necrosis.[56] Further research in this area has unfortunately been hampered by extremely high costs of producing various oxides of cholesterol in sufficient quantities for feeding experiments. The question, of course, is of immense public health importance because oxidation products of cholesterol have been isolated from various food products containing egg yolk as an ingredient.[61]

On the other hand, recent investigations have questioned the validity of recommendations to lower serum cholesterol for persons 55 or older.[2,19,73] Available data suggest that restrictions to lower serum cholesterol may be too late to help those in late maturity and old age. In fact, serum cholesterol does not seem to be a good biochemical marker for CHD after the sixth decade of life.

It is obvious that if the disease is to be conquered to any significant degree preventive measures will have to be instituted at a much younger age. There is no doubt that the modern man smokes more cigarettes, drinks more, and eats more animal fat, is more obese, subject to more tensions, and exercises less than at any time in human history. He also suffers more frequently from atherosclerosis. The results of the most comprehensive studies of this debilitating disease ever undertaken, the Framingham Study, are now available. They suggest that our way of life and most prominent way of dying are interrelated.[57]

In summary, all these so-called risk factors (hypercholesterolemia, hypertension, diabetes mellitus, obesity, etc.) are the interrelated intrinsic body processes which act together for many years with the "environmental factors" (cigarette smoking, psychosocial stress, dietary and sedentary habits) to produce this age-related degenerative disease called atherosclerosis.[67]

ACKNOWLEDGEMENT

The authors wish to express their gratitude to Messrs. Frank Cerra, Arlen Sager and Ms. Joan Shields and Deborah French for help during the preparation of this manuscript.

REFERENCES

(1) Abramzon, M.N. and Malkiel, R.Y. 1973. Lipolytic activity of the small intestine and fat absorption in patients with atherosclerosis. *Kardiologiia* 13(10):90-94.
(2) Ahrens, E.H., Jr. 1976. The management of hyperlipidemia: Whether, rather than how. *Ann. Intern. Med.* 85:87-93.
(3) Eldersberg, D. and Schaefer, L.E. 1959. The interplay of heredity and environment in the regulation of circulating lipids and in atherogenesis. *Am. J. Med.* 26:1-7.
(4) Angel, A. and Farkas, J. 1970. Cholesterol storage in white adipose tissue. *Circulation* 42 (Suppl. III):III-1 (Abst.).
(5) Anitschkow, N.N. 1934. Pathologische Anatomie und Allgemeine Pathologie der Arteriosklerose. In *Comp. Rend de la Deuxieme Conf. de Pathol. Geographique*, pp. 44-101. Utrecht.
(6) Arteriosclerosis—A report by the National Heart and Lung Institute Task Force on Arteriosclerosis. 1971. 2:72-219.
(7) Aschoff, L. 1924. Arteriosclerosis. In *Lectures on Pathology*, pp. 131-168. Paul B. Hoeber, New York.
(8) Bagdade, J.D., Bierman, E.L. and Porte, D., Jr. Significance of insulin levels in the evaluation of the insulin response to glucose in diabetic and non-diabetic subjects. *J. Clin. Invest.* 46:1549-1557.
(9) Becker, G.H., Meyer, J. and Necheles, H. 1950. Fat absorption in young and old age. *Gastroenterology* 14:80-92.
(10) Benditt, E.P. 1977. The origin of atherosclerosis. *Scient. Am.* 236:74-85.
(11) Benjamin, W., Gellhorn, A., Wagner, M. and Kundel, H. 1961. Effect of aging on lipid composition and metabolism in the adipose tissues of the rat. *Am. J. Physiol.* 201:540-546.
(12) Berger, G.M.B. 1978. High-density lipoproteins in the prevention of atherosclerotic heart disease. *S. Afr. Med. J.* 54:693-697.
(13) Bertelsen, S. 1963. The role of ground substance, collagen, and elastic fibers in the genesis of atherosclerosis. In M. Sandler and G.H. Bourne (Eds.) *Atherosclerosis and Its Origin*, pp. 119-165. Academic Press, New York.
(14) Bhattathiry, E.P.M. and Siperstein, M.D. 1963. Feedback control of cholesterol synthesis in men. *J. Clin. Invest.* 42:1613-1618.
(15) Bierman, E.L. 1976. Obesity, carbohydrate and lipid interactions in the elderly. In M. Winick (Ed.) *Nutrition and Aging*, pp. 171-188. John Wiley & Sons, New York.
(16) Bierman, E.L. 1973. Fat metabolism, atherosclerosis and aging in man: A review. *Mechan. Aging and Develop.* 2:315-332.

(17) Bierman, E.L. and Porte, D., Jr. 1968. Carbohydrate intolerance and lipemia. *Ann. Intern. Med.* 68:926-933.
(18) Bierman, E.L. and Ross, R. 1977. Aging and atherosclerosis. *Atherosclerosis Rev.* 2:79-111.
(19) Bilheimer, D.W. 1977. Needed: New therapy for hypercholesterolemia. *New Engl. J. Med.* 296:508-510.
(20) Brunner, D., Weisbort, J., Lobel, K., Schwartz, S., Altman, S., Bearman, J.E. and Levin, S. 1978. Serum cholesterol and high density lipoprotein cholesterol in coronary patient and healthy persons. *Atherosclerosis* 33:9-16.
(21) Cady, L.D., Jr., Gertler, M.M., Gottsch, L.G. and Woodbury, M.A. 1961. The factor structure of variables concerned with coronary artery disease. *Behav. Sci.* 6:37-41.
(22) Carlson, L.A. and Bottiger, L.E. 1972. Ischemic heart disease in relation to fasting values of plasma triglycerides and cholesterol. *Lancet* 1:865-868.
(23) Cholesterol metabolism—A review. 1979. *Dairy Council Digest* 50:1-6.
(24) Christie, D. 1974. Mortality from cardiovascular disease. *Med. J. Aust.* 11:390-393.
(25) Civin, W.H. 1972. Serum cholesterol in health and disease. *Ann. of Clin. Lab. Sci.* 2:367-375.
(26) Clarkson, T.B., Lofland, H.B., Bullock, B.C. and Goodman, H.O. 1971. Genetic control of plasma cholesterol—Studies on squirrel monkeys. *Arch. Pathol.* 92:37-45.
(27) Cohen, L.S., Berman, M., Bulkley, B.H., Josephson, M.E., Manchester, J.H., Moser, M., Mullins, C.B., Ross, A.M., Schlant, R.C. and Zipes, D.P. 1979. Cardiovascular Diseases. Medical knowledge self assessment program. 5:159-191.
(28) Crouse, R.R., Grundy, S.M. and Ahrens, E.H., Jr. 1972. Cholesterol distribution in the bulk tissues of man: variation with age. *J. Clin. Invest.* 51:1292-1296.
(29) Daly, M.M., Deming, Q.B., Raeff, V.M. and Brun, L.M. 1963. Cholesterol concentration and cholesterol synthesis in aortas of rats with renal hypertension. *J. Clin. Invest.* 42:1606-1608.
(30) Das, B.C., Bhattacharya, S.K. 1961. Variation in lipoprotein level with changes in age, weight and cholesterol ester. *Gerentologia* 5:25-39.
(31) Dietschy, J.M. and Wilson, J.D. 1970. Regulation of cholesterol metabolism. *New Eng. J. Med.* 282:1128-1138.
(32) Dietschy, J.M. and Wilson, J.D. 1970. Regulation of cholesterol metabolism. *New Eng. J. Med.* 282:1179-1183.
(33) Eisenberg, S. and Levy, R.I. 1975. Lipoprotein metabolism. *Adv. Lipid Res.* 13:1-89.
(34) Eisenberg, S., Stein, Y. and Stein, O. 1969. Phospholipases in arterial tissue. IV. The role of phosphatide acyl hydrolase, lysophosphatide acyl hydrolase, and sphingomyelin choline phosphohydrolase in the regulation of phospholipid composition in the normal human aorta with age. *J. Clin. Invest.* 48:2320-2329.
(35) Enos, W.J., Jr., Beyer, J.C. and Holmes, R.H. 1955. Pathogenesis of Coronary disease in American soldiers killed in Korea. *J. Am. Med. Assn. 158:912-914.*
(36) Epstein, F.H. 1967. Hyperglycemia: A risk factor in coronary heart disease. *Circulation* 36:609-619.
(37) Epstein, F.H. 1965. The epidemiology of coronary heart disease. A review. *J. Chron. Dis.* 18:735-774.

(38) Erickson, B.A., Coots, R.H., Mattson, F.H. and Kligman, A.M. 1964. The effect of partial hydrogenation of dietary fats, of the ratio of polyunsaturated to saturated fatty acids, and of dietary cholesterol upon plasma lipids in man. *J. Clin. Invest.* 43:2017-2025.
(39) Ernst, N. and Levy, R.I. 1980. Diet, hyperlipidemia and atherosclerosis. In Robert S. Goodhart and Maurice E. Shils (Eds.) *Modern Nutrition in Health and Disease*, pp. 1045-1070. Lea and Febiger, Philadelphia.
(40) Frantz, I.M., Dawson, E.A., Kuba, K., Brewer, E.R., Gatewood, L.C. and Bartsch, G.E. 1975. The Minnesota coronary survey: Effect of diet on cardiovascular events and deaths. *Circulation* 52:II-4 (Abst.)
(41) Fujiwara, T., Hirono, H. and Arakawa, T. 1965. Idiopathic hypercholesterolemia; demonstration of an impaired feedback control of cholesterol synthesis *in vivo*. *Tohoku J. Exp. Med.* 87:155-162.
(42) Garrison, R.J., Kannel, W.B., Feinleib, M., Castelli, W.P., McNamara, P.M. and Padget, J.J. 1978. Cigarette smoking and HDL cholesterol— The Framingham offspring study. *Atherosclerosis* 30:17-25.
(43) Geer, J.C., McGill, H.C., Jr., Robertson, W.B. and Strong, J.P. 1968. Histologic characteristics of coronary artery fatty streaks. *Lab. Invest.* 18:565-570.
(44) Goldstein, J.L., Hazzard, W.R., Schrott, H.G., Bierman, E.L. and Motulsky, A.G. 1973. Hyperlipidemia in coronary heart disease. I. Lipid levels in 500 survivors of myocardial infarction. *J. Clin. Invest.* 52: 1533-1543.
(45) Gordon, T., Casteli, W.P., Hjortland, M.C., Kannel, W.B. and Dawber, T.R. 1977. High density lipoprotein as a protective factor against coronary heart disease. The Framingham Study. *Am. J. Med.* 62: 707-714.
(46) Grad, B. and Hoffman, M.M. 1955. Thyroxine secretion rates and plasma cholesterol levels of young and old rats. *Am. J. Physiol.* 182:497-512.
(47) Grundy, S.M. and Ahrens, E.H., Jr. 1970. The effects of unsaturated dietary fats on absorption, excretion, synthesis, and distribution of cholesterol in man. *J. Clin. Invest.* 49:1135-1152.
(48) Haimovici, H. 1977. Atherogenesis: Recent biological concepts and clinical implications. *Am. J. Surg.* 134:173-178.
(49) Hollander, D. and Morgan, D. 1978. Increase in cholesterol intestinal absorption with aging in the rat. *Exp. Gerontol.* 14:201-206.
(50) Hollman, R.L., McGill, H.C., Jr., Strong, J.P. and Greer, H.C. 1958. The natural history of atherosclerosis. The early aortic lesions as seen in New Orleans in the middle of 20th century. *Am. J. Pathol.* 35:209-235.
(51) Hruza, Z. 1971. Effect of endocrine factors on cholesterol turnover in young and old rats. *Exp. Gerontol.* 6:199-204.
(52) Hruza, Z. and Jelinkova, M. 1965. Carbohydrate metabolism after epinephrine. *Exp. Gerontol.* 1:139-145.
(53) Hruza, Z. and Wachtlova, M. 1969. Decrease of cholesterol turnover in old rats. *Exp. Gerontol.* 4:245-250.
(54) Hruza, Z. and Zbuzkova, V. 1975. Cholesterol turnover in plasma, aorta, muscles and erythrocytes in young and old rats. *Mech. of Aging and Devel.* 4:169-179.
(55) Hruza, Z. and Zbuzkova, V. 1972. Decrease of excretion of cholesterol during aging. *Exp. Gerontol.* 8:29-37.
(56) Imai, H., Werthessen, N.T., Taylor, C.B. and Lee, K.T. 1976. Angiotoxicity and arteriosclerosis due to contaminants of USP cholesterol. *Arch. Pathol. and Lab. Med.* 100:565-572.

(57) Kannel, W.B. 1972. The disease of living. *Nutrition Today* 6:2-11.
(58) Kannel, W.B. and Dawber, T.R. 1972. Atherosclerosis as a pediatric problem. *J. Ped.* 80:544-554.
(59) Kent, S. 1978. Lipoprotein metabolism and atherosclerosis. *Geriatrics* 33:93-100.
(60) Keys, A. 1963. The role of the diet in human atherosclerosis and its complications. In M. Sandler and G.H. Bourne (Eds.) *Atherosclerosis and Its Origin*, pp. 263-299. Academic Press, New York.
(61) Khan, M.A. 1980. Unpublished data.
(62) Khan, M.A., Earl, F.L., Farber, T.M., Miller, E., Husain, M.M., Nelson, E., Gertz, S.D., Forbes, M.S., Rennels, M.L. and Heald, F.P. 1977. Elevation of serum cholesterol and increased fatty streaking in egg yolk:lard fed castrated miniature-pigs. *Exp. Mol. Pathol.* 26:63-74.
(63) Kinsell, L.W., Michaels, G.D., Partridge, J.W., Boling, L.A., Balch, H.E. and Cochrane, G.C. 1953. Effect upon serum cholesterol and phospholipid of diets containing large amounts of vegetable fats. *J. Clin. Nutr.* 1:224-231.
(64) Kipshidze, N.N. 1978. Atherosclerosis in long living subjects. In Lars A. Carlson, R. Paoletti, C.R. Sirtori and G. Weber (Eds.). *International Conference on Atherosclerosis*, pp. 365-370. Raven Press, New York.
(65) Kirk, J.E. 1963. Intermediary metabolism of human arterial tissue and its changes with age and atherosclerosis. In M. Sandler and G.H. Bourne (Eds.) *Atherosclerosis and Its Origin*, pp. 67-117. Academic Press, New York.
(66) Kirstein, P. and Olsson, A.G. 1979. HDL-cholesterol is low in young and increases with age in male claudicators. *Atherosclerosis* 33:145-148.
(67) Klimenko, E.D., Martsevich, M.S., Mukhina, A.P. and Shenkman, N.S. 1975. Interaction between the cardiovascular and digestive systems in the genesis of experimental atherosclerosis. *Bull. Exp. Biol. Med.* 80:1164-1167.
(68) Kritchevsky, D. 1979. Diet, metabolism and aging. *Fed. Proc.* 38:2001-2006.
(69) Kwiterovich, P.O., Jr., Chase, G.A. and Bachorik, P.S. 1978. The Columbia population study. I. Plasma cholesterol and triglyceride levels. *The Johns Hopkins Med. J.* 143:32-42.
(70) Lee, K.T., Jarmolyeh, J., Kim, D.B., Grant, C., Krasney, J.A., Thomas, W.A. and Bruno, A.M. 1971. Coronary atherosclerosis, myocardial infarction and "sudden death" in swine. *Exp. Mol. Pathol.* 15:170-190.
(71) Lewis, L.A. and Naito, H.K. 1978. Relation of hypertension lipids and lipoproteins to atherosclerosis. *Clinical Chem.* 24:2081-2098.
(72) Lundbaek, K. 1974. The special role of diabetic angiopathy. *Hormones and Metab. Res.* (Suppl. 4):158-163.
(73) Mann, G.V. 1977. Diet-Heart: End of an era. *New Engl. J. Med.* 297:644:650.
(74) Mann, G.V. 1977. A factor in yogurt which lowers cholesterolemia in man. *Atherosclerosis* 26:335-337.
(75) Martin, G.M. and Sprague, C.A. 1973. Symposium on *in vitro* studies related to atherogenesis: life histories of hyperplastoid cell lines from aorta and skin. *Exp. Mol. Pathol.* 18:125-141.
(76) Mathur, K.S., Patney, N.L. and Kumar, V. 1961. Atherosclerosis in India. An autopsy study of the aorta and the coronary, cerebral, renal and pulmonary arteries. *Circulation* 24:68-75.

(77) McGill, H.C., Jr., Geer, J.C. and Strong, J.P. 1963. Natural history of human atherosclerotic lesions. In M. Sandler and G.H. Bourne (Eds.). *Atherosclerosis and Its Origin*, pp. 39-66. Academic Press, New York.
(78) McGill, H.C. 1968. Introduction to the geographic pathology of atherosclerosis. *Lab. Invest.* 18:465-467.
(79) McNamara, S.S., Molot, M.A., Stremple, J.F. and Cutting, R.T. 1971. Coronary artery disease in combat casualties in Vietnam. *J. Am. Med. Assn.* 216:1185-1187.
(80) Miettinen, T.A. 1974. Current views on cholesterol metabolism. *Hormones Metab. Res.* 6:37-44.
(81) Miettinen, T.A. 1971. Cholesterol production in obesity. *Circulation* 44:842-850.
(82) Montgomery, R., Dryer, R.L., Conway, T.W., Spector, A.A. 1977. *Biochemistry—A Case Oriented Approach*, 2nd Edition, pp. 470-517. C.V. Mosby Company, St. Louis, MO.
(83) Morris, J.N., Chave, S.P.W. and Adams, C. 1973. Vigorous exercise in leisure time and the incidence of coronary heart disease. *Lancet* 1:333-339.
(84) Morton, N.E. 1976. Genetic markers in atherosclerosis: A review. *J. Med. Genetics* 131:81-90.
(85) Nestel, P.J., Whyte, H.M. and Goodman, D.S. 1969. Distribution and turnover of cholesterol in humans. *J. Clin. Invest.* 48:982-991.
(86) Nikkila, E.A. 1969. Control of plasma and liver triglyceride kinetics by carbohydrate metabolism and insulin. *Adv. Lipid Res.* 7:63-134.
(87) Nikkila, E.A. 1978. Metabolic regulation of plasma high density lipoprotein concentrations. *Europ. J. Clin. Invest.* 8:111-113.
(88) Pawliger, D.F. and Shipp, J.C. 1968. Familial hypercholesterolemia: Effects of exogenous cholesterol on cholesterol biosynthesis in vivo and by liver in vitro. *Clin. Res.* 16:51 (Abstr.).
(89) Quintao, E., Grundy, S.M. and Ahrens, E.H., Jr. 1971. An evaluation of four methods for measuring cholesterol absorption by the intestine in man. *J. Lipid Res.* 12:221-232.
(90) Report of Inter-Society Commission for Heart Disease Resources. *Circulation* 42:1-43.
(91) Rhoads, G.G., Gulbransen, C.L. and Kagan, A. 1976. Serum lipoproteins and coronary heart disease in a population study of Hawaiian-Japanese men. *New Eng. J. Med.* 294:293-298.
(92) Rouser, G. and Solomon, R.D. 1969. Changes in phospholipid composition of human aorta with age. *Lipids* 4:232-234.
(93) Schilling, R.J., Christakis, G., Orbach, A. and Becker, W.H. 1969. Serum cholesterol and triglyceride. An epidemiological and pathogenetic interpretation. *Am. J. Clin. Nutr.* 22:133-138.
(94) Smith, L.E. and Shock, V.W. 1949. Intravenous glucose tolerance test in the aged males. *J. Gerontol.* 4:27-33.
(95) Spain, D.M., Natham, D.J. and Gellis, M. 1963. Weight, body type and the prevalence of coronary atherosclerotic heart disease in males. *Am. J. Med. Sci.* 245:63-69.
(96) Spector, W.S. 1961. *Handbook of Biological Data*, pp. 584. W.B. Saunders Co., Philadelphia.
(97) Spina, M. and Garbin, G. 1976. Age-related chemical changes in human elastins from non-atherosclerotic areas of thoracid aorta. *Atherosclerosis* 24:267-279.
(98) Stamler, J. and Epstein, F.H. 1972. Coronary heart disease: Risk factors or guides to preventive action. *Prev. Med.* 1:27-48.

(99) Stein, Y. and Stein, O. 1973. Lipid synthesis and degradation and lipoprotein transport in mammalian aorta. In *Atherogenesis: Initiating Factors*, pp. 165-179, Ciba Foundation Symposium 12. Elsevier, Amsterdam.
(100) Story, J.A. and Kritchevsky, D. 1976. Dietary fiber and lipid metabolism. In G.A. Spiller and R.J. Amen (Eds.). *Fiber in Human Nutrition*. pp. 171-184. Plenum Press, New York.
(101) Stout, R.W., Bierman, E.L. and Ross, R. 1975. Effect of insulin on the proliferation of cultured primate arterial smooth-muscle cells. *Circul. Res.* 36:319-332.
(102) Stout, R.W. and Vallance-Owen, J. 1969. Insulin and atheroma. *Lancet* 1:1078-1080.
(103) Strong, J.P. and McGill, H.C., Jr. 1962. The natural history of coronary atherosclerosis. *Am. J. Pathol.* 40:37-49.
(104) Strong, W.B., Rao, P.S. and Steinbaugh, M. Primary prevention of atherosclerosis: A challenge to the physician caring for children. *South. Med. J.* 68:319-327.
(105) Sylven, C. and Borgstrom, B. 1969. Intestinal absorption and lymphatic transport of cholesterol in the rat: Influence of the fatty acid chain length of the carrier triglyceride. *J. Lipid Res.* 10:351-355.
(106) Thomas, C.B., Ross, D.C. and Duszynski, K.R. 1975. Youthful hypercholesterolemia associated characteristics and role in premature myocardial infarction. *The Johns Hopkins Med. J.* 136:193-208.
(107) Vavrik, M., Priddle, W.W. and Lie, S.F. 1979. Serum cholesterol concentration and atherosclerotic cardiovascular disease in the aged. *J. Am. Geriat. Soc.* 12:56-61.
(108) Vesselinovitch, D., Getz, G.S., Hughes, R.H. and Wissler, R.W. 1974. Atherosclerosis in the rhesus monkey fed three food fats. *Atherosclerosis* 20:303-321.
(109) Watts, H.F. 1971. Basic aspects of the pathogenesis of human atherosclerosis. *Human Pathol.* 2:31-55.
(110) Weber, G., Fabbrini, P., Resi, L., Jones, R., Vesselinovitch, D. and Wissler, R.W. 1977. Regression of arteriosclerotic lesions in rhesus monkey aortas after regression diet: Scanning and electron microscopic observations of the endothelium. *Atherosclerosis* 26:535-547.
(111) Werner, M., Tolls, R.E., Hultin, J.V. and Mellecker, J. 1970. Influence of sex and age on the normal range of eleven serum constituents. *Z. Klin. Chem. Klin. Biochem.* 8:105-115.
(112) Wilson, J.D. 1972. The relation between cholesterol and cholesterol synthesis in baboon. *J. Clin. Invest.* 51:1450-1458.
(113) Wolinsky, H. 1972. Long-term effects of hypertension on the rat aortic wall and their relation to concurrent aging changes. *Circ. Res.* 30:301-309.
(114) Wolinsky, H. and Fowler, S. 1978. Participation of lysosomes in atherosclerosis. *New Engl. J. Med.* 299:1173-1178.
(115) Wood, P., Haskel, W., Klein, H., Lewis, S., Stern, M.P. and Farquhar, J.W. 1976. The distribution of plasma lipoproteins in middle-aged male runners. *Metabolism* 25:1249-1257.
(116) Yamamoto, M. and Yamamura, Y. 1971. Changes in cholesterol metabolism in the aging rat. *Atherosclerosis* 13:365-374.
(117) Zemplenyi, T. and Grafnetter, D. The lipolytic activity of the aorta: Its relation to aging and to atherosclerosis. *Gerontologia* 3:55-64.

–7–
Carbohydrate Nutrition and Aging

Sheldon Reiser
and
Judith Hallfrisch

INTRODUCTION

Since controlled studies of aging in humans are very difficult and costly, most studies have used laboratory animals. Most studies of longevity have measured effects of caloric restriction or of protein restriction. A few studies have measured the effects of the type of dietary carbohydrate on various physiological and biochemical parameters associated with aging. These studies will be discussed in comparative detail. The effects of type of carbohydrate on longevity, maternal and postweaning diets, and on metabolic risk factors associated with degenerative diseases that afflict the aged in this country will be discussed.

General nutritional status of the elderly, although not a direct study of the aging process, gives an indication of the effects of aging on nutritional processes. Physiological and sociological changes during aging which affect carbohydrate nutrition in the elderly will be discussed. Study of the type of diet, including carbohydrates, of three population groups who apparently enjoy very long lives, is included. Nutritional intakes of elderly in this country may indicate nutrients which would be supplemented and types of foods to be emphasized to improve their health and longevity.

ANIMAL STUDIES

Longevity

The restriction of total food intake of laboratory animals increased life expectancy.[4,36,37,40,41,42,53,54,55,57,64] Since no other known environmental condition has as yet been found to be as effective in increasing longevity, these findings illustrate the importance of dietary factors on health and well-being.

The influences of the type and the amount of dietary carbohydrate on the life expectancy of laboratory animals have received little attention. Durand et al[23] determined the effect on the longevity of two strains of rats when fed nutritionally adequate diets containing 39% by weight of either sucrose, cornstarch or glucose. Table 1 summarizes their results. The life span of BHE rats was significantly lower when they consumed the sucrose diet rather than the starch and glucose diet. Kidney disease was the major cause of death in BHE rats. In contrast, type of dietary carbohydrate did not affect longevity of Wistar rats and lung disease was the major cause of death in the Wistar rats. Dietary intakes and body weights of the rats were not reported.

Table 1: Effect of Dietary Carbohydrate (39% by Weight) on Age of Death of BHE and Wistar Rats

Dietary Carbohydrate	..... BHE Rat		.... Wistar Rat	
	Number	Age of Death (days)	Number	Age of Death (days)
Sucrose	38	444±24*	32	583±40
Cornstarch	13	595±34	7	636±43
Glucose	15	543±48	8	565±48

*Mean±SEM.

Source: Adapted from Reference No. 23.

Dalderup and Visser[17] investigated the effect of the replacement of 15% of starch calories by sucrose on the longevity of male and female Wistar rats (Table 2). The average life span of male rats was significantly less ($P < 0.05$) on the sucrose than on the starch diet. The trend was in the same direction in females, but was not significant. Both dietary intake (Table 2) and body weight (not shown) were depressed in the male rats fed the sucrose diet. Thus, differences in either amount of diet consumed early in life or in body weight[58] could not explain the decreased longevity of the sucrose-fed male rats.

Ross and Bras[56] determined life expectancies of COBS-SD male rats fed three diets in which the proportion of casein and sucrose were varied to provide 10, 22, and 51% protein and 70.5, 58.5, and 29.5% carbohydrate, respectively. The diets were fed in a restricted pattern or ad libitum. Table 3 summarizes the longevity of the rats as a function of diet. With each feeding pattern, life expectancy increased as the casein content increased and the sucrose content decreased. Since both the protein and carbohydrate contents of the diets were changing, it is difficult to attribute the effect on longevity to either of these dietary components. However, it appears that the determining factor in longevity might be the carbohydrate component, since an inverse relationship between carbohydrate intake and life expectancy irrespective of the intake of protein or the caloric level has been reported.[57] In view of the other reports specifically implicating sucrose as compared to other carbohydrates as a factor in decreased longevity,[17,23] it would be of interest to determine whether the inverse relationship between dietary carbohydrate intake and life expectancy would be found using carbohydrates other than sucrose.

Table 2: Age at Death and Food Consumption of 22 Male and 22 Female Wistar Rats Consuming Diets in Which 15% of the Starch Calories Were Replaced by Sucrose

Diet	Average Lifespan (days)		Total Food Consumption During the First 6 Weeks (grams)	
	Males	Females	Males	Females
Starch	566	607	473	416
Sucrose	486*	582	410	402

*Significantly less than corresponding starch-fed males ($P < 0.05$).

Source: Adapted from Reference No. 17.

Table 3: Influence of the Proportion of Dietary Protein (Casein) and Dietary Carbohydrate (Sucrose) on the Longevity of Male Rats Fed in a Restricted Pattern or ad Libitum

Diet Casein/Sucrose (%)	Life Expectancy (days)	
	Restricted	Ad Libitum
10/70.5	692	540
22/58.5	838	585
51/29.5	934	614

Source: Adapted from Reference No. 56.

Early Nutrition

The composition of the maternal diet profoundly affects the

growth and metabolism of the progeny. Restriction of dietary energy[11,12] or protein[35,61,72] intakes retarded growth and development in the offspring. A few studies focused on the effects of type of dietary carbohydrate fed pregnant laboratory mammals on various physiological and metabolic parameters in the progeny.

Davis et al[19,20] maintained pregnant mice on either chow or a mixture of 65 parts of chow and 35 parts of sucrose during gestation and lactation. After weaning, the offspring were fed undiluted chow. The body weights and food consumptions of the progeny from dams fed the two different diets are shown in Table 4. At 45 weeks of age, body weight was significantly greater in both male and female offspring of the dams fed the added sucrose.[19] Despite the greater body weight, the progeny from sucrose-fed mothers did not consume more diet per unit body weight, suggesting that they had greater food efficiency. Offspring were given an intraperitoneal glucose tolerance test (1.25 mg/g body weight); blood glucose responses are shown in Table 5. The progeny from the sucrose-fed dams showed significantly elevated fasting blood glucose and glucose response at 2 and 3 hours and a significantly greater lag in the decline of these levels after 6 hours.[19] The serum lipoprotein pattern of the newborn from sucrose-fed mothers resembled that of type IV hyperlipoproteinemia with 50% pre-beta lipoprotein as compared to 5% from the chow-fed mothers.[20] The pre-beta lipoproteins were still elevated at 2 months of age but declined significantly by 4 months of age. Serum cholesterol from the progeny of sucrose-fed dams was also apparently higher than from chow-fed dams.[20]

Table 4: Body Weight and Food Consumption of 45-Week-Old Offspring of Mice Fed Chow Either With or Without Sucrose

Diet	Number	Sex	Body Weight (grams)		Food Consumed per Gram Body Weight	
Sucrose	14	Male	51.8*		0.124	
				$P < 0.001$		NS**
Chow	12	Male	45.2		0.137	
Sucrose	10	Female	50.2		0.124	
				$P < 0.001$		$P < 0.01$
Chow	11	Female	36.9		0.140	

*Mean from number of mice indicated.
**NS = Not Significant.

Source: Adapted from Reference No. 19.

Table 5: Blood Glucose Levels Before and After Progeny from Mice Fed Chow With or Without Sucrose Were Given an Intraperitoneal Glucose Tolerance Test (1.25 mg/g Body Weight)

Diet	Number	Blood Glucose (mg/100 ml)				
		Fasting	1 Hour	2 Hours	3 Hours	6 Hours
Sucrose	28	113*	201	152*	127*	101*
Chow	22	100	184	127	111	90

*Significantly greater ($P < 0.02$) than corresponding chow value.

Source: Adapted from Reference No. 19.

Similar studies of the effect of maternal carbohydrate nutrition have also been carried out.[3,25] Berdanier[3] fed pregnant rats diets containing either 65% sucrose or cornstarch during gestation and lactation. Male pups from each dietary group were then maintained on either the sucrose or cornstarch diets until they were 142 days of age when the effects on various metabolic parameters were determined. Table 6 summarizes some of the results for pups maintained on the cornstarch diet. In contrast to the results with mice,[19,20] sucrose intake during gestation and lactation did not increase weight gain or fasting blood glucose levels in the offspring. Maternal sucrose intake significantly lowered serum triglycerides and significantly increased hepatic lipid. Irrespective of maternal diet, progeny fed sucrose had significantly higher weight gain, serum triglycerides and hepatic glucose-6-phosphate dehydrogenase (G6PD) activity[3] than progeny fed starch. Evers et al[25] also fed female rats diets containing either 65% sucrose or 65% cornstarch during gestation and lactation. Male offspring were then fed the 65% cornstarch diet until they were 60 or 120 days of age. Effects of maternal carbohydrate on body weight, serum triglycerides and cholesterol, hepatic total lipid and G6PD activity, and on blood glucose response to an intraperitoneal glucose tolerance test were determined in the progeny. The only significant difference was that progeny from sucrose-fed dams had higher levels of serum cholesterol at 120 days of age.

The effects of sucrose feeding to pregnant rodents on subsequent weight gain and glucose levels in the progeny differed in mice and rats. In the studies with mice the dilution of the chow diet with sucrose may have lowered the levels of other dietary components that could have influenced weight gain and glucose tolerance. It is also possible that the difference in effect of maternal sucrose feeding is due to species differences.

There is some evidence that metabolic responses to the amount and type of dietary carbohydrate differ between young and older animals.[24] Rats aged 3 weeks, 8 weeks, or 52 weeks were fed diets containing either 40% glucose (control), 68% glucose, 68% fructose or 68% sucrose for 3 days. The activities of the hepatic glycolytic enzymes fructose-1-phosphate aldolase, fructose-1,6-diphosphate aldolase and pyruvic kinase were then determined. The results showed: (1) for each enzyme, increase in enzyme activity was greatest at 3 weeks and was least at 72 weeks, (2) at all ages, enzyme activity was greater for the 68% carbohydrate diets than for the 40% glucose diet, and (3) enzyme activity was greater for animals fed fructose and sucrose than for animals fed a comparable quantity of glucose.

Table 6: Effect of Sucrose or Starch Fed to Pregnant Rats During Gestation and Lactation on Physiological and Metabolic Parameters of 142-Day-Old Male Progeny Fed Starch

Maternal Diet	Progeny Wt Gain (grams)	Serum Glucose (mg/100 ml)	Serum Triglycerides (mg/100 ml)	Serum Cholesterol (mg/100 ml)	Hepatic Lipid (mg/g)	Hepatic G6PD (U/100 g body wt)
Sucrose	333[a]	152[a]	69[a]	82[a]	91.0[a]	3.9[a]
Starch	338[a]	169[a]	180[b]	82[a]	70.4[b]	5.5[a]

Values are mean of six rats.
Values having different superscript letters are significantly different ($P < 0.05$).

Source: Adapted from Reference No. 3.

Moser and Berdanier[44] determined whether the kind of carbohydrate fed rats during the initial postweaning period had long-lasting effects on metabolic parameters after a different carbohydrate was fed for extended times. Weanling rats were fed either 65% sucrose or 65% starch diets until 50 days of age. They were then either switched to the alternate diet or remained on their initial diet until they were 142 days of age. After 142 days, the rats fed sucrose during the initial postweaning period had higher fasting serum insulin and triglycerides and greater hepatic α-glycerol phosphate dehydrogenase than those fed starch initially. Rats initially fed sucrose and then switched to starch had higher liver lipid and cholesterol levels than those fed starch continually. It was concluded that in rats the kind of carbohydrate fed during the initial postweaning period exerted long-lasting metabolic effects even after that carbohydrate had not been fed for as long as 92 days. In contrast, Macdonald et al[38] found no support for the view that the amount or type of dietary carbohydrate consumed by a young rat would result in permanent metabolic changes throughout the remainder of the animal's life. Possibly, differences in the strain of the rats used and in experimental conditions employed (e.g., time of fast before killing) explain these conflicting results.

Risk Factors in Degenerative Diseases

Glucose Tolerance: Glucose tolerance in humans[18] and rats[6] is impaired during aging. In addition, fructose and galactose utilization is impaired in elderly humans.[9] It appears that impaired tissue sensitivity to insulin is the primary factor responsible for the decrease in glucose tolerance with advancing age.[21] Decreases in hormone receptor levels may constitute a common manifestation of the aging process.[59] A few studies have examined the effect of the type of dietary carbohydrate on parameters of glucose tolerance when fed during a considerable portion of the life span of the experimental animal. The results of these studies will be presented in this section.

Cohen et al[13] fed male rats initially weighing 60 to 70 g either standard laboratory chow or synthetic diets containing 72% cornstarch or sucrose. The chow was fed for 8 months and the synthetic diets fed for 12 months. A glucose tolerance test (350 mg/100 g body weight) was then administered to rats fasted overnight. Blood samples were taken from the tail vein at fasting and 60 minutes after the glucose load. Fasting blood glucose was not affected by diet. However, glucose tolerance deteriorated in sucrose-fed rats (149 mg % at 60 minutes, n = 9) as compared to either chow-fed rats (109 mg %, n = 6) or starch-fed rats (106 mg %, n = 9). The activities of two liver gluconeogenic enzymes were also determined in this study. Glucose-6-phosphatase activity was significantly higher in sucrose-fed rats than in either chow-fed or starch-fed rats. In contrast, phosphoenolpyruvate carboxykinase activity was not significantly affected by diet.

Table 7: Effect of Diet (Cornstarch or Cornstarch plus Fructose) and Age on Fasting Serum Insulin and the Insulin Response to an Oral Glucose Tolerance Test (250 mg Glucose/100 g Body Weight) of Rats

Age* (months)	Fasting Insulin		Insulin Response (Σ ½, 1, 2 and 3 hours)	
	54% Starch	39% Starch + 15% Fructose**	54% Starch	39% Starch + 15% Fructose**
3	16***	20	124	192
5	22	30	170	286
7	44	62	387	455
9	56	72	466	550
15†	75	98	449	591

*Significant increase in insulin response with age (ANOVA, $P < 0.001$).
**Significantly higher than starch-fed rats (ANOVA, $P < 0.03$).
***Mean insulin value (μunits/ml serum) from at least eight rats.
†2-hour response value estimated.

Source: Adapted from Reference No. 5.

Blakely et al[5] fed weanling, male Wistar rats diets containing either 54% cornstarch or 39% cornstarch plus 15% fructose for periods up to 15 months of age. Oral glucose tolerance tests (250 mg glucose/100 g body weight) were performed at 3, 5, 7, 9, and 15 months. Insulin and glucose were determined from tail vein blood at fasting and ½, 1, 2, and 3 hours after the glucose load. Body weight and food intake were not significantly different between the two diet groups. Table 7 summarizes the effects of diet and age on fasting insulin and the insulin response to the glucose load. Both fasting serum insulin and the insulin response to a glucose load increased significantly as the age of the rats increased. In addition, rats fed the cornstarch plus fructose had significantly higher insulin values than did rats fed only the starch. Fasting serum glucose decreased with age in starch-fed rats but increased with age in cornstarch plus fructose-fed rats. Serum glucose response after the glucose load tended to be greater in the rats fed fructose. Adverse effects of fructose feeding on glucose tolerance were previously reported in rats[14] and humans.[2]

Blood Lipids and Lipogenesis: Elevated levels of total blood cholesterol[46] and fasting blood triglycerides[46,67] are considered to be risk factors in the etiology of heart disease. In the United States, blood cholesterol levels progressively increase from age 20 and peak at about age 65.[45] Blood triglycerides show a similar increase with age, peaking at 55 in males and 65 in females.[45] Long-term studies with experimental animals showed that the type of dietary carbohydrate can differentially influence the levels of blood lipids and lipogenesis.

Taylor et al[65] fed two strains of weanling rats diets containing either 39% sucrose, cornstarch or glucose. Fasted rats were killed at either 150 or 350 days of age. Blood cholesterol increased with age but was not affected by dietary carbohydrate. Noncholesterol blood lipids also increased with age and were higher in 150-day-old rats fed sucrose than in rats fed either starch or glucose. At 350 days of age, the increase in noncholesterol blood lipids due to sucrose had virtually disappeared. The most striking effect of diet was on liver lipids; at both ages, both cholesterol and noncholesterol lipids were higher in rats fed sucrose than those fed starch or glucose.

Qureshi et al[51] fed weanling rats for 20 weeks diets containing either 60% sucrose or cornstarch. The rats were then transferred to an atherogenic diet (16% hydrogenated coconut oil, 1% cholesterol, 1% cholic acid) containing 47% of their respective carbohydrates. Fasting blood lipids and liver fat were determined after 100 and 180 days on diet. After 100 days (age of rats, 261 days), cholesterol, triglycerides, and liver fat were 36%, 82%, and 12% higher, respectively, when the rats were fed sucrose than when they were fed starch. By 180 days on diet (age of rats, 341 days), the levels of the metabolic parameters had fallen but still were higher in the sucrose-fed rats as compared to the starch-fed rats.

The activity of the hepatic lipogenic enzyme G6PD and of the amount of liver lipid were significantly higher in two strains of rats at ages 3, 6, and 9 months when the diet contained 50% sucrose rather than starch or glucose.[10] Both of these parameters were based on relative body weight and tended to decrease in the sucrose-fed rats as a function of age.

The effects of the long-term feeding of chow, 72% starch or 72% sucrose diets to weanling rats on the activity of various hepatic lipogenic enzymes are summarized in Table 8.[13] Sucrose-fed rats had 5.1 to 6.3 times more enzyme activity than did chow-fed rats and 1.8 to 2.7 times more enzyme activity than did starch-fed rats.

Table 8: Hepatic Lipogenic Enzyme Activity in Rats Fed Different Carbohydrates for Extended Times

Diet	Time on Diet (months)	Number	Enzyme (nmols substrate metabolized/mg soluble protein)		
			G6PD	Malic Enzyme	Acetyl CoA Carboxylase
Chow	8	6	34	26	13*
72% Starch	12	9	85*	49*	40*
72% Sucrose	12	9	214*,**	132*,**	74*,**

*Significantly greater than chow ($P < 0.05$).
**Significantly greater than starch ($P < 0.05$).

Source: Adapted from Reference No. 13.

St. Clair et al[62] fed 2-month-old miniature swine cholesterol-containing diets with either 41% sucrose or cornstarch for 2 years. Fasting blood was collected every 2 months for determination of cholesterol and triglycerides and the extent of atherosclerosis ascertained by examination of the aorta and coronary arteries. There were no significant differences in serum cholesterol due to diet for up to 2 years on study. Similarly, serum triglycerides were not affected by diet. The extremely low levels of triglycerides in the serum (less than 30 mg % after 100 days of age) are probably due to the minor lipogenic capacity of liver as compared to adipose tissue in swine.[48] Aortic atherosclerosis was not affected by diet. However, the coronary arteries of the sucrose-fed swine were significantly more diseased than the arteries of starch-fed swine after 1 year. This effect was not present after 2 years.

HUMAN STUDIES

Factors Affecting Nutritional Status

Dentures: Over 32 million Americans are over 60 years of age.[68]

More than 20,000,000 of these are 65 or over. One-half of all Americans aged 65 have no teeth.[8] By 75 years of age about two-thirds of them have no teeth. The incidence of periodontal disease increased with age to 79% in nearly every subgroup studied in the Ten State Nutrition Survey.[66] Poorly fitting dentures can cause nutritional problems. Although high fiber foods are recommended to avoid constipation, they may be difficult to chew and therefore may be avoided by the elderly, thus aggravating another problem of old age. A British nutrition survey of elderly subjects[47] found that 87% of obese subjects had poor masticatory performance. The study concluded that loss of teeth or ill-fitting dentures might lead to selection of foods that might be high in fats and carbohydrates. The percentage of energy from protein was lower for those with inefficient mastication. Average daily meat intake was 2.9 oz for those with no chewing problems and 2.3 oz for inefficient masticators. This study involved 50 elderly subjects; 40 wore dentures and 10 had a few remaining teeth. A nutritional follow-up survey of old people in Dalby, Sweden[16] also concluded that missing teeth, poorly functioning dentures, and dental decay could explain their avoidance of high fiber foods.

Taste and Smell: Changes in taste and smell are associated with aging. There is a decline in taste sensitivity with age which may be due to a reduction in the number of taste buds per papilla. Taste buds that detect sweet and salt go first and then those that detect bitter and sour.[8] Therefore, more salt and sugar would be required to achieve the same sensations of sweet or salty tastes received at an earlier age. This may also explain reports of sour or bitter tasting food by some older people.[60] The ability to identify foods blindfolded declines with age as does the ability to detect odors. Schiffman[60] studied 256 "normal" elderly people (average age, 70.8 years) and found one-fourth slightly diminished in olfaction, and almost one-half markedly diminished in olfaction. Only 32% of the elderly could detect odors as well as young adults. Cohen and Gitman[15] found that 38.9% of elderly females and 25.8% of elderly males had a reduced satisfaction with taste and smell. George et al[27] reported a higher intake of sugar and salt in elderly men than elderly women. This may be due either to reduced sense of taste or to denture problems.

Intestinal and Renal Factors: Because intestinal muscle tone diminishes with age, the elderly are more prone to constipation than younger individuals. Impaired glucose tolerance also appears to be a serious problem in the elderly. The intake of high-fiber foods promotes proper bowel function[31] and improves glucose tolerance.[52]

Intestinal absorption and efficiency of digestion decrease with advancing age.[70] The secretions of hydrochloric acid and of the digestive enzymes decrease with age. Calcium absorption decreases after age 65 and xylose absorption decreases after age 80.

The reduced secretion of hydrochloric acid may indicate an increased need for nutrients whose absorption is enhanced by it such as iron and vitamin B_{12}. The combination of reduced calcium absorption and bone resorption might indicate an increased requirement for calcium or a change in both calcium and phosphorus requirements. Osteoporosis is a common debilitating disease of the aged which may be a result of chronic calcium deficiency. Winick[70] believes that although this relationship has not been proven, calcium-rich foods should be recommended throughout life.

Other diseases of the gastrointestinal tract that are associated with aging include diverticulosis, gall bladder disease, and colon cancer. Low-fiber diets might be associated with these diseases.[7] Increasing fiber in the diet decreases transit time and prevents constipation, a prevalent complaint of older people. Asp et al[1] found that aged male pensioners ate about one-half the amount of unextracted carbohydrate as men of working age. Women pensioners ate less than women of working age, but the difference was not significant. These results suggest that elderly subjects eat low-fiber diets. It therefore appears that the elderly would benefit from recommendations to modify the present American diet that would result in additional fiber intake.[22] Although most of the physiological effects of fiber would be considered beneficial, or at least not harmful, the effect on mineral balance poses a potential nutritional risk. Table 9 summarizes the results of Kelsay et al[32,33] on the mineral balance of 12 men receiving either high-fiber (20 g neutral detergent) or low-fiber (3 to 6 g neutral detergent) diets for 26 days. The high-fiber diet significantly decreased the mean balances for calcium, magnesium, zinc, and copper; balances for these minerals were negative on the high-fiber diet and positive on the low-fiber diet. These findings illustrate the need for definitive studies to determine the levels and sources of dietary fiber that would produce satisfactory laxation and bowel function and improve glucose tolerance without adversely affecting mineral balances.

Table 9: Mineral Balances of 12 Men During Last 7 Days of 26-Day High- and Low-Fiber Dietary Periods

Mineral	Balance (mg/day)		Significance
	High Fiber	Low Fiber	
Calcium	−122	72	$P < 0.01$
Phosphorus	292	361	NS
Magnesium	−32	25	$P < 0.01$
Iron	4.6	3.8	NS
Zinc	−1.3	3.1	$P < 0.001$
Copper	−0.44	0.18	$P < 0.005$

NS = Not Significant.

Source: Adapted from References 32 and 33.

A number of renal changes related to aging can affect nutrition.[71] Renal blood flow and glomerular filtration rate decline. There is a loss of functioning nephrons. There are also changes in the ability to form urine as age increases. These changes would tend to decrease the efficiency of the kidneys, reducing their ability to reabsorb soluble nutrients such as glucose, amino acids, vitamin C, B vitamins, and some minerals. A metabolite of vitamin D, $1,25(OH)_2$, vitamin D_3, is formed in the kidney. Reduced kidney function may affect vitamin D metabolism in old age. The decrease with age in nephrons may be associated with bone loss occurring in old age. Diuretics, which are used to treat hypertension, cause increased loss of potassium in the urine. They reduce blood pressure, but increase renal blood flow.

Drug Treatment: Many of the elderly are affected with diseases that require chronic drug treatment. Biological half-lives of many drugs seem to be longer in elderly than in younger patients.[28] This could result in greater incidence of adverse reactions in elderly patients. Digitoxin, commonly given to elderly patients, was higher in blood of elderly than in younger patients probably due to reduced kidney function.[26] Anorexia and nausea are two side effects associated with digitoxin treatment. Antipyrine, a vasoconstrictor analgesic and antipyretic, has an extended half-life in older patients due to decreased metabolism by microsomes of the liver.[49] Nausea and vomiting were reported in patients receiving average doses. Combinations of drugs given to elderly patients may also affect their appetites and thus their intakes.

Social Factors: A number of social factors can affect nutrition in the aged. In the British survey of the elderly,[47] of 824 subjects, 225 lived alone, 417 with spouses, 163 with relatives, 19 in other ways. Elderly people living in institutions were not considered. Men aged over 75 years living alone had the poorest nutrient intakes of all groups. Many men aged 65 to 74 years lived alone, but appeared to eat adequately. Women between 65 and 74 years had the best nutritional status of the four groups.

The subjects in the British study also were divided into three economic groups. An increase in meat consumption with social class was seen. Grotkowski and Sims[29] found a positive correlation between nutritional adequacy and socioeconomic status. Self-evaluation of nutritional knowledge was also correlated with nutritional adequacy. A correlation between nutritional attitudes and adequate intake has been reported.[30]

Longevity

Diet has long been considered to influence longevity. There are three areas of the world where people apparently have longer than normal life spans. These three groups of people live in the Caucasus Mountains of Soviet Georgia, Vilacabamba, Ecuador, and the Karahurum Range of Kashmir. All three areas are rural and mountainous. The people lead active lives and work well into old age. The diets of these people are of interest to nutritionists although environment and genetics must also play an important role in their long lives.

Although ages of the Russians[43] and the South Americans[39] may be exaggerated, all three groups do appear to have a high proportion of elderly. Diets in the three regions vary, but there are a few similarities. In all three groups, obesity is rare, consumption of meat is low, and of carbohydrate is high, but the carbohydrate is generally complex and includes a lot of fiber. Leaf[34] has made a study of all three groups searching for the secret of their long lives.

In the Caucasus, the Abkhazians eat meat one or two times a week. They prefer chicken and mutton to fish. Goat cheese, yogurt, and about two glasses of buttermilk are eaten daily. They eat fresh fruits and vegetables including grapes, green onions, lima beans, cabbage, cucumbers, tomatoes and lots of garlic. They drink large quantities of wine with high alcohol content, but no coffee or tea. Only a few of them smoke. They eat cornmeal mush instead of bread. They believe that being overweight is dangerous and that physical activity is necessary. Of 15,000 individuals over 80 years of age, 60% still worked and 70% were still very active. The incidence of atherosclerosis of those in the Caucasus Mountains was one-half that of those living in the sea-level villages. The fact that these people are a combination of Russians, Jews, Armenians, and Turks indicates that genetic factors alone do not explain their long lives.[34]

In Vilacabamba, the average daily intake is 1,200 kilocalories. The residents usually lose their teeth in their early 20s, but their gums become very hard and they can eat fruits and vegetables. The intake of meat and fat is low. They eat a lot of dairy products including one similar to cottage cheese. Dried beans, raw eggs, and Yucca root starch are also included in their diets. Most of the Vilacabambans smoke and drink locally brewed rum. The population is genetically quite homogeneous.[34]

The Hunzukuts of Kashmir have been reported to eat an average of 1,923 kilocalories daily. Carbohydrate intake was 354 g, or over 70% of the total calories. Protein intake was 50 g and fat only 36 g. Like the other two groups of old folks, they eat a low-fat, high-carbohydrate, high-fiber diet.

Nutritional Intake

In this country, recent studies indicated that the elderly may be at risk for deficiencies of certain nutrients. In a preliminary report of the USDA 1977-78 Nationwide Food Consumption Survey,[50] average daily intake for men over 65 was 1,921 kilocalories and for women, 1,418 kilocalories. This is slightly less than intake averages obtained in the 1965 Survey.[69] A survey of nursing home subjects reported that men consume 1,720 kilocalories daily and women consume 1,333 kilocalories daily.[63] In a study of senior citizens groups, elderly men were reported to consume 1,801 kilocalories while women consumed 1,363 kilocalories per day. In the 1977-78 survey, elderly men and women consumed 42 and 46% of their calories from carbohydrates, slightly more than young adults.[50] Total carbohydrate intake was higher in 1977 than in 1965 in women 51 to 64 and over 65 and in men over 65. Stiedemann et al[63] found that elderly men and women consume 48.4 and 49.2% of their calories as carbohydrate. Unfortunately, the type of carbohydrate consumed was not determined in these studies.

In both elderly men and women, calcium intake was higher in the 1977 household survey[50] than in the 1965-66 survey.[69] Both surveys reported caloric intakes below RDAs, especially in women. The vitamin A and vitamin C contents of diets of the elderly increased considerably in the last 12 years. Pao[50] concluded that the elderly may be eating more foods high in vitamin A such as dark green vegetables. Increased fortification of foods with vitamin C probably accounts for the increase in that vitamin in the 1977 survey. Levels of thiamin and niacin were 67 and 64% of RDA for elderly women in nursing homes.[63] Iron intake was 80% of RDA. Calcium intake of the women in nursing homes was low with 43% taking in only about 500 mg.

SUMMARY AND CONCLUSION

Comparatively few studies have measured the effects of the type of dietary carbohydrate on various physiological and biochemical parameters associated with aging. The longevity of rats was decreased when sucrose replaced either starch or casein in the diet, an effect influenced by the strain and sex of the rat. The feeding of sucrose instead of starch to pregnant mice, but not rats, increased blood glucose and triglycerides in the adult offspring. Long-lasting metabolic effects due to the feeding of different carbohydrates for short periods of time immediately after weaning have not been consistently observed in the rat. The long-term feeding of either sucrose or fructose as compared to starch appears to produce undesirable effects on parameters of glucose tolerance in rats. Blood lipids and hepatic lipogenesis were generally greater in rats fed diets containing sucrose as compared to starch over long periods of time. In human population groups with increased longevity, the diets are usually high in complex carbohydrate and fiber and low in meat products. Since constipation and glucose intolerance are very common in the elderly, the intake of foods rich in complex carbohydrate may also be advisable in view of the reported beneficial effects of fiber on laxation and glucose tolerance. However, care should be taken to avoid the decrease in mineral balance reported on high-fiber diets. Proper dentition also appears to be an important factor in the nutrition of the elderly since poorly fitting dentures or loss of teeth appear to promote the intake of diets high in fat and simple sugars and low in fiber. The studies reported in this section favor the view that the intake of complex carbohydrate as compared to simple sugars over a considerable portion of the life span promotes a more desirable metabolic environment for health and well-being.

REFERENCES

(1) Asp, N., Carlstedt, I., Dahlqvist, A., Johansson, C., and Paulsson, M. 1979. Dietary fibre. In B. Borgstrom, A. Norden, B. Akesson, M. Abdulla, and M. Jagerstad (eds.), *Nutrition and Old Age*, chapter 12, pp. 128-137. *Scand. J. Gastroent.* 14, Suppl. 52.

(2) Beck-Nielsen, H., Pedersen, O., and Linkskov, H.O. 1980. Impaired cellular insulin binding and insulin sensitivity induced by high fructose feeding in normal subjects. *Am. J. Clin. Nutr.* 33:273-278.

(3) Berdanier, C.D. 1975. Effect of maternal sucrose intake on the metabolic patterns of mature rat progeny. *Am. J. Clin. Nutr.* 28:1416-1421.

(4) Berg, B.N. and Simms, H.S. 1961. Nutrition and longevity in the rat. II. Longevity and onset of disease with different levels of food intake. *J. Nutr.* 71:255-263.

(5) Blakely, S.R., Hallfrisch, J. and Reiser, S. 1981. Long-term effects of moderate fructose feeding on glucose tolerance parameters in rats. *J. Nutr.* 111:307-314.

(6) Bracho-Romero, E. and Reaven, G.M. 1977. Effect of age and weight on plasma glucose and insulin responses in the rat. *J. Am. Geriat. Soc.* 25:299-302.

(7) Burkitt, D.P., Walker, A.R.P. and Painter, N.S. 1974. Dietary fiber and disease. *J. Am. Med. Assoc.* 299:1068-1074.

(8) Busse, E.W. 1979. Eating in late life: physiologic and psychologic factors. *Contemp. Nutr.* 4(11): November.

(9) Calloway, N.O. and Kujak, R. 1971. Age and the kinetics of responses to sugars and insulin. *J. Am. Geriat. Soc.* 19:122-130.

(10) Chang, M.L.W., Lee, J.A., Schuster, E.M. and Trout, D.L. 1971. Metabolic adaptation to dietary carbohydrate in two strains of rats at three ages. *J. Nutr.* 101:323-329.

(11) Chow, B.F. and Lee, C.J. 1964. Effect of dietary restriction of pregnant rats on body weight gain in the offspring. *J. Nutr.* 82:10-18.

(12) Chow, B.F. and Stephan, J.K. 1971. Fetal undernourishment and growth potential. *Nutr. Rep. Int.* 4:245-255.

(13) Cohen, A.M., Briller, S. and Shafrir, E. 1972. Effect of long-term sucrose feeding on the activity of some enzymes regulating glycolysis, lipogenesis, and gluconeogenesis in the rat liver and adipose tissue. *Biochim. Biophys. Acta* 279:129-138.

(14) Cohen, A.M., Teitelbaum, A. and Rosenmann, E. 1977. Diabetes induced by a high fructose diet. *Metabolism* 26:17-24.

(15) Cohen, T. and Gitman, L. 1959. Oral complaints and taste perception in the aged. *J. Gerontol.* 14:294-298.

(16) Dahlqvist, A. and Asp., N. 1979. Carbohydrates. In B. Borgstrom, A. Norden, B. Akesson, M. Abdulla and M. Jagerstad (eds.), *Nutrition and Old Age*, chapter 11, pp. 121-127 *Scand. J. Gastroent.* 14, Suppl. 52.

(17) Dalderup, L.M. and Visser, W. 1969. Influence of extra sucrose in the daily food on the life span of Wistar albino rats. *Nature* 222:1050-1052.
(18) Davidson, M.B. 1979. The effect of aging on carbohydrate metabolism: a review of the English literature and a practical approach to the diagnosis of diabetes mellitus in the elderly. *Metabolism* 28:688-705.
(19) Davis, R.L., Hargen, S.M. and Chow, B.F. 1972. The effect of maternal diet on the growth and metabolic patterns of progeny (mice). *Nutr. Rep. Int.* 6:1-7.
(20) Davis, R.L., Hargen, S.M. Yeomans, F.M. and Chow, B.F. 1973. Long-term effects of alterations of maternal diet in mice. *Nutr. Rep. Int.* 7:463-473.
(21) DeFronzo, R.A. 1979. Glucose intolerance and aging. Evidence for tissue insensitivity. *Diabetes* 28:1095-1101.
(22) Dietary Goals for the United States (2nd ed.). 1977. Select Committee on Nutrition and Human Needs, United States Senate. U.S. Government Printing Office, Washington, D.C.
(23) Durand, A.M.A., Fisher, M. and Adams, M.A. 1968. The influence of type of dietary carbohydrate. Effect on histological findings in two strains of rats. *Arch. Path.* 85:318-324.
(24) Espinoza, J. and Rosensweig, N.S. 1973. Effect of aging on the response of rat hepatic glycolytic enzyme activities to dietary sugars. *Am. J. Clin. Nutr.* 26:608-611.
(25) Evers, W.D., Chenoweth, W.L. and Bennink, M.R. 1977. Effect on offspring of feeding a sucrose diet during gestation and lactation in rats. *Nutr. Rep. Int.* 15:391-396.
(26) Ewy, G.A., Kapadia, G.C., Yao, L., Lullin, M. and Marcus, F.I. 1969. Digitoxin metabolism in the elderly. *Circulation* 39:449-453.
(27) George, R., Krondl, M. and Lau, D. 1980. Sex as a risk factor in nutrition behavior of adolescent and elderly anglophones. *Fed. Proc.* 39:653.
(28) Gillette, J.R. 1979. Biotransformation of drugs during aging. *Fed. Proc.* 38:1900-1909.
(29) Grotkowski, M.L. and Sims, L.S. 1978. Nutritional knowledge, attitudes and dietary practices of the elderly. *JADA* 73:499-506.
(30) Health Practices and Opinions. 1972. Food and Drug Administration, Washington, D.C.
(31) Kelsay, J.L. 1978. A review of research on effects of fiber intake on man. *Am. J. Clin. Nutr.* 31:142-159.
(32) Kelsay, J.L., Behall, K.M. and Prather, E.S. 1979. Effect of fiber from fruits and vegetables on metabolic responses of human subjects. II. Calcium, magnesium, iron, and silicon balances. *Am. J. Clin. Nutr.* 32:1876-1880.
(33) Kelsay, J.L., Jacob, R.A. and Prather, E.S. 1979. Effect of fiber from fruits and vegetables on metabolic responses of human subjects. III. Zinc, copper, and phosphorus balances. *Am. J. Clin. Nutr.* 32:2307-2311.
(34) Leaf, A. 1975. *Youth in Old Age.* McGraw-Hill, New York.
(35) Lee, C.J. and Chow, B.F. 1965. Protein metabolism in the offspring of underfed mother rats. *J. Nutr.* 87:439-443.
(36) Leto, S., Kokkonen, G.C. and Barrows, C.H. 1976. Dietary protein, life-span and biochemical variables in female mice. *J. Gerontol.* 31:144-148.
(37) Leveille, G.A. 1972. The long-term effects of meal-eating on lipogenesis, enzyme activity, and longevity in the rat. *J. Nutr.* 102:549-556.

(38) Macdonald, I., Rebello, T. and Keyser, A. 1975. An investigation in rats, into the metabolic consequences of early ingestion of dietary carbohydrates. *Nutr. Metab.* 18:283-293.
(39) Mazess, R. 1978. Conference on the Longevous Population of Vilacabamba, Ecuador. National Institute on Aging, Fogerty International Center.
(40) McCay, C.M., Crowell, M.F. and Maynard, L.A. 1935. The effect of retarded growth upon the length of life span and upon ultimate body size. *J. Nutr.* 10:63-79.
(41) McCay, C.M., Maynard, L.A., Sperling, G. and Barnes, L.L. 1939. Retarded growth, life span, ultimate body size and age changes in the albino rat after feeding diets restricted in calories. *J. Nutr.* 18:1-13.
(42) McCay, C.M., Sperling, G. and Barnes, L.L. 1943. Growth, aging, chronic diseases and life span in rats. *Arch. Biochem.* 2:469-479.
(43) Medvedev, Z.A. 1974. A Caucasus and Altay longevity: A biological or social problem? *Gerontologist* 14:381-384.
(44) Moser, P.B. and Berdanier, C.D. 1974. Effect of early sucrose feeding on the metabolic patterns of mature rats. *J. Nutr.* 104:687-694.
(45) Nicholson, J., Gartside, P.S., Siegel, M., Spencer, W., Steiner, P.M. and Glueck, C.J. 1979. Lipid and lipoprotein distributions in octo- and nonagenarians. *Metabolism* 28:51-55.
(46) Norum, K.R. 1978. Some present concepts concerning diet and prevention of coronary heart disease. *Nutr. Metab.* 22:1-7.
(47) Nutrition Study of the Elderly. Report by the Panel on Nutrition of the Elderly. 1972. Dept. Health and Social Security. Report on Health and Social Subjects No. 3, Her Majesty's Stationery Office, London.
(48) O'Hea, E.K. and Leveille, G.A. 1969. Influence of fasting and refeeding on lipogenesis and enzymatic activity of pig adipose tissue. *J. Nutr.* 99:345-352.
(49) O'Malley, K., Crooks, J., Duke, E. and Stevenson, I.H. 1971. Effect of age and sex on human drug metabolism. *Brit. Med. J.* 3:607-609.
(50) Pao, E.M. 1979. Nutrient Consumption Patterns of Individuals in 1977 and 1965. Agricultural Outlook Conference, Session 11, Washington, D.C.
(51) Qureshi, R.U., Akinyanju, P.A. and Yudkin, J. 1970. The effect of an 'atherogenic' diet containing starch or sucrose upon carcass composition and plasma lipids in the rat. *Nutr. Metab.* 12:347-357.
(52) Reiser, S. 1979. Effect of dietary fiber on parameters of glucose tolerance in humans. In G.E. Inglett and S.I. Falkenhag (eds.), *Dietary Fibers: Chemistry and Nutrition*, pp. 173-191. Academic Press, New York.
(53) Riesen, W.H., Herbst, E.J., Walliker, C. and Elvehjem, C.A. 1947. The effect of restricted caloric intake on the longevity of rats. *Am. J. Physiol.* 148:614-617.
(54) Ross, M.H. 1959. Protein, calories and life expectancy. *Fed. Proc.* 18:1190-1207.
(55) Ross, M.H. 1972. Length of life and caloric intake. *Am. J. Clin. Nutr.* 25:834-838.
(56) Ross, M.H. and Bras, G. 1973. Influence of protein under- and overnutrition on spontaneous tumor prevalence in the rat. *J. Nutr.* 103:944-963.
(57) Ross, M.H. 1976. In M. Winick (ed.), *Nutrition and Aging*, Chapter 3, pp. 43-57. John Wiley and Sons, New York.

(58) Ross, M.H. 1977. Dietary behavior and longevity. *Nutr. Rev.* 35:257-265.
(59) Roth, G.S. 1979. Hormone receptor changes during adulthood and senescence: significance for aging research. *Fed. Proc.* 38:1910-1914.
(60) Schiffman, S.S. 1978. Changes in taste and smell in old persons. *Advances in Research*, Vol. 2, pp. 1-6. Duke University Center for the Study of Aging and Human Development.
(61) Srivastava, U., Vu, M. and Goswami, T. 1974. Maternal dietary deficiency and cellular development of progeny in the rat. *J. Nutr.* 104:512-520.
(62) St. Clair, R.W., Bullock, B.C., Lehner, N.D.M., Clarkson, T.B. and Lofland, H.B., Jr. 1971. Long-term effects of dietary sucrose and starch on serum lipids and atherosclerosis in miniature swine. *Exp. Mol. Pathol.* 15:21-33.
(63) Stiedemann, M., Jansen, D. and Harrell, I. 1978. Nutritional states of elderly men and women. *JADA* 73:132-139.
(64) Stuchlikova, E., Juricava-Horakova, M. and Deyl, Z. 1975. New aspects of the dietary effect of life prolongation in rodents. What is the role of obesity in aging? *Exp. Gerontol.* 10:141-144.
(65) Taylor, D.D., Conway, E.S., Schuster, E.M. and Adams, M. 1967. Influence of dietary carbohydrates on liver content and on serum lipids in relation to age and strain of rat. *J. Nutr.* 91:275-282.
(66) Ten State Nutrition Survey. 1972. Department of Health, Education and Welfare, Center for Disease Control.
(67) Tzagournis, M. 1978. Triglycerides in clinical medicine. A review. *Am. J. Clin. Nutr.* 31:1437-1452.
(68) U.S. Bureau of Census. 1977. Current Population Reports, Series P-25, No. 643. U.S. Government Printing Office, Washington, D.C.
(69) U.S. Department of Agriculture. 1972. Household Food Consumption Survey, 1965-66. Report No. 17, U.S. Government Printing Office, Washington, D.C.
(70) Winick, M. 1976. Nutrition and Aging. In *Current Concepts in Nutrition*, vol. 4. John Wiley and Sons, New York.
(71) Winick, M. 1977. Nutrition and aging. *Contemp. Nutr.* 2(6):June.
(72) Zeman, F.J. 1967. Effect on the young of maternal protein restriction. *J. Nutr.* 93:167-173.

The Current Status of Ascorbic Acid, Vitamin B_6, Folic Acid and Vitamin B_{12} in the Elderly

Jeng M. Hsu

INTRODUCTION

Vitamins are chemical compounds essential for the maintenance of life. Because the human organism cannot synthesize these substances, a continuous and adequate supply of such micronutrients in the diet is indispensable, either in the active form or as inactive precursors.

From a theoretical standpoint one would believe that the required nutrients for old people should be lower than for younger people. With advancing age the need for caloric intake decreases because of reduced activity and body mass.[56] However, there are other complex factors involved in the formulation of nutritional requirements for the elderly. The physiological demand for B-vitamins is related to the amount of ingested calories, and any reduction in caloric intake should correlate to a corresponding decrease in the body's need for vitamins. This concept, however, is often challenged by dietitians. With aging, the body is exposed to various insults which result in a marked physiological deterioration in the gastrointestinal system. For example, the ability of parietal cells to secrete HCl declines with age.[118] There is a relative insufficiency in the capacity to digest protein,[195] a general reduction in the secretion of digestive juice,[118,195] and a decrease of calcium absorption[28] in old age. These conditions may have a marked effect in the nutritional requirement of older people. In addition, a variety of psychlological, ethnic, sociological and economic factors affect food habits. Thus, it would not be surprising to discover that many older individuals increase their vitamin intake. Whether such increased nutrient intake has any effect on the subject's health or longevity has not been conclusively established.

The purpose of this report is to update the current knowledge and to present a general review of four water-soluble vitamins: ascorbic acid, pyridoxine, folic acid, and cobalamin. Emphasis will be given to their involvement in the aging process.

ASCORBIC ACID

General Consideration

Ascorbic acid (vitamin C) participates in several reactions and is considered to be of importance in the aging process. This vitamin is concerned fundamentally with the formation of intercellular substances, including the collagen of fibrous tissue structure, the matrices of bone, cartilage and dentin, and all nonepithelial cement substances including that of the vascular endothelium. In the absence of the protection afforded by this vitamin, the condition known as scurvy develops.

Ascorbic acid is mainly synthesized from glucose through the intermediate formation of glucuronic acid and gulonic acid. Species such as man, monkey, and guinea pig lack the liver microsomal enzyme, L-gulonolactone oxidase, which converts L-gulonolactone into ascorbic acid. It is this missing enzyme that necessitates the dietary requirement of vitamin C. On ingestion, ascorbic acid is easily absorbed in the normal human individual from the upper part of the small intestine. The exact mechanism is not clear; it may be absorbed by simple diffusion or by active sodium dependent transport. Following absorption into the portal blood, it passes into the tissues and maintains an equilibrium between the reduced form, ascorbic acid, and the oxidized form, dehydroascorbic acid, with only a small fraction in the latter state. Both forms are antiscorbutic. According to Horning,[91] ascorbic acid and not dehydroascorbic acid is the preferred form of transport of the vitamin in pituitary, adrenals, lung, liver, kidneys, bone, skin, and nasal mucosa. On the other hand, dehydroascorbic acid and not ascorbic acid is the preferred form of uptake by the neutrophils, erythrocytes and lymphocytes.[15] In the erythrocytes, dehydroascorbate is enzymatically reduced to ascorbate by the reduced glutathione-dependent dehydroascorbic acid-reducing enzyme.[91,98] Dehydroascorbate reduction also occurs chemically in erythrocytes. In the leukocytes the reduction of dehydroascorbate to ascorbate is achieved through coupled NADPH oxidation.[183] In man, the ingested vitamin is excreted in the form of urinary ascorbic acid, dehydroascorbic acid, oxalate and a number of other metabolites, of which only ascorbate-2-sulfate has been identified.[3] The compound has been shown to be present in the liver, spleen, adrenals, bile, lymph, and urine of the rat.[92,144] The biological role of ascorbate sulfate is of some interest since this metabolite has been shown to cure scorbutic symptoms in fish[73] but not in guinea pigs.[120] It appears that the antiscorbutic activity of ascorbate sulfate may be species dependent, and at present time it would be premature to make suggestions concerning its biological activity in primates or humans. Radioisotopic studies in rats[31] and experiments with cultured human fibroblasts[18] suggest that ascorbate sulfate may be involved in sulfation reactions.

Studies in human subjects have indicated that respiratory exchange is not a significant excretory route for ascorbate catabolism.[49,5,77] Less than one percent of the ingested ^{14}C labeled ascorbic acid was converted to respiratory $^{14}CO_2$.[6]

Vitamin C Deficiency in the Elderly

The elderly, in general, are reputed to be in poor vitamin C status.[70,154,190] The evidence is based on the observation of the decreased ascorbic acid levels in whole blood,[116,117] serum,[133] plasma,[20,111,197] leukocytes,[1,20,105,111,127] and in the cerebrospinal fluid.[23] Impaired absorption and poor utilization are associated with low blood ascorbic acid levels in a few cases.[117,154] Inadequate dietary intake, however, is considered to be the principal cause of these findings.[114,149,155]

In a report for an aging population Steinkamp and coworkers[184] indicated that the mean intakes of nutrients generally satisfied the Recommended Daily Allowances (RDA); however, dietary intake of calcium, vitamin A and vitamin C were below the recommendation. A similar trend was noted in the diet of 100 elderly people who lived alone in New York State.[108] Of those surveyed, 59% were deficient in vitamin A and 40% in vitamin C. A 1956 survey of 100 low-income elderly indicated that one quarter of the residents were dietary deficient in vitamin C, riboflavin, calcium, and iron.[131] The reduced intake of vitamin C (71% RDA) was also observed by LeBovit[122] in a survey of 283 recipients in Rochester.

By using radioactive labeled compounds, L-(1-^{14}C) ascorbic acid and L-(4-^3H) ascorbic acid, Baker et al[4,5] found that when each of nine volunteers received 75 mg of ascorbic acid daily, the body pool sizes of this vitamin were remarkably constant ranging from 1,486–1,542 (mean 1,500) mg despite substantial variation in age, height and body weight. The mean value fell to 300 mg after 55 days on an ascorbic acid-free diet. These findings are comparable to those of Hellman and Burns[77] who reported body ascorbic acid pools of three patients to be 26, 21 and 19 mg/kg/day.

Plasma level of ascorbic acid is affected by dietary intake and is about 12 mg/ℓ in a diet containing 100 mg. The value below 1 mg/ℓ is an indication of a high-risk level for scurvy.[103]

Kirk and Chieffi[116] indicated that whole blood ascorbic acid levels were significantly decreased with age in men, but showed no major change in women. Similarly, Roderuck et al[165] reported no significant changes with age in whole blood, plasma and serum ascorbic acid concentration in women. On the other hand, Brook and Grimshaw[25] observed a significant decline in the plasma vitamin C concentration with increasing age in both men and women.

The human cell for which there is the most information about vitamin C concentration is the leukocyte. Its ascorbic acid content reflects a better index of vitamin C storage and depletion. The normal level is about 150 mg/ℓ, and a value below 75 mg/ℓ indicates a high risk of developing scurvy.

According to Milne et al[139] the leukocyte ascorbic acid concentration decreased significantly with aged women but not in aged men. Loh and Wilson[128] also observed that the fall in leukocyte ascorbic acid was greater in elderly women than in elderly men. Thus, some effects of both age and sex on ascorbic acid metabolism are indicated.

The highest concentrations of ascorbic acid in human beings are reported in the adrenal and pituitary glands.[115,173] Other organs including the brain, pancreas, spleen, eye lens, kidneys, liver, lung, heart, muscle and thymus also contain appreciable amounts of ascorbic acid. The effect of age on ascorbic acid contents in various organs obtained from autopsies is shown in Table 1. The figures are averages and show very high concentrations in early life and steady fall with increasing age.[200] Whether these changes are related to the findings of diseases, dietary intakes and plasma concentrations in life is not clear. However, additional evidence indicating that the ascorbic acid contents of the tissues decrease with advancing age has been demonstrated in rats[142] and in chick embryo.[202]

Table 1: Effect of Age on Vitamin C Content in Human Organs[200]

Age:	1-30 Days	1-10 Years	11-45 Years	46-77 Years
No. of Subjects:	11	11	17	19
Adrenal	581*	550	393	230
Brain	460	433	–	110
Pancreas	365	225	152	95
Liver	149	163	135	64
Spleen	153	157	127	81
Kidney	153	98	98	47
Lung	126	58	65	45
Heart	76	42	42	21
Thymus	304	190	–	46

*Values represent means for the number of mg found in each kg of tissue.

Hazards of Excessive Vitamin C

Millions of people in the United States are routinely taking large quantities of vitamin C in the belief that this may protect them from the common cold, prevent cancer and improve their general health. In contrast, evidence from various workers indicates that massive doses (over 250 mg per day on a continuing basis) do not appear to be beneficial and may even be harmful to some people.

Schrauzer and Rhead[175] reported that persons taking 1 gram or more per day of vitamin C for extended periods show signs of "systemic conditioning" characterized by lower plasma and red cell ascorbic levels and high urinary ascorbic acid excretion. This systematic conditioning appears to have no detrimental effects as long as the high intake of vitamin C is maintained, but sudden cessation of the excessive intake may lead to development of mild scurvy. Herbert and Jacob[85] have recently noted that megavitamin C (100 to 500 mg), when added to a typical Veterans Administration hospital diet, could destroy 50% of vitamin B_{12} in the diet. They recommended regular evaluation of vitamin B_{12} status in anyone taking more than 0.5 gram of vitamin C daily. Other possible hazards occurring with large doses include oxalic aciduria, uric acidemia, increased urinary calcium and urinary sodium in humans.[76] Furthermore, use of vitamin C at high dosage levels increases susceptability of red cells to hemolysis.[137] In one case, it was apparently involved as a causative factor producing death in a patient with glucose-6-phosphate dehydrogenase deficiency.[3]

Requirement

Human requirements for vitamin C have been investigated extensively. For many years it has been recognized that 10 mg of ascorbic acid per day will prevent the development of scurvy in adult man. Considerable evidence indicates that the desirable intake for the maintenance of good health is much larger than this. The preferable amount has been equated with maintenance of plasma concentration in excess of 0.6 mg/100 ml which required an intake of at least 60 mg or more of ascorbic acid for a 70 kg man.[75] Now, the recent work of Baker et al[4,5] on ascorbic acid metabolism in man provides evidence for lowering the RDA for ascorbic acid in adult man from 60 to 45 mg/day. No specific recommendation was given regarding the daily requirement of ascorbic acid intake for the aged.

Estimates of vitamin C requirements for the elderly have been based on studies of blood and urinary responses to varying intake. Bowers and Kubik[20] studied 100 men and women, 56 to 92 years old, and concluded that vitamin C requirements for the aged are not greater than those of young people. Roderuck et al[165] found that an ascorbic acid intake of 1.1 mg or more per day per kg of body weight from self-selected diets provides the elderly woman with apparently satisfactory amounts of this vitamin. In another study, Morse and her associates[143] investigated the amount of ascorbic acid required for tissue saturation. These workers stated that an intake of 57 mg of ascorbic acid a day resulted in saturation of the leukocytes in both younger and older women. On the other hand, Gander and Niederberger,[60] on the basis of the results of urinary ascorbic acid levels after oral doses, concluded that elderly people require about 50% more vitamin C than do young people. This concept has received support from the work of Kirchmann[113] and Rafsky and Newman.[155]

Finally, it is important to emphasize that the RDA is not a

scientifically precise value. Individual variations and other factors such as environmental temperature,[145] physical exertion,[146] tobacco smoking,[29] wound healing[9] and infection[52] are all associated with the requirement of vitamin C. An individual receiving an intake below the RDA cannot be regarded as having an inadequate intake for ascorbic acid unless clinical and biochemical manifestations are obtained to support this conclusion. More accurate measurement of vitamin C requirements must await further research to obtain a better understanding of the biochemical role of this vitamin. Thus, a correct judgment will be formulated which reflects human needs more accurately.

VITAMIN B_6

General Consideration

Vitamin B_6 is a group name and consists of three pyridine-related compounds: pyridoxine, pyridoxal and pyridoxamine. These may appear in tissues and foodstuffs in the free form or combined with phosphate or with phosphate and protein. The physiological active form of vitamin B_6 is the coenzyme, pyridoxal phosphate, which can be formed from any of the three compounds. The coenzyme of vitamin B_6 is necessary for transamination, racemization, decarboxylation and desulfuration. Pyridoxal phosphate is also required in the conversion of tryptophan to niacin; for the activity of glycogen phosphorylase, an enzyme by which glycogen is broken down to glucose; for the formation of antibodies; for the synthesis of delta-aminolevulinic acid, a precursor of the porphyrin ring and possibly for the conversion of linolein to arachidonic acid. Additional information on the biochemical roles of vitamin B_6 is available in the cited reviews.[26,78,187]

Absorption, Transport and Metabolism

Absorption of vitamin B_6 occurs in the jejunum and the ileum but primarily in the upper small intestine. In experimental animals and in man there is a linear relationship between the intake of vitamin B_6 and urinary excretion of 4-pyridoxic acid.[19,125] These observations indicate that vitamin B_6 is absorbed by diffusion. Absorption from the colon was very slight[22] and passive transport was suggested by studies *in vitro* in the animal intestine.[177]

In liver, ingested pyridoxine is phosphorylated by a specific kinase and then oxidized to pyridoxal phosphate by a specific flavoprotein.

Vitamin B_6 is found in blood and tissues principally in bound form, either pyridoxal or pyridoxamine phosphate. In adult human subjects, concentration of pyridoxine phosphate in blood is less than 10 mμg/ml; in leukocytes, 1 to 2 μg/g; in liver, 6 to 9 μg/g; in muscle, 2 to 5 µg/g; and in nerve tissue, 0.5 to 2 µg/g.[64] The vitamin B_6 level in skin is approximately 1 µg/g when measured by *Saccharomyces carlsbergensis* suggesting that most, if not all, of the tissue vitamin B_6 is present in the form of the coenzyme.

Approximately 90% of the pyridoxine administered to man is oxidized to an inactive metabolite, 4-pyridoxic acid, and excreted in this form. Small amounts of pyridoxal, pyridoxamine and pyridoxine and their phosphorated derivatives are also found in the urine.

Biochemical Assessment

A number of biochemical procedures have been developed and used in the evaluation of vitamin B_6 nutritional status.[171]

Plasma levels of pyridoxine-5'-phosphate, the biologically active form of vitamin B_6, have been shown to be a reliable index of vitamin B_6 nutrition.[171] Hamfelt,[74] employing the tyrosine decarboxylase method to measure pyridoxine phosphate, found a striking difference in plasma pyridoxine phosphate with age. Supplementation of pyridoxine markedly increased the plasma pyridoxine phosphate levels and corrected the biochemical alterations. Labadarios et al[121] recently observed that 22 out 31 patients with decompensated cirrhosis or subacute hepatic necrosis had a significant decrease of plasma pyridoxine phosphate levels.

Tryptophan load test is based on the requirement for pyridoxine-5'-phosphate in the conversion of tryptophan metabolite, 3-hydroxykyurenine to 3-hydroxyanthranilic acid. In the absence of pyridoxine, 3-hydroxykyurenine is converted to xanthurenic acid, the excess amount of which is excreted in the urine. The increased amounts of urinary xanthurenic acid have been observed in vitamin B_6 deficient dogs[57] and rats.[124] This led to the demonstration by Greenberg et al[68] that a dietary restriction of vitamin B_6 in man caused an increased excretion of xanthurenic acid following a load with tryptophan. Subsequently, this load test has become a useful technique for detecting or evaluating vitamin B_6 deficiency. It is important to mention that high xanthurenic acid levels in human urine have been reported in patients with riboflavin deficiency[44] in toxemia of pregnancy[193] and in elderly subjects.[157]

Urinary excretion of 4-pyridoxic acid has also been used as an indicator of vitamin B_6 status. It is the principal excretion product from vitamin B_6 ingestion in man; however, isotopic studies[107] as well as other investigations[171] have demonstrated that only 40 to 50% of vitamin B_6 ingested in a normal diet is excreted in the urine as 4-pyridoxic acid. This method is, therefore, not a totally reliable indicator of vitamin B_6 status.

The measurements of transaminase activities in erythrocytes, leukocytes and plasma have frequently been suggested as a criterion of vitamin B_6 nutritional status. This measurement represents a biochemical function test that reflects the state of deficiency or the degree of depletion of vitamin B_6 storage. The transaminase system is based on the dependence of pyridoxine phosphate of the alanine and aspartic transaminase in the body fluid. Deficiency of vitamin B_6 resulted in decreased enzyme activities in erythrocytes, leukocytes and serum.[14,41,125] The alanine enzyme appears to be more sensitive.[2,24] Plasma contains considerably less transaminase activities than do the erythrocytes. In addition, plasma transaminase activities observed in normal subjects have a wide range.[171] Chenney et al[36] have studied transaminase activities in human erythrocyte and were in agreement with others that erythrocyte activity provided a much closer reflection of vitamin B_6 intake than plasma enzyme activity. Thus, the measurements of these enzyme activities in erythrocytes appear to be a better method for detecting and evaluating vitamin B_6 deficiency in humans.

Vitamin B_6 Deficiency in the Aged

The existence of a mild state of vitamin B_6 deficiency among the aged has been suggested.

Boxer et al[21] found that the content of pyridoxal phosphate in leukocytes from children was significantly higher than in adults but maintained that the hypothesis of a regression of the pyridoxal phosphate content with age needed more extensive testing in various age groups.

When serum glutamic-oxaloacetic transaminase (GOT) was taken as a measure of the state of vitamin B_6 nutriture, Ranke and her associates[157] found that the enzyme content was lower in sera from 54 old healthy individuals than from 60 young ones, but it increased after oral supplementation with vitamin B_6. These workers also noted that elderly individuals excreted significantly more xanthurenic acid in the urine after a tryptophan load test than did young adults. Similar alteration of vitamin B_6 metabolism with advancing age was also reported by Hamfelt.[74] He observed a striking decrease of plasma pyridoxine phosphate and an increased urinary xanthurenic acid excretion after tryptophan load test in elderly subjects. In addition, he found that an increase in serum GOT level after pyridoxal phosphate incubation was much more pronounced in the aged group than in the younger subjects.[74]

Since stimulation of erythrocyte GOT and GPT (glutamic-pyruvic transaminase) *in vitro* is frequently considered a better index of vitamin B_6 status than the basic enzyme activity,[156] this technique was employed by Jacobs et al[104] to determine vitamin B_6 status in the aged. They found that there was a decline of erythrocyte GPT content with age, both when basic activity was measured and when pyridoxine phosphate stimulation was used. The ratio of basic to stimulated GPT activity was not correlated in men, but in women there was a slight decrease with age. In neither sex was there a correlation of basic or stimulated GOT activity with age. Stimulation of erythrocyte GPT is a measure of the amount of apoenzyme present; its decline with increasing age suggests that the older individuals not only have reduced amounts of coenzyme but that this has resulted in a reduced synthesis of apoenzyme.

Since pyridoxine deficiency in animals[69,163] develops arteriosclerosis, and since this disease primarily occurs in the aged, the relationship between age and vascular tissue pyridoxine concentrations was investigated by Kheim and Kirk.[112] Their findings indicate that lower pyridoxine concentrations were generally recorded for arterial specimens from children than from adults. No statistically significant changes with age were found for vascular specimens derived from 20 to 84 year old individuals, but a tendency was noted for the pyridoxine level of aortic tissue to increase with age and for the vena cava to decrease. No correlation was observed between assayed pyridoxine concentration and arterial GOT and GPT activities.

Driskell[49] has recently studied the vitamin B_6 status of 17 males and 20 females, above age 60, by measuring the activity of erythrocyte alanine aminotransferase after coenzyme stimulation and concluded that one third of the subjects had an inadequate vitamin B_6 status.

The reason for this increasing vitamin B_6 deficiency with age is not apparent. It is unlikely that dietary inadequacy was responsible for these metabolic changes since the subjects used by various investigators[74,104,157] were known to be healthy eaters. It is possible that the older individuals have a reduced absorption for vitamin B_6, a defect in the synthesis of pyridoxine phosphokinase, or an increased excretion of pyridoxic acid in the urine. Further studies are clearly needed to determine these possibilities.

Requirement

The need for vitamin B_6 is proportional to the amount of protein ingested. The recommended allowance for men and women is 2.0 mg daily.[55] This provides a reasonable margin of safety and permits a protein intake of 100 grams or more. The requirement may be higher than 2.0 grams when the protein intake exceeds 150 grams. An intake of 0.76 mg of vitamin B_6 was not adequate for men even with protein intakes as low as 30 g/day.[126]

FOLIC ACID

General Consideration

Folic acid (pteroylglutamic acid, folate, folacin) is the nutritionally essential precursor of a large family of compounds that serve as coenzymes for one carbon transfer reactions. These coenzymes are found in practically all living systems ranging from viruses and bacteria to man. The molecule structure contains a pteridine derivative, para-aminobenzoic acid and glutamic acid. The modifications of any part of structural units such as the reduction of a one carbon fragment in N^5 and/or N^{10} can form the folate coenzyme. Baugh and

Krumdieck[12] estimated that there could be at least 150 different folacins if the poly-gamma-glutamyl side chain does not contain more than six residues. The chemistry of the folacins has been recently reviewed in detail by Rodriguez[166] and Krumdieck et al.[119]

Folic acid is converted to its coenzyme form by a pyrimidine nucleotide-dependent enzyme, dihydrofolate reductase. The folacin coenzymes are involved in the transfer of one carbon unit throughout the body. They are required for the synthesis of pyrine and pyrimidine bases, for the interconversion of glycine and serine, for the biosynthesis of methionine methyl group and for the degradation of histidine. In addition, the vitamin has been shown to increase jejunal glycolytic enzymes[168] and to be involved in the desaturation and hydroxylation of long-chain fatty acids in the brain.[137] Recent research in animals indicates that folacin coenzymes are required in a number of different reactions in the brain. The discovery of folacin in mouse synaptic region[135] suggests that the vitamin might represent an important role in neural metabolism and function. Fehling et al[53] observed that the central nervous system of rats is resistant to systemic folate depletion whereas the peripheral nerves are depleted to the same degree as the extra neural tissues. During aging, physiological changes take place in the central nervous system.[192] It is not clear whether these changes will alter the utilization or requirement for the nutrients involved in brain metabolism. Our recent knowledge on enzyme reactions involving folacin coenzymes has been reviewed by Stokstad and Koch[186] and Rader and Huennekens.[153] A deficiency of the vitamin is clinically manifested as megaloblastic anemia.[16]

Absorption, Transport and Storage

Current evidence indicates that food folate is absorbed in the proximal jejunum, although it is capable of being absorbed from entire length of the small intestine. Folate in food is present in polyglutamate forms. These polyglutamates must be deconjugated by folic deconjugase (gamma glutamylcarboxypeptidase) to mono-diglutamates for absorption to occur.[167] The absorption of food folate is thus controlled by the deconjugating mechanism which may, in turn, be affected by conjugase inhibitors in some food, such as yeast. Studies in the isolated segments of dog intestine indicate that the rate of absorption of conjugated folates appears to be inversely proportional to the length of the gamma glutamyl side chain.[13] These observations, if true in the human, will emphasize the importance of complete enzymatic digestion of the folate conjugates in various foods so that a good source of folate can be properly classified.

Like vitamin B_{12}, folate has two separate and distinct absorption mechanisms. The active mechanism is energy-dependent absorption. The passive mechanism is probably diffusion and accounts for absorption of a small quantity of free folate in the intestine.

Folate, bound to a protein, is transported to bone marrow cells, reticulocytes and other tissues by both an energy-dependent active manner and passive diffusion. Methyl-folate (N^5-methyl-tetrahydrofolate) appears to be the main form of the vitamin in serum, liver and other tissues.[81] Normal total folate stores are in the range of 5 to 10 mg of which approximately half is in the liver.[84] Normal serum values in adults range from 3.0 to 12.0 ng/ml with an average of around 6.0 ng/ml.[194] Approximately 0.1 mg of folate is normally excreted in the bile daily.[79]

Biochemical Assessment

The megaloblastic anemia is a manifestation resulting from advanced deficiency of either folic acid or vitamin B_{12}. Regardless of the causes, the morphological changes observed in the blood and bone marrow are identical. Thus, the hemotological sign alone cannot be used to distinguish between these two vitamin deficiencies. For this reason, a number of biochemical measurements have been established to detect the specific nutritional state of folacin in the individual subject.

Of the procedures indicated, measurement of serum (plasma) folic acid is most commonly used. The predominant form of folate in plasma is N^5-methyl-tetrahydrofolic acid[86] no polyglutamate being present. Plasma folate levels were widely measured by microbiological method using *Lactobacillus casei*,[194] the only organism which responds to N^5-methyl-tetrahydrofolic acid. In recent years, radioisotopic assays for serum folate have been developed.[169,170] The radioassay for methyl tetrahydrofolic acid using various folate binders and various radioactive labeled forms is sensitive, versatile and is not influenced by drugs such as folate antagonists or antibiotics which might affect the microbiological assay.[169]

The folate level in red blood cells has been regarded as a more accurate and less variable quantitative index than serum folate of the severity of folacin deficiency in patients. Red cells contain polyglutamyl derivatives which require degradation by conjugase prior to bioassay. It should be mentioned that red cell folate levels do not distinguish between megaloblastic anemia due to B_{12} deficiency and that due to a folic deficiency.[90] The low folate values in both serum and red cells provide a strong indication of folic acid deficiency which is usually accompanied by morphological and hematological changes in bone marrow.

Formiminoglutamic acid (FIGLU) and urocanic acid are degradation metabolites of histidine. These compounds are normally converted to glutamic acid through the participation of formiminotransferase and the acceptor, tetrahydrofolic acid. In folic acid deficiency, the conversion is inhibited and the amounts of FIGLU and urocanic in the urine are increased. The urinary excretion of these compounds are further increased in the folic acid deficient subject after an oral dose of 15 grams of L-histidine.[129] The measurement of urinary FIGLU or of urinary FIGLU plus urocanic acid has thus become a useful tool for the detection of folate deficiency.

Newer tests including the deoxyuridine suppression test,[87] the estimation of ^{14}C-formate conversion into serine by lymphocytes,[181] the determination of $^{14}CO_2$ production from serine-3-^{14}C[48] appear promising and may prove useful in research studies.

Folate Status in the Elderly

It is only within recent years that a satisfactory method for estimating the serum folate level has become available and, since this test was introduced, a number of investigators have reported the occurrence of nutritional megaloblastic anemia due to folic acid deficiency in the elderly. Elwood et al[51] in a survey of 533 subjects aged 65 years and over in South Wales, found that 8% had reduced red cell folate levels. Read et al[159] reported that 40 of the 51 elderly males and females admitted to an old people's home had a significant lower serum folate level as compared with the control subjects, a group of apparently healthy individuals of similar age. Further studies by Batata et al[11] and Read et al[159] showed that low serum folate levels in geriatric patients resulted most frequently from a poor intake of folate containing foods such as milk and fresh uncooked fruits and vegetables. A direct correlation between low dietary folate intake and low levels of serum folate in geriatric patients was reported by Hurdle and associates.[100,101,102] They found the mean serum and red cell folacin values of those patients consuming less than 80 µg free folacin activity per day were 4.2 and 119 ng/ml respectively. The mean values for those who consumed more than 79 µg per day were 6.1 and 202 ng/ml. In another study Hurdle[99] measured folate intake and again found that geriatric patients consumed less folate daily than normal controls. In addition to low dietary intake of folate, other factors have been suggested to cause folate deficiency in man.[82] These include (a) destruction of folate by processing and cooking of food, (b) inadequate absorption, (c) inadequate utilization, (d) increased excretion of folate, and (e) increased requirement of folate.

Requirement

Folates are present in food of both plant and animal origin. The vitamin is light-sensitive and heat-labile, and losses occur during cooking. Data thus from surveys are less accurate and only provide general guidelines for estimating adequate amounts of dietary folacin. Estimates of the folacin content of the diet of healthy adults vary with the locale and the assay method.

The Recommended Dietary Allowances (RDA) by the National Academy of Sciences National Research Council,[55] designed for the maintenance of good nutrition of practically all healthy people in the United States, is 400 µg of folate for all males and females aged

11 years or older. These allowances are set at a level above the normal requirement so as to allow a margin of safety to encompass such factors as variable absorption of folate from different foodstuffs, variable conjugase inhibitors in foodstuffs, internal microbes and interrelationship of folate utilization with other nutrients (iron, ascorbic acid, vitamin B_{12} and methionine). No specific value is recommended for the elderly.

VITAMIN B_{12}

General Consideration

The discovery of liver therapy by Minot and Murphy[140] in 1926 for pernicious anemia first drew attention to the possible nutritional aspects of this disease. Since half a pound of liver had to be consumed daily, Castle[32] began to suspect that some digestive process of the normal stomach was concerned in the etiology of the condition. The results of his clinical observations[33,34] suggested the presence of an extrinsic factor in the beef muscle and an intrinsic factor in normal human gastric juice. The isolation of a red crystalline compound designated vitamin B_{12} from liver was achieved independently in 1948 by a pharmaceutical industry research team of Merck scientists[162] working in the United States, and by a research team at Glaxo Laboratories in England.[128] This new vitamin was found to be clinically active[196] and identical with extrinsic factor of Castle[33] which is essential for normal bone marrow erythropoietic activity in man. In 1964 Hodgkin and coworkers[89] delineated the structure formula of vitamin B_{12} by x-ray crystallography.

The chemical structure of vitamin B_{12} was established as an extremely complex nitrogenous compound containing two major portions, the corrin nucleus which includes cobalt and the attached nucleotide. The nucleotide (5,6-dimethylbenzimidazole) is connected to ribose by an alpha-glycoside linkage. A second bond between the two major parts of the molecule is the coordinate linkage of the cobalt atom to one of the nitrogen atoms of the nucleotide. The naturally occurring forms of this vitamin include cyanocobalamin (vitamin B_{12}), hydroxycobalamin (vitamin $B_{12}a$), aquocobalamin (vitamin $B_{12}b$) and nitrocobalamin (vitamin $B_{12}c$).

Coenzyme B_{12} (5-deoxyadenosylcobalamin or adenosylcobalamin) and methylcobalamin (methyl-B_{12}) are the two vitamin B_{12} coenzymes (cobamides) known to be metabolically active in man and constitute the dominant forms of B_{12} in mammalian tissues. Adenosylcobalamin is required for the conversion of methylmalonyl coenzyme A to succinyl coenzyme A. Lack of this cofactor tends to increase the urinary excretion of methylmalonate which has been used as a diagnostic test of vitamin B_{12} deficiency. Methylcobalamin, an essential cofactor in the conversion of homocysteine to methionine, regenerates from N^5-methyltetrahydrofolic ($N^5CH_3FH_4$) which is converted to tetrahydrofolic acid (FH$_4$). FH$_4$ is required to form N^5,N^{10}-methylene-FH$_4$, the cofactor essential in the conversion of deoxyuridylate (dUMP) to deoxythymidylate (dTMP), for DNA synthesis. Cobamides are thus involved in folic acid metabolism. Since N^5-methyltetrahydrofolic acid is the dominant form of folate in human tissues and since this methyl compound may only return to the body's folate pool by the presence of methylcobalamin, deficiency of vitamin B$_{12}$ will result in an accumulation of methylfolate which is metabolically useless. The folate trap hypothesis[88] has been claimed as the primary mechanism underlying the defects in DNA synthesis when deficiency of vitamin B$_{12}$ or folic acid occurs singly or jointly.

Absorption, Transport and Storage

There are two separate and distinct mechanisms for the absorption of vitamin B$_{12}$.[81] The first one involves the following steps: (a) ingested B$_{12}$ is freed from polypeptide linkages to food by gastric acid and gastric, pancreatic, and intestinal enzymes; (b) the free vitamin B$_{12}$ is attached to gastric intrinsic factor (IF) to form a complex; (c) this complex is carried down to specific receptors on the mucosal brush border of the distal ileum where calcium ion and a pH above 6 exist; (d) in a manner not fully understood, vitamin B$_{12}$ is released from its complex and is transported across the cell into the portal venous blood, but the IF is either hydrolyzed or returned to the intestinal luman.[199] In the blood circulation, vitamin B$_{12}$ is bound to a vitamin B$_{12}$-binding-beta-globulin (transcobalamin II, MW 35,000) which delivers the vitamin to liver, bone marrow cells, reticulocytes and possibly other tissues as well. A second vitamin B$_{12}$-binding alpha-globulin (transcobalamin I, MW 121,000) carries most of the vitamin B$_{12}$ in plasma. It has been suggested that transcobalamin I serves as a storage function since it binds vitamin B$_{12}$ more tightly than II does. A third vitamin B$_{12}$-binding protein is present in serum, normally in smaller quantity and with different electrophoretic mobility from the other two.[72,80]

The second mechanism of vitamin B$_{12}$ absorption is simple diffusion and occurs throughout the small intestine. This is the mechanism which makes possible oral therapy of vitamin B$_{12}$ deficiency due to vitamin B$_{12}$ malabsorption.

In addition to intrinsic factor, digestive enzymes and a receptor site in the epithelial cells of the ileum, a number of other factors are also associated with vitamin B$_{12}$ absorption. Deficiency of pyridoxine[95] or iron[201] will reduce B$_{12}$ absorption whereas deficiency of folic acid will increase it.[97] The deficiency of other vitamins in the B-group have essentially no effect. Prefeeding of the test subjects, shortly before testing, will cause an elevation of absorption.[94] The absorption is also greater if the vitamin B$_{12}$ is present in three meals than if it is all provided in a single meal. However, as the quantity of the vitamin increases in the diet, the percent absorbed decreases.

Normal storage of vitamin B_{12} in human tissues is around 1 to 10 mg,[158] of which 50 to 90% is in the liver. There is no evidence for significant catabolism of vitamin B_{12} by man. This vitamin is excreted mainly in the bile and reabsorbed in ileum.[148] The whole body turnover of vitamin B_{12} is between 0.1 and 0.2% daily regardless of whether body stores are normal or reduced.

Pernicious anemia results not from inadequate ingestion of vitamin B_{12}, but from a defect in production by the parietal cells of the stomach or one or more mucoproteins (intrinsic factor) essential for the normal absorption of dietary vitamin B_{12} from the intestine. Purified IF from human gastric juice (MW 50,000) binds one mol of vitamin B_{12} per mol of complex. The protein contains about 13% carbohydrate including glucose, mannose, fucose, N-acetylglucosamine, N-acetylgalactosamine and a sialic acid. Biological activity is rapidly destroyed by incubation with neuraminidase.

Biochemical Assessment

The biochemical procedures employed to evaluate vitamin B_{12} status are designed to establish whether a deficiency exists and, if so, whether the deficiency is due to an impaired absorption of vitamin B_{12}.

The most common procedure useful for the diagnosis of vitamin B_{12} deficiency has been the determination of the serum or plasma vitamin B_{12} level. As a general rule, a serum B_{12} level below 100 pg per ml is diagnostic for vitamin B_{12} deficiency.[83] Microbiological assay methods are commonly used for this purpose.[7,67,172] Of several organisms, *L. leichmannii* (ATCC 7830) is the least troublesome and appears to be most widely employed in the assay for serum vitamin B_{12}.

Serum vitamin B_{12} levels can also be determined with the radioactive vitamin B_{12} isotope technique.[71] This method is equally reliable and reproducable.

The urinary excretion of methylmalonic acid (MMA) is often increased in patients with a vitamin B_{12} deficiency.[8,10,43,65,198] This elevated excretion probably results from a decreased activity in methylmalonyl CoA mutase, a vitamin B_{12} coenzyme linked enzyme catalyzing the isomerization of methylmalonyl CoA to succinyl CoA.[50,54,123,134,180,182,185]

Normal subjects excreted less than 12 mg of methylmalonic acid in 24 hours,[42,61,66,110,130] while patients with B_{12} deficiency may excrete more than 300 mg in 24 hours.[61] Since folate deficiency is not accompanied by an increased urinary excretion of methylmalonic acid,[27,65] this measurement can be used to distinguish a vitamin B_{12} deficiency from a folate deficiency. However, the amount of methylmalonic acid in the urine does not necessarily correlate with the severity of vitamin B_{12} deficiency.[147] Additionally, methylmalonic aciduria also exists as a result of an inborn metabolic error.[42,65,93,132]

For these reasons, plus the fact that 25% of individuals with B_{12} deficiency show no methylmalonic aciduria,[35] its quantative measurement in urine is not routinely used as a practical index for the evaluation of vitamin B_{12} status.[65]

According to Herbert et al[45,87] and others,[46,138] the dU suppression test is of particular value in biochemical assessment of B_{12} status. In this test, human bone marrow cells[87,138] or peripheral blood lymphocytes[46] are incubated with a radioactive precursor of DNA. By the amount of radioactivity incorporation into cellular DNA in the presence or absence of vitamin B_{12} one can determine the nutritional status of vitamin B_{12}. Recent findings[45] further claim that this test can detect a past B_{12} deficiency even after the start of treatment because the resting lymphocytes reflect the nutritional status of the patient at the time the lymphocytes were produced.

The Schilling test[174] is employed primarily to determine whether vitamin B_{12} deficiency is a result of (a) lack of intrinsic factor (pernicious anemia), (b) other forms of malabsorption, or (c) dietary deficiency of vitamin B_{12}. Pernicious anemia can be readily diagnosed by determining the amount of radioactivity excreted in the urine after an oral dose of vitamin $B_{12}{}^{60}Co$ or ^{57}Co given with and without intrinsic factor.[150,151,152,160] Normal subjects absorb 45 to 80% of the administered vitamin B_{12} in comparison to 0 to 17% by patients with pernicious anemia. If vitamin B_{12} deficiency is the result of malabsorption other than that of the pernicious anemia such as sprue[164] and ileitis,[109] the administration of intrinsic factor will have little or no improvement in the absorption of the orally-administered vitamin $B_{12}{}^{60}Co$.

Tissue Distribution

Because serum B_{12} level is normally maintained at the expense of body stores, an evaluation of the magnitude of vitamin B_{12} storage in the liver and other vital organs in disease may help to explain the mechanism of elevation or depression of serum vitamin B_{12} observed. The distribution of vitamin B_{12} in human tissues is shown in Table 2.[96]

Table 2: Concentration of Total Cobalamin in Human Organs[96]

Tissue	No. of Samples	Total Cobalamin (ng/g wet wt)
Liver	6	653±385*
Kidney	6	337±330
Adrenal	6	161±51
Heart	6	119±41
Pancreas	3	116±100
Spleen	5	55±50
Lung	3	50±53
Brain	2	27±13

*Mean ± S.D.

The tissues are ranked in order of diminishing concentration of the vitamin, which was greatest in the liver. Considerable variations between replicate tissue sample concentrations are indicated by the magnitude of the standard deviations listed. The relatively high concentration of the vitamin in the adrenals in all the subjects examined suggests that this vitamin may be involved in the metabolism of steroid compounds or the maintenance of the normal function of these glands. The data in Table 3[96] show that there is an upward trend in hepatic vitamin B_{12} concentration associated with increased age of the subjects studied.

Table 3: Liver Total Cobalamin Concentration at Different Ages[96]

Age	No. of Cases	Total Cobalamin (ng/g wet wt)
Stillborn	5	56±28*
4-48 hours	3	108±16
0.5-2 years	2	383±120
20-40 years	5	310±111
40-60 years	6	598±168
60-85 years	6	775±511

*Mean ± S.D.

Swendseid et al[188] found the average value of vitamin B_{12} in liver tissue obtained at autopsy from 132 individuals was approximately 0.7 µg per gram of wet tissue. In cirrhosis, the average value fell to 0.26 µg per gram. It was concluded that there are no apparent differences in vitamin B_{12} stores in the various age groups. The reduction of liver vitamin B_{12} content in the patients with cirrhosis was also reported by Hsu et al.[96]

Vitamin B_{12} Deficiency

The literature on the behavior of serum vitamin B_{12} in old age contains a great deal of confusion which, in part, is related to the experimental condition and diet.

Mollin and Ross[141] first reported that a group of elderly patients had a significantly lower mean serum B_{12} level than the healthy hospital staff under 40 years of age. Since then, reduction of serum vitamin B_{12} with advancing age has been found by several investigators.[17,40,58] The most convincing data came from the work of Gaffney and associates.[58] Their study involved 528 males and females in five groups, 12 to 94 years of age, living under substantially different conditions and revealed significant agewise regresssions of serum B_{12} level for each of the groups. The agreement in regression lines between the several groups widely different in background and living conditions suggests that the B_{12} level as measured by the amount of *Lactobacillus leichmannii*[178] is relatively unaffected by these conditions. In a recent study, Jagerstad et al,[106] using the radioisotope method[191] for the determination of serum B_{12}, also observed that mean values of serum vitamin B_{12} levels for male and female pensioners decrease as age advances. On the other hand, several groups[11,188] have not found any reduction with age. In fact, Meindok and Dvorsky[136] reported higher serum levels of vitamin B_{12} in older age groups compared with young controls. These controversial findings indicate that factors other than age may be important in the regulation of serum B_{12} concentration.

One physiologic change observed in older persons is a decreased capacity of the gastric juice to bind vitamin B_{12}.[38,39,40] Achlorhydria and hypochlorhydria also are found in the aged and it is thus possible that marginal B_{12} differences due to poor absorption may occur with some frequency in the aged. Glass and associates[62] have shown by means of their hepatic uptake technique that the absorption of orally-administered vitamin B_{12} is poor in subjects with achlorhydria. However, these authors[63] did not observe any significant difference in the amount of radioactive vitamin B_{12} retained by the liver of the young and old individuals after oral administration of vitamin B_{12}. Gaffney et al[59] also reported that there was no demonstrable age differences in urinary excretion of orally-administered $^{60}CoB_{12}$. Davis et al[47] and Tauber et al[189] found no impairment of B_{12} absorption in the aged population.

Requirement

The Food and Nutrition Board of the National Research Council recommends 3 µg of vitamin B_{12} per day for adults.[55] No specific value is recommended for people over 51 years of age. Pernicious anemia patients who have been treated to replenish their stores will meet their body needs with 1.5 mg B_{12} daily, given parenterally. The joint FAO/WHO Expert Group[161] recommends a daily intake of 2 µg of vitamin B_{12} for the normal adult.

CONCLUDING REMARKS

Despite an abundant food supply in most Western developing countries, it is evident that certain nutritional deficiencies may occur in elderly populations as a result of individual behavioral, sociological and cultural factors. In this review, specific nutrient deficiencies such as vitamin C, pyridoxine, folic acid and vitamin B_{12} have been identified in a significant proportion of the aged. Whether these observations are associated with the method of evaluation, poor eating habits, or the basic physiological changes during aging are not apparent. Obviously much more clinical research is needed to determine these possibilities. In the meantime, it may be advisable to administer selected vitamins to the elderly as a prophylactic measure, particularly the B-complex vitamins and ascorbic acid, which are often found in abnormally low levels in their blood and urine.

REFERENCES

(1) Andrews, J. and Brook, M. 1966. Leucocyte-vitamin-C content and clinical signs in the elderly. *Lancet* 1:1350-1351.
(2) Babcock, M.J., Brush, M. and Sostman, E. 1960. Evaluation of vitamin B_6 nutrition. *J. Nutr.* 70:369-376.
(3) Baker, E.M., III, Hammer, D.C., March, S.C., Tolbert, B.M. and Canham, J.E. 1971. Ascorbate sulfate: A urinary metabolite of ascorbic acid in man. *Science* 173:826-837.
(4) Baker, E.M., Hodges, R.E., Hood, J., Sauberlich, H.E. and March, S.C. 1969. Metabolism of ascorbic-1-^{14}C acid in experimental human scurvy. *Am. J. Clin. Nutr.* 22:549-558.
(5) Baker, E.M., Hodges, R.E., Hood, J., Sauberlich, H.E., March, S.C. and Canham, J.E. 1971. Metabolism of ^{14}C and ^{3}H-labeled L-ascorbic acid in human scurvy. *Am. J. Clin. Nutr.* 24:444-454.
(6) Baker, E.M., Levandoski, N.G. and Sauberlich, H.E. 1963. Respiratory catabolism in man of the degradative intermediates of L-ascorbic-1-^{14}C acid. *Proc. Soc. Exp. Biol. Med.* 113:379-383.
(7) Baker, S.J. 1967. Human vitamin B_{12} deficiency. *World Rev. Nutr. Diet.* 8:62-78.
(8) Barness, L.A., Young, D., Mellman, W.J., Kahn, S.B. and Williams, W.J. 1963. Methylmalonate excretion in a patient with pernicious anemia. *N. Eng. J. Med.* 268:144-146.
(9) Bartlett, M.K., Jones, C.M. and Ryan, A.E. 1940. Vitamin C studies on surgical patients. *Ann. Surg.* 111:1-26.
(10) Bashir, H.V., Hinterberger, H. and Jones, B.P. 1966. Methylmalonic acid excretion in vitamin B_{12} deficiency. *Br. J. Haematol.* 12:704-711.
(11) Batata, M., Spray, G.H., Bolton, F.G., Higgins, G. and Wollner, L. 1967. Blood and bone marrow changes in elderly patients with special reference to folic acid, vitamin B_{12}, iron and ascorbic acid. *Br. Med. J.* 2:667-669.
(12) Baugh, C.M. and Krumdieck, C.L. 1971. Naturally occurring folates. *Ann. N.Y. Acad. Sci.* 186:7-28.
(13) Baugh, C.M., Krumdieck, C.L., Baker, H.J. and Butterworth, C.E., Jr. 1971. Studies on the absorption and metabolism of folic acid. I. Folate absorption in the dog after exposure of isolated intestinal segments to synthetic pterolylpolyglutamates of various chain lengths. *J. Clin. Invest.* 50:2009-2021.
(14) Baysal, A., Johnson, B.A. and Linkswiller, H. 1966. Vitamin B depletion in man: blood vitamin B_6, plasma pyridoxal-phosphate, serum cholesterol, serum transminases and urinary vitamin B_6 and 4-pyridoxic acid. *J. Nutr.* 89:19-23.
(15) Bigley, R.H. and Stankova, L. 1974. Uptake and reduction of oxidized and reduced ascorbate by human leukocytes. *J. Exper. Med.* 139:1084-1092.
(16) Blakley, R.L. 1969. *The Biochemistry of Folic Acid and Related Pteridines, Frontiers of Biology*, vol. 13, pp 29-32, North-Holland Pub. Co., Amsterdam-London.
(17) Boger, W.P., Wright, L.D., Strickland, S.C., Gylfe, J.S. and Ciminera, T.L. 1955. Vitamin B_{12} correlation of serum concentration and age. *Proc. Soc. Exp. Biol. Med.* 89:375-378.
(18) Bond, A.D. 1975. Ascorbate-2-sulfate metabolism in human fibroblasts. *Ann. N.Y. Acad. Sci.* 258:307-313.

(19) Booth, C.C. and Brain, M.C. 1962. The absorption of tritium-labeled pyridoxine hydrochloride in the rat. *J. Physiol.* (London): 164:282-294.
(20) Bowers, E.F. and Kubik, M.M. 1965. Vitamin C levels in old people and the response to ascorbic acid and to the juice of the acerola *(Malpighia punicifolia L.). Br. J. Clin. Prac.* 19:141-147.
(21) Boxer, G.E., Pruss, M.P. and Goodhart, R.S. 1957. Pyridoxal-5-phosphoric acid in whole blood and isolated leukocytes of man and animals. *J. Nutr.* 63:623-636.
(22) Brain, M.C. and Booth, C.C. 1964. The absorption of tritium-labeled pyridoxi HCl in control subjects and in patients with intestinal malabsorption. *Gut*, 241-247.
(23) Bertolini, A.M. 1969. Vitamins and old age. In *Gerontologic Metabolism*, edited by A.M. Bertolini, pp 370-389. Charles Thomas, Springfield, IL.
(24) Brin, M., Tai, M., Ostrashever, A.S. and Kalinsky, H. 1960. The relative effects of pyridoxine deficiency on two plasma transaminase in the growing and in the adult rat. *J. Nutr.* 71:416-420.
(25) Brook, M. and Grimshaw, J.J. 1968. Vitamin C concentration of plasma and leukocytes as related to smoking habit, age and sex of human. *Am. J. Clin. Nutr.* 21:1254-1258.
(26) Broquist, H.P. and Trupin, J.S. 1966. Amino acid metabolism. *Ann. Review Biochem.* 35:231-274.
(27) Brozovic, M., Hoffbrand, A.V., Dimitriadore, A. and Mollin, D.L. 1967. The excretion of methylmalonic acid and succinic acid in vitamin B_{12} and folate deficiency. *Br. J. Haematol.* 13:1021-1032.
(28) Bullamore, J.R., Wilkinson, R., Gallagher, J.C., Nordim, B.F.C. and Marshall, P. 1970. Effect of age on calcium absorption. *Lancet* 2:535-537.
(29) Calder, J.H., Curtis, R.C. and Fore, H. 1963. Comparison of vitamin C in plasma and leucocytes of smokers and nonsmokers. *Lancet* 1:556.
(30) Campbell, G.D., Jr., Steinberg, M.H. and Bower, J.D. 1975. Ascorbic acid-induced hemolysis in G-6-D deficiency. *Ann. Int. Med.* 82:810.
(31) Campeau, J.D. and March, S.C. 1972. Distribution of sulfur-35 in subcellular fraction from rat tissues following incubation with ascorbate-3-$^{35}SO_4$. *Fed. Proc.* 31:705 (Abstract).
(32) Castle, W.B. 1929. Observations on the etiologic relationship of achylia gastrica to pernicious anemia. I. The effect of the administration to patients with pernicious anemia of the contents of the normal human stomach recovered after the ingestion of beef muscle. *Am. J. Med. Sci.* 178:748-764.
(33) Castle, W.B. 1953. Development of knowledge concerning the gastric intrinsic factor and its relation to pernicious anemia. *N. Eng. J. Med.* 249:603-614.
(34) Castle, W.B. and Townsend, W.C. 1929. Observations on the etiologic relationships of *Achylia gastica* to pernicious anemia. II. The effect of the administration to patients with pernicious anemia of beef muscle after incubation with normal human gastric juice. *Am. J. Med. Sci.* 178:764-777.
(35) Chanarin, I., England, J.M., Mollin, C. and Perry, S. 1973. Methylmalonic acid excretion studies. *Br. J. Haematol.* 25:45-53.
(36) Chenney, M.C., Sabry, Z.I. and Beaton, G.H. 1965. Erythrocyte glutamic-pyruvic transaminase activity in man. *Am. J. Clin. Nutr.* 16:337-338.

(37) Chida, N., Hirona, H. and Arakawa, T. 1972. Effects of dietary folate deficiency on fatty acid composition of myelin cerebroside in growing rats. *Tohoku, J. Exp. Med.* 108:219-224.
(38) Chow, B.F. 1954. Disturbances in the metabolism of vitamin B_{12} in diabetes and their significance. In *Newer Concept of Causes and Treatment of Diabetes Mellitus*, pp 105-114. National Vitamin Found., Inc., New York.
(39) Chow, B.F., Rosen, D.A. and Lang, C.A. 1954. Vitamin B_{12} serum levels and diabetic retinopathy. *Proc. Soc. Exp. Biol. Med.* 87:38-39.
(40) Chow, B.F., Wood, R., Horonick, A. and Okuda, K. 1956. Agewise variation of vitamin B_{12} serum levels. *J. Geront.* 11:142-146.
(41) Cinnamon, A.D. and Beaton, J.R. 1970. Biochemical assessment of vitamin B_6 in man. *Am. J. Clin. Nutr.* 23:696-702.
(42) Contreras, E. and Giorgio, A.J. 1972. Leukocyte methylmalonyl-CoA mutase. I. Vitamin B_{12} deficiency. *Am. J. Clin. Nutr.* 25:695-702.
(43) Cox, E.V. and White, A.M. 1962. Methylmalonic acid excretion: An index of vitamin B_{12} deficiency. *Lancet* 2:853-856.
(44) Dalgleish, C.E. 1956. Interrelationships of tryptophan, nicotinic acid and other B vitamins. *Br. Med. Bulletin* 12:49-51.
(45) Das, K.C. and Herbert, V. 1976. Use of the lymphocyte deoxyuridine (dU) suppression test and lymphocyte chromosome study for retrospective diagnosis of vitamin B_{12} and/or folate deficiency despite "shotgun" treatment. *Clin. Res.* 24:480A.
(46) Das, K.C. and Hoffbrand, A.V. 1970. Lymphocyte transformation in megaloblastic anemia. Morphology and DNA synthesis. *Br. J. Haematol.* 19:459-468.
(47) Davis, R.L., Lawton, A.H., Prouty, R. and Chow, B.F. 1965. The absorption of oral vitamin B_{12} in an aged population. *J. Geront.* 20:169-172.
(48) Degrazia, J.A., Fish, M.B., Pollycove, M., Wallerstein, R.O. and Hollander, L. 1972. The oxidation of the beta carbon of serine in human folate and vitamin B_{12} deficiency. *J. Lab. Clin. Med.* 80:395-404.
(49) Driskell, J.A. 1978. Vitamin B_6 status of the elderly. In *Human Vitamin B_6 Requirements*, pp 252-256. National Academy of Sciences, Washington, DC.
(50) Eggerer, H., Overath, P. and Lunen, F. 1960. On the mechanisms of the cobamide coenzyme dependent isomerization of methylmalonyl CoA to succinyl CoA. *J. Am. Chem. Soc.* 82:2643.
(51) Elwood, P.C., Shinton, N.K., Wilson, C.I.D., Sweetnam, P. and Frazer, A.C. 1971. Haemoglobin, vitamin B_{12} and folate levels in the elderly. *Br. J. Haematol.* 21:557-563.
(52) Faulkner, J.M. and Taylor, F.H.L. 1937. Vitamin C and infection. *Ann. Intern. Med.* 10:1867-1873.
(53) Fehling, C., Jagerstad, M., Lindstrand, K. and Elmquist, D. 1976. Reduction of folate levels in the rat: Difference in depletion between the central and the peripheral nervous system. *Z. Ernahrungswiss.* 15:1-8.
(54) Flavin, M. and Ochoa, S. 1957. Metabolism of propionic acid in animal tissue. *J. Biol. Chem.* 229:965-979.
(55) FNB (Food and Nutrition Board, National Research Council) 1974. *Recommended Dietary Allowances*, 8th edition, p 29. National Academy of Science, Washington, DC.
(56) Forbes, G.B. and Reina, J.C. 1970. Adult lean body mass declines with age: Some longitudinal observations. *Metabolism* 191:653-663.

(57) Fouts, P.J. and Lepkovsky, S. 1942. A green pigment-producing compound in urine of pyridoxine-deficient dogs. *Proc. Soc. Exptl. Biol. Med.* 50:221-222.

(58) Gaffney, G.W., Horonick, A., Okuda, K., Meier, P., Chow, B.F. and Shock, N.W. 1957. Vitamin B_{12} serum concentrations in 528 apparently healthy human subjects of ages 12-94. *J. Geront.* 12:32-38.

(59) Gaffney, G.W., Watkin, D.M. and Chow, B.F. 1959. Vitamin B_{12} absorption: Relationship between oral administration and urinary excretion of cobalt-60 labeled cyanocobalamin following a parenteral dose. *J. Lab. Clin. Med.* 53:525-534.

(60) Gander, J. and Niederberger, W. 1936. Uebenden vitamin C-bedarf alter leute. *Munchener Med. Wochenschr.* 83:1386-1389.

(61) Giorgio, A.J. and Paunt, G.W.E. 1965. A method for the colorimetric determination of urinary methylmalonic acid in pernicious anemia. *J. Lab. Clin. Med.* 66:667-676.

(62) Glass, G.J.B., Boyd, L.J., Gellin, G.A. Stephanson, L. 1954. Uptake of radioactive vitamin B_{12} by the liver in humans: Test for measurement of intestinal absorption of vitamin B_{12} and intrinsic factor activity. *Fed. Proc.* 13:54.

(63) Glass, G.B.J., Goldbloom, A., Boyd, L.J., Laughton, R., Rosen, S. and Rich, M. 1956. Intestinal absorption and hepatic uptake of radioactive vitamin B_{12} in various age groups and the effect of intrinsic factor preparations. *Am. J. Clin. Nutr.* 4:124-133.

(64) Goldsmith, G.A. 1964. Vitamins and antivitamins. In *Disease of Metabolism*, edited by G.G. Duncan, pp 567-663. W.B. Saunders Co., Philadelphia.

(65) Gompertz, D. and Hoffbrand, A.V. 1970. Methylmalonic aciduria. *Br. J. Haematol.* 18:377-381.

(66) Gompertz, D., Jones, J.H. and Knowles, J.P. 1967. Metabolic precursors of methylmalonic acid in vitamin B_{12} deficiency anemia. *Clin. Chim. Acta* 18:197-204.

(67) Grasbeck, R. 1960. Physiology and pathology of vitamin B_{12} absorption distribution and excretion. *Adv. Clin. Chem.* 3:299-366.

(68) Greenberg, L.D., Bohr, D.F., McGrath, H. and Rinehart, J.F. 1949. Xanthurenic acid excretion in the human subject on a pyridoxine-deficient diet. *Arch. Biochem.* 21:237-239.

(69) Greenberg, L.D., McIvor, B. and Moon, H.D. 1958. Experimental arteriosclerosis in pyridoxine-deficient rhesus monkeys. *Am. J. Clin. Nutr.* 6:635-637.

(70) Griffiths, L.L., Brocklehurst, J.C., Scott, D.L., Marks, J. and Blackley, J. 1967. Thiamine and ascorbic acid levels in the elderly. *Geront. Clin.* 9:1-10.

(71) Grossowicz, N., Sulitzeanu, D. and Merzbach, D. 1962. Isotopic determination of vitamin B_{12} binding capacity and concentration. *Proc. Soc. Exp. Biol. Med.* 109:604-608.

(72) Hall, C.A. 1971. Vitamin B_{12} binding protein of man. *Ann. Int. Med.* 75:297-301.

(73) Halver, J.E., Smith, R.R., Tolbert, B.M. and Baker, E.M. 1975. Utilization of ascorbic acid in fish. *Ann. N.Y. Acad. Sci.* 258:81-101.

(74) Hamfelt, A. 1964. Age variation of vitamin B_6 metabolism in man. *Clin. Chem. Acta.* 10:48-54.

(75) Harper, A.E. 1975. The recommended dietary allowances for ascorbic acid. *Ann. N.Y. Acad. Sci.* 258:491-494.

(76) Hayes, K.C. and Hegsted, D.M. 1973. Toxicity of the vitamins. In *Toxicants Occurring Naturally in Food*, 2nd edition pp 235-253. National Academy of Sciences, Washington, DC.

(77) Hellman, I. and Burns, J.J. 1958. Metabolism of L-ascorbic acid-1-^{14}C in man. *J. Biol. Chem.* 230:923-930.

(78) Henderson, L.M. and Hulse, J.D. 1978. Vitamin B_6 relationship in tryptophan metabolism. In *Human Vitamin B_6 Requirement* pp 21-36. National Academy of Sciences, Washington, DC.

(79) Herbert, V. 1965. Excretion of folic acid in bile. *Lancet* 1:913.

(80) Herbert, V. 1968. Diagnostic and prognostic values of measurement of serum vitamin B_{12}-binding proteins. *Blood* 32:305-312.

(81) Herbert, V. 1973. Folic acid and vitamin B_{12}. In *Modern Nutrition in Health and Disease*, edited by R.S. Goodhart and M.E. Shils, pp 221-244. Lea and Febiger, Philadelphia.

(82) Herbert, V. 1973. The five possible causes of all nutrient deficiency: illustrated by deficiencies of vitamin B_{12} and folic acid. *Am. J. Clin. Nutr.* 26:77-88.

(83) Herbert, V. 1979. Megaloblastic anemias. In *Textbook of Medicine*, edited by P.B. Beeson, W. McDermott and J.B. Wyngaarden, pp 1719-1729. W.B. Saunders, Philadelphia.

(84) Herbert, V., Cunneen, N., Jaskiet, L. and Kapff, C. 1962. Mineral daily adult folate requirements. *Arch. Intern. Med.* 110:649-652.

(85) Herbert, V. and Jacob, E. 1974. Destruction of vitamin B_{12} by ascorbic acid. *JAMA* 230:241-242.

(86) Herbert, V., Larrabee, A.R. and Buchanan, J.M. 1962. Studies on the identification of a folate compound in human serum. *J. Clin. Invest.* 41:1134-1138.

(87) Herbert, V., Tisman, G., Go, L.T. and Brenner, L. 1973. The dU suppression test using ^{125}I-UdR to define biochemical megaloblastosis. *Br. J. Haematol.* 24:713-723.

(88) Herbert, V. and Zalusky, R. 1962. Interrelations of vitamin B_{12} and folic acid metabolism: Folic acid clearance studies. *J. Clin. Invest.* 41:1263-1276.

(89) Hodgkin, D.C. 1964. Vitamin B_{12} and the porphyrins. *Fed. Proc.* 23:592-598.

(90) Hoffbrand, A.V., Newcombe, B.F.A. and Mollin, D.L. 1966. Method of assay of red cell folate activity and the value of the assay as a test for folate deficiency. *J. Clin. Pathol.* 19:17-28.

(91) Horning, D. 1975: Distribution of amino acid metabolites and analogues in man and animals. *Ann. N.Y. Acad. Sci.* 258:103-117.

(92) Horning, D., Gallo-Torres, H.E. and Weiser, H. 1973. A biliary metabolite of ascorbic acid in normal and hypophysectomized rats. *Biochim. Biophys. Acta.* 320:549-556.

(93) Hsia, Y.E., Lilljeqvist, A.C. and Rosenberg, L.E. 1970. Vitamin B_{12} dependent methylmalonicaciduria: Amino acid toxicity, long chain ketonuria, and protective effect of vitamin B_{12}. *Pediatrics* 46:497-507.

(94) Hsu, J.M. 1963. Effect of deficiencies of certain B-vitamins and ascorbic acid on absorption of vitamin B_{12}. *Am. J. Clin. Nutr.* 12:170-179.

(95) Hsu, J.M. and Chow, B.F. 1957. Effect of pyridoxine deficiency on the absorption of vitamin B_{12}. *Arch. Biochem.* 72:322-330.

(96) Hsu, J.M., Kawin, B., Minor, P. and Mitchell, J.A. 1966. Vitamin B_{12} concentrations in human tissues. *Nature* 210:1264-1265.

(97) Hsu, J.M., Tantengco, V. and Chow, B.F. 1962. Effect of folic acid deficiency on absorption of Co^{60} labeled vitamin B_{12} by rats. *J. Clin. Invest.* 41:532-536.
(98) Hughes, R.E. 1964. Reduction of dehydroascorbic acid by animal tissues. *Nature* (London) 203:1068-1069.
(99) Hurdle, A.D.F. 1968. The influence of hospital food on the folic-acid status of long-stay elderly patients. *Med. J. Aust.* 2:104-110.
(100) Hurdle, A.D.F. 1968. An assessment of the folate intake of elderly patients in hospital. *Med. J. Aust.* 2:101-104.
(101) Hurdle, A.D.F. 1973. The assay of folate in food. *Nutrition* 27:12-16.
(102) Hurdle, A.D.F. and Williams, P. 1966. Folic-acid deficiency in elderly patients admitted to hospital. *Br. Med. J.* 2:202-205.
(103) Irwin, M.I. and Hutchins, B.K. 1976. A conspectus of research on vitamin C requirements of man. *J. Nutr.* 106:823-879.
(104) Jacobs, A., Cavill, I.A.J. and Hughes, J.N.P. 1968. Erythrocyte transaminase activity. *Am. J. Clin. Nutr.* 21:502-507.
(105) Jacobs, A., Greenman, D., Owen, E. and Cavill, I. 1971. Ascorbic acid status in iron deficiency anemia. *J. Clin. Path.* 24:694-697.
(106) Jagerstad, M., Kristinkvist, A.N. and Westesson, A.K. 1978. Vitamin B_{12}. In *Nutrition and Old Age*, edited by B. Borgstrom, A. Norden, B. Akesson, M. Abdulla and M. Jagerstad, pp 191-195. Carl Bloms Boktryckeri AB Lund, Sweden.
(107) Johansson, S., Lindstedt, S., Register, U. and Wadstrom, L. 1966. Studies on the metabolism of labeled pyridoxine in man. *Am. Clin. J. Nutr.* 18:185-196.
(108) Jordan, M., Kepes, M., Hayes, R.B. and Hammon, W. 1954. Dietary habits of persons living alone. *Geriatrics* 9:230-232.
(109) Kahn, S.B. 1970. Recent advances in the nutritional anemias. *Med. Clin. of N. Amer.* 54:631-645.
(110) Kahn, S.B., Williams, W.J., Barness, L.A., Young, D., Shafer, B., Vivacqua, R.J. and Beaupre, E.M. 1965. Methylmalonic acid excretion: A sensitive indicator of vitamin B_{12} deficiency in man. *J. Lab. Clin. Med.* 66:75-83.
(111) Kataria, M.S., Rao, D.B. and Curtis, R.C. 1965. Vitamin C levels in the elderly. *Geront. Clin.* 7:189-190.
(112) Kheim, T. and Kirk, J.E. 1967. Vitamin B_6 content of human arterial and venous tissue. *Am. J. Clin. Nutr.* 20:702-707.
(113) Kirchmann, L.L. 1930. Über die bedeutung des vitamin C für die klinische medizin. *Ergebn. Inn. Med. Kinderheelk* 56:101-153.
(114) Kirk, J.E. 1954. Blood and urine vitamin levels in the aged. Symposium on Problems of Gerontology, *Nutrition Symposium Series* 9:73-94, Natl. Vit. Found. Inc., New York.
(115) Kirk, J.E. 1962. Variations with age in the tissue content. *Vitamins and Hormones* 20:83-92.
(116) Kirk, J.E. and Chieffi, M. 1953. Vitamin studies in middle-aged and old individuals. XI. The concentration of total ascorbic acid in whole blood. *J. Gerontol.* 8:301-304.
(117) Kirk, J.E. and Chieffi, M. 1953. Vitamin studies in middle-aged and old individuals. XII. Hypovitaminemia C effects of ascorbic acid administration on the blood ascorbic acid concentrations. *J. Gerontol.* 8:305-311.
(118) Kohn, R.K. 1971. *Principles of Mammalian Aging.* Prentice-Hall, Englewood Cliffs, NJ.

(119) Krumdieck, L., Cornwell, P.E., Thompson, R.W. and White, W.E., Jr. 1975. Studies on the biological role of folic acid polyglutamates. In *Folic Acid*, pp 25-42. National Academy of Science, Washington, DC.

(120) Kuenzig, W., Avenia, R. and Kamm, J.J. 1974. Studies on the antiscorbutic acitivity of ascorbate-2-sulfate in the guinea pig. *J. Nutr.* 104:952-956.

(121) Labadarios, D., Rossouw, J.E., McConnell, J.B., Davis, M. and Williams, R. 1971. Vitamin B_6 deficiency in chronic liver disease—evidence for increased degradation of pyridoxal-5'-phosphate. *Gut* 18:23-27.

(122) LeBovit, C. 1965. The food of older persons living alone. *J. Amer. Dietetic Assn.* 46:285-289.

(123) Lengyel, P., Mazunder, R. and Ochoa, S. 1970. Metabolism of propionic acid in animal tissue. *Proc. Nat. Acad. Sci.* 46:1312-1318.

(124) Lepkovsky, S. and Nielson, E. 1942. A green pigment-producing compound in urine of pyridocine-deficient rats. *J. Biol. Chem.* 144:135-138.

(125) Linkswiler, H. 1967. Biochemical and physiological changes in vitamin B_6 deficiency. *Am. J. Clin. Nutr.* 20:547-557.

(126) Linkswiler, H.M. 1978. Vitamin B_6 requirement of med. In *Human Vitamin B_6 Requirement*, pp 279-290. National Academy of Science, Washington, DC.

(127) Loh, H.S. and Wilson, C.W.M. 1971. Relationship between leukocyte ascorbic acid and hemoglobin levels at different ages. *Int. J. Vit. Nutr. Res.* 41:259-267.

(128) Loh, H.S. and Wilson, C.W.M. 1971. Relationship of human ascorbic acid metabolism to ovulation. *Lancet* 1:110-112.

(129) Luhby, A.L. and Cooperman, J.M. 1964. Folic acid deficiency in man and its interrelationship with vitamin B_{12} metabolism. *Adv. Metab. Disorders* 1:263-334.

(130) Luhby, A.L., Cooperman, J.M., Lopez, R. and Giorgio, A.J. 1969. Vitamin B_{12} metabolism in thalassemia major. *Ann. N.Y. Acad. Sci.* 165:443-460.

(131) Lyons, J.S. and Trulson, M.F. 1956. Food practices of older people living at home. *J. Geront.* 11:66-72.

(132) Mahoney, M.J. and Rosenberg, L.E. 1970. Inherited defects of B_{12} metabolism. *Am. J. Med.* 48:584-593.

(133) Mäkilä, E. 1970. The vitamin status of elderly denture wearers. *Int. Z. Vetaminforsch.* 40:81-89.

(134) Mazumder, R., Sasakawa, T. and Ochoa, S. 1963. Metabolism of propionic acid in animal tissue. X. Methylmalonyl coenzyme A mutase holoenzyme. *J. Biol. Chem.* 238:50-53.

(135) McCkaub, L.D., Carl, G.F. and Bridgers, W.F. 1975. Distribution of folic acid coenzymes and folate dependent enzymes in mouse brain. *J. Neurochem.* 24:719-722.

(136) Meindok, H. and Dvorsky, R. 1970. Serum folate and vitamin B_{12} levels in the elderly. *J. Am. Geriat. Soc.* 18:317-326.

(137) Mengel, C.E. and Greene, H.L., Jr. 1976. Ascorbic acid effects on erytrocytes. *Ann. Int. Med.* 84:490.

(138) Metz, J., Kelly, A., Swett, V.C., Waxman, S. and Herbert, V. 1968. Deranged DNA synthesis by bone marrow from vitamin B_{12} deficient humans. *Br. J. Haematol.* 14:575-592.

(139) Milne, J.S., Longergan, M.E., Williamson, J., Moore, F.M.L., McMaster, R. and Percy, N. 1971. Leucocyte ascorbic acid levels and vitamin C intake in older people. *Brit. Med. J.* 4:383-385.
(140) Minot, G.R. and Murphy, W.P. 1926. Treatment of pernicious anemia by special diet. *J. Amer. Med. Assn.* 87:470-476.
(141) Mollin, D.L. and Ross, G.I.M. 1952. The vitamin B_{12} concentrations of serum and urine of normals and patients with megaloblastic anemias and other diseases. *J. Clin. Path.* 5:129-139.
(142) Morehouse, A.L. and Guerrant, N.B. 1952. Influence of age and of sex of the albino rat on hepatic ascorbic acid. *J. Nutr.* 46:551-564.
(143) Morse, E.H., Potgieter, M. and Walker, G.R. 1956. Ascorbic acid utilization by women. Response of blood serum and white blood cells to increasing levels of intake. *J. Nutr.* 58:291-298.
(144) Mumma, R.O. and Verlangieri, A.J. 1972. The isolation of ascorbic acid 2-sulfate from selected rat organs. *Biochim. Biophys. Acta.* 273:249-253.
(145) Nakamura, M., Kawagoe, T., Ogino, Y., Nishiyama, K., Ichikawa, H. and Sugahara, K. 1967. Experimental study on the effect of vitamin C on the basal metabolism and resistance to cold in human aging. *Tokyo J. Exp. Med.* 92:207-219.
(146) Namyslowski, L. 1956. Spostrzezenia nad zapotrzebowaniem na witamine C u sportówcow w zależności od Wysilku fizycznego. *Rocz. Państwow. Zakl. Hig.* 7:97-122.
(147) Neumann, E. 1970. Methylmalonsaureausscheidung und megaloblastische Blutbildungsstorung. *Wiener Klinische Wochenschrift.* 82:373-375.
(148) Okuda, K., Grasbeck, R. and Chow, B.F. 1958. Bile and vitamin B_{12} absorption. *J. Lab. Clin. Med.* 51:17-23.
(149) O'Sullivan, D.H., Callaghan, N., Ferriss, J.B., Finucane, J.F. and Hegarty, M. 1968. Ascorbic acid deficiency in the elderly. *Irish J. Med. Sci.* 1(4):151-156.
(150) Oxenhorn, S., Estres, S., Wasserman, L.R. and Adlersberg, D. 1958. Malabsorption syndrome-intestinal absorption of vitamin B_{12}. *Ann. Int. Med.* 48:30-38.
(151) Pollycover, M. and Apt, L. 1956. Absorption and elimination and excretion of orally administered vitamin B_{12} in normal subjects and in patients with pernicious anemia. *N. Eng. J. Med.* 255:207-212.
(152) Rabiner, F.F., Lichtman, H.C., Messite, J., Watson, J., Ginsberg, U., Ellenbogen, L. and Williams, W.L. 1956. Urinary excretion test in diagnosis of addisonian pernicious anemia. *Ann. Int. Med.* 44:437-445.
(153) Rader, J.I. and Huennekens, F.M. 1973. Folate coenzyme-mediated transfer of one-carbon groups. In *Enzymes*, edited by P.D. Boyer, pp 197-223. Academic Press, New York.
(154) Rafsky, H.A. and Newman, B. 1947. Nutritional aspects of aging. *Geriatrics* 2:101-104.
(155) Rafsky, H.A. and Newman, B. 1948. A quantitative study of diet in the aged. *Geriatrics* 3:267-272.
(156) Raica, N. and Sauberlich, H.E. 1964. Blood cell transaminase activity in human vitamin B_6 deficiency. *Am. J. Clin. Nutr.* 15:67-72.
(157) Ranke, E., Tauber, A.S.A., Horonick, A., Ranke, B., Goodhart, R.S. and Chow, B.F. 1960. Vitamin B_6 deficiency in the aged. *J. Gerontol.* 15:41 44.

(158) Rappazzo, M.E., Salmi, H.A. and Hall, C.A. 1970. The content of vitamin B_{12} in the adult or foetal tissue: a comparative study. *Br. J. Haematol.* 18:425-433.
(159) Read, A.E., Gough, K.R., Pardoe, J.L. and Nicholas, A. 1965. Nutritional studies on the entrants to an old people's home with particular reference to folic acid deficiency. *Br. Med. J.* 11:834-848.
(160) Reisner, E.H., Gilbert, J.P., Rosenblum, C. and Morgan, M.C. 1956. Applications of urinary tracer test (of Schilling) as an index of vitamin B_{12} absorption. *Am. J. Clin. Nutr.* 4:134-141.
(161) *Requirement of Ascorbic Acid, Vitamin D, Vitamin B_{12}, and Folate and Iron.* 1970. World Health Organization Technical Report Series No. 452.
(162) Rickes, E.L., Brink, N.G., Koniuszy, F.R., Wood, T.R. and Folker, K. 1948. Crystalline vitamin B_{12}. *Science* 107:396-397.
(163) Rinehart, J.F. and Greenberg, L.D. 1949. Arteriosclerotic lesions in pyridoxine-deficient monkeys. *Am. J. Pathol.* 25:481-490.
(164) Rivera, J.V., de La Obra, F.R. and Maldonado, M.M. 1966. Anemia due to vitamin B_{12} deficiency after treatment with folic acid in tropical sprue. *Am. J. Clin. Nutr.* 18:110-115.
(165) Roderuck, C., Burrill, L., Campbell, L.J., Brakke, B.E., Childs, M.T., Leverton, R., Chaloupka, M., Jebe, E.H. and Swanson, P.P. 1958. Estimated dietary intake, urinary excretion and blood vitamin C in women of different age. *J. Nutr.* 66:15-27.
(166) Rodriguez, M.S. 1978. A conspectus of research on folacin requirements of man. *J. Nutr.* 108:1983-2075.
(167) Rosenberg, I.H., Streiff, R.R., Godwin, H.A. and Castle, W.B. 1969. Absorption of polyglutamic folate: participation of deconjugating enzymes of the intestinal mucosa. *N. Eng. J. Med.* 280:985-988.
(168) Rosensweig, N.S. 1975. Diet and intestinal enzyme adaptions: Implications for gastrointestinal disorders. *Am. J. Clin. Nutr.* 28:648-655.
(169) Rothenberg, S.P. 1973. Application of competitive ligand binding for the radioassay of vitamin B_{12} and folic acid. *Metabolism* 22:1075-1082.
(170) Rothenberg, S.P., daCosta, M. and Rosenberg, Z. 1972. A radioassay for serum folate: use of a two-phase sequential-incubation, ligand binding system. *N. Eng. J. Med.* 286:1355-1359.
(171) Sauberlich, H.E., Canham, J.E., Baker, E.M., Ralca, N., Jr. and Herman, Y.F. 1972. Biochemical assessment of the nutritional status of vitamin B_6 in the human. *Am. J. Clin. Nutr.* 25:629-642.
(172) Sauberlich, H.E., Dowdy, R.P. and Skala, J.H. 1974. *Laboratory Tests for the Assessment of Nutritional Status.* CRC Press, Inc., Cleveland.
(173) Schaus, R. 1957. The ascorbic acid content of human pituitary, cerebral cortex, heart and skeletal muscle and its relation to age. *Am. J. Clin. Nutr.* 5:39-41.
(174) Schilling, R.F. 1953. Intrinsic factor studies. II. The effects of gastric juice on the urinary excretion of radioactivity after the oral administration of radioactive vitamin B_{12}. *J. Lab. Clin. Med.* 42:860-866.
(175) Schrauzer, G.N. and Rhead, W.J. 1973. Ascorbic acid abuse: effects of long term ingestion of excessive amounts on blood levels and urinary excretion. *Int. J. Vit. Nutr. Res.* 43:201-211.
(176) Scudi, J.V., Unna, K. and Antopol, W. 1940. A study of the urinary excretion of vitamin B_6 by a colorimetric method. *J. Biol. Chem.* 135:371-376.

(177) Serebro, H.A., Solomon, H.M., Johnson, J.H. and Hendrix, T.R. 1966. The intestinal absorption of vitamin B_6 compounds by the rat and hamster. *Bull. Johns Hopkins Hosp.* 119:166-171.
(178) Skeggs, H.R., Nepple, H.M., Valentik, I.O., Huff, J.W. and Wright, L.D. 1950. Observation on the use of *Lactobacillus leichmannii* 4797 in the microbiological assay of vitamin B_{12}. *J. Biol. Chem.* 184:211-224.
(179) Smith, E.L. and Parker, L.F.J. 1948. Purification of anti-pernicious anemia factor. *Biochem. J.* 43:viii-ix.
(180) Smith, R.M. and Monty, K.J. 1959. Vitamin B_{12} and propionate metabolism. *Biochem. Biophys. Res. Commun.* 1:105-109.
(181) Solling, H., Ellegaard, J. and Esmann, V.A. 1973. A clinical evaluation of a rapid method for determination of folate deficiency by ^{14}C-formate incorporation into serine of lymphocytes. *Scandinav. J. Clin. Lab. Invest.* 31:453-457.
(182) Stadtman, E.R., Overath, P., Eggerer, H. and Lynen, F. 1960. The role of biotin and vitamin B_{12} coenzyme in propionate metabolism. *Biochem. Biophys. Res. Commun.* 2:1-7.
(183) Stankova, L., Rigas, D.A. and Bigley, R.H. 1975. Dehydroascorbate uptake and reduction by human blood neutrophils, erythrocytes and lymphocytes. *Ann. N.Y. Acad. Sci.* 258:238-242.
(184) Steinkamp, R., Cohen, N. and Walsh, H. 1965. Resurvey of an aging population—fourteen year follow-up. *J. Amer. Dietetic Assn.* 46:103-110.
(185) Stern, J.R. and Friedman, D.L. 1960. Vitamin B_{12} and methylmalonyl CoA isomerase. *Biochem. Biophys. Res. Commun.* 2:82-87.
(186) Stokstad, E.L.R. and Koch, J. 1967. Folic acid metabolism. *Physiol. Rev.* 47:83-116.
(187) Sturman, J.A. 1978. Vitamin B_6 and the metabolism of sulfur amino acid. In *Human Vitamin B_6 Requirements*, pp 37-60. National Academy of Science, Washington, DC.
(188) Swendseid, M.E., Hvolboll, E., Schick, G. and Halsted, J.A. 1957. The vitamin B_{12} content of human liver tissue and its nutritional significance. *Blood* 12:24-28.
(189) Tauber, S.A., Goodhart, R.S., Hsu, J.M., Blumberg, N., Kassab, J. and Chow, B.F. 1957. Vitamin B_{12} deficiency in the aged. *Geriatrics* 12:368-374.
(190) Taylor, G. 1966. Diet of elderly women. *Lancet* 1:926.
(191) Tibbling, G. 1969. A method for determination of vitamin B_{12} serum by radioassay. *Clin. Chim. Acta.* 23:209-218.
(192) Timiras, P.S. and Vernadakis, A. 1972. Structural, biochemical, and functional aging of the nervous system. In *Developmental Physiology and Aging*, edited by P.S. Timiras, pp 502-526. Macmillan Co., New York.
(193) Wachstein, M. and Gudaitis, A. 1952. Disturbance of vitamin B_6 metabolism. *J. Lab. Clin. Med.* 40:550-557.
(194) Waters, A.H. and Mollin, D.L. 1961. Studies on the folic acid activity of human serum. *J. Clin. Path.* 14:335-344.
(195) Watkin, D. 1964. Protein metabolism and requirement in the elderly. In *Mammalian Protein Metabolism*, vol 2, edited by H.N. Munro and J.B. Allison, pp 247-263. Academic Press, New York.
(196) West, R. 1948. Activity of vitamin B_{12} in addisonian pernicious anemia. *Science* 107-398.

(197) Westergaard, F. 1940. Staseproven og dens kleniske betydning. E. Munksgaard, Copenhagen. Cited by Kirk, J.E., and Chieffi, M. 1953. Vitamin studies in middle-aged and old individuals. XI. The concentration of total ascorbic acid in whole blood. *J. Geront.* 8:301-304.
(198) White, A.M. and Cox, E.V. 1964. Methylmalonic acid excretion and vitamin B_{12} deficiency in the human. *Ann. N.Y. Acad. Sci.* 112:915-921.
(199) Wilson, T.H. 1965. Intrinsic factor and vitamin B_{12} absorption—a problem in cell biology. *Nutr. Rev.* 23:33-35.
(200) Yavorsky, M., Almaden, P. and King, C.G. 1934. The vitamin C content of human tissues. *J. Biol. Chem.* 106:525-529.
(201) Yeh, S.D.J., Chow, B.F., Velasco, C. and Gabriel, R.R. 1961. In *Second European Symposium on Vitamin B_{12} and Intrinsic Factor*, edited by H.C. Heinrich, pp 69. Verlag Hans Huber, Bern und Stuttgart, Federal Republic of Germany.
(202) Yew, Man-Li S. 1975. Biological variation in ascorbic acid needs. *Ann. N.Y. Acad. Sci.* 258:451-456.

−9−
Vitamin D Metabolism in Aging

John F. Aloia

INTRODUCTION

The last decade has seen a virtual knowledge explosion in the field of vitamin D metabolism. The active form of vitamin D has been characterized as a hormone, and the relationship of vitamin D metabolites to other hormonal systems has been investigated. Sensitive assays for the measurement of vitamin D metabolites in body fluids have been developed and some of these are commercially available for clinical use, such as the assay for 25-hydroxy vitamin D (25OHD). Vitamin D metabolites have been synthesized and used in clinical trials for treatment of several disorders of calcium homeostasis (e.g., hypoparathyroidism, osteomalacia and uremic osteodystrophy). The advance from the basic science laboratory to clinical management of humans has been extremely rapid in this field.

It is the purpose of this chapter to briefly review these advances, to examine the changes in vitamin D metabolism with aging, and to relate these to the prevention and management of two metabolic bone disorders that are common in the elderly, osteoporosis and ostemalacia.

BACKGROUND

Vitamin D

Sources of Vitamin D: Vitamin D has become a vitamin only since progress in civilization has resulted in inadequate exposure to sunlight, making ingestion of vitamin D necessary.[24] The epidermis contains a significant quantity of 7-dehydrocholesterol (Figure 1) which, when exposed to ultraviolet light undergoes photolysis pro-

Figure 1: Vitamin D Metabolism

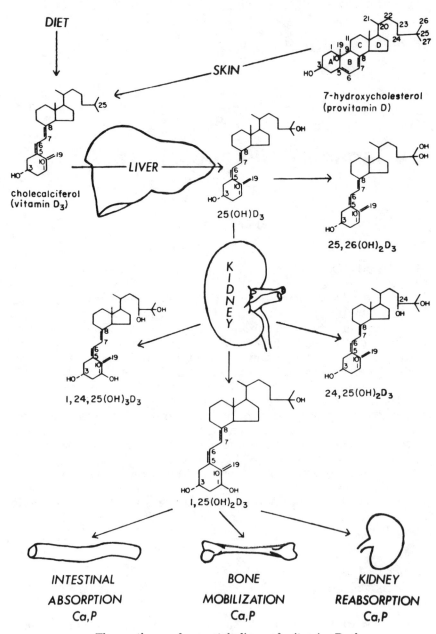

The pathway for metabolism of vitamin D derived from the diet and ultraviolet light is depicted. Following hydroxylation in the liver, when appropriate stimuli are present the active form of vitamin D [$1,25(OH)_2D_3$] is formed.

ducing previtamin D_3. The previtamin D_3 undergoes a temperature-dependent conversion to vitamin D_3.[24] Dietary sources of vitamin D are found in high quantities in the livers of sea animals.[85] Although vitamin D is not prevalent in the plant world, the plant sterol, ergosterol, is artifically converted to vitamin D_2 by ultraviolet light and is used widely in fortification of foods, such as milk, bread and butter.

Absorption and Storage: Vitamin D is absorbed in the small intestine and requires the presence of bile salts.[84] Whether the absorption is most active in the proximal or distal small intestine remains controversial. Since vitamin D is fat-soluble, it is probably absorbed via the lymphatics and it is stored in fat deposits.[5]

Hepatic Conversion to $25OHD_3$: Within an hour of ingestion, vitamin D accumulates in the liver and is hydroxylated to its major circulating form, $25OHD_3$.[71,74] The hepatic enzyme controlling this hydroxylation is regulated by $25OHD_3$ levels (feedback inhibition).[9] The $25OHD_3$ is several times more active than vitamin D_3 in intestinal calcium and phosphate absorption, and in the cure of rickets.[92] It is generally considered an intermediary in the formation of $1,25(OH)_2D_3$ although there is some evidence that $25OHD_3$ may play a physiologic role in calcium homeostasis.[76]

Renal Metabolism of $25OHD_3$: The $25OHD_3$ is converted to the most active metabolite of vitamin D, $1,25(OH)_2D_3$ (10 times the potency of vitamin D_3) in the renal tubules.[29] The $25OHD_3$-1-hydroxylase is a mitochondrial enzyme; the reaction is dependent on cytochrome P-450.[36] Since $1,25(OH)_2D_3$ is synthesized in the kidney and acts on the intestine and bone, it fits the criteria for designation as a hormone. The major stimuli for increased synthesis of $1,25(OH)_2D_3$ are hypophosphatemia, hypocalcemia, and parathyroid hormone (Figures 2 and 3).[23,33] In addition, a number of other hormones may play a role in control of the renal hydroxylase (e.g., growth hormone, prolactin, sex steroids, glucocorticoids).[27,89]

Under conditions where there is suppression of the activity of the 25OHD-1-hydroxylase, such as hypercalcemia and hyperphosphatemia, a 24-hydroxylase becomes active which results in the formation of $24,25(OH)_2D_3$.[41] The 24-hydroxylase can also act on $1,25(OH)_2D_3$ to produce $1,24,25(OH)_3D_3$.[40] Although much must be learned about these reactions, it is currently believed that they represent a step in the inactivation and excretion of vitamin D metabolites. In addition to hydroxylation to $1,24,25(OH)_3D_3$, side chain oxidation is a major pathway of $1,25(OH)_2D_3$ metabolism. Most of the vitamin D metabolites are excreted in the bile, with a minor amount excreted in the urine.

Analogues of Vitamin D: A variety of vitamin D analogues have been synthesized for biologic testing and possible clinical use.[42,68] Methods were developed for introduction of the 1-hydroxyl function into vitamin D precursors leading to the synthesis of 1α-OHD_3. This was the most important analogue clinically, it is about one-half as

Figure 2: The Hypocalcemic Signal for 1,25(OH)$_2$D$_3$ Synthesis

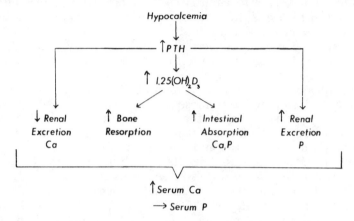

Hypocalcemia results in increased release of parathyroid hormone which increases renal synthesis of 1,25(OH)$_2$D$_3$. This active metabolite then results in normalization of serum calcium. Serum phosphate does not increase inordinately because in the presence of parathyroid hormone there is increased renal excretion of phosphate. Thus, serum calcium may be returned to the normal range with no change in serum phosphate.

Figure 3: The Hypophosphatemic Signal for 1,25(OH)$_2$D$_3$ Synthesis

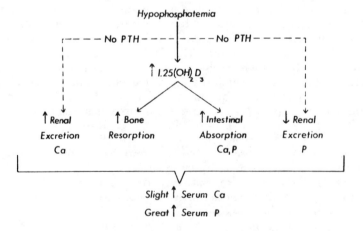

Hypophosphatemia directly results in increased synthesis of 1,25(OH)$_2$D$_3$. In this instance there is no increased secretion of parathyroid hormone. Therefore, serum phosphate increases into the normal range. As a result of the lack of parathyroid hormone activity, the increase in serum calcium is minimal.

active as $1,25(OH)_2D_3$. Although $1\alpha\text{-}OHD_3$ has been used in a variety of clinical situations, it is likely that it will be replaced by $1,25\text{-}(OH)_2D_3$ now that the latter is readily synthesized and available for clinical use.

Osteoporosis and Osteomalacia

Osteoporosis: Osteopenia is a preferable term to osteoporosis, when the term is used to connote a reduction in bone mass. Osteoporosis, then, may be defined as osteopenia of sufficient magnitude to have resulted in structural failure (fracture of the femur, wrist or crush fracture of the spine). The definition of osteoporosis must also include reference to the lack of pathologic findings on bone biopsy; i.e., the bone is normally mineralized with a reduced amount of normal bone per unit volume. Osteopenia is present in many elderly individuals, since loss of bone mass occurs with aging. There are about one million fractures in the elderly each year in the United States. It is immediately apparent that osteoporosis is of great public health importance.

Involutional bone loss appears to be a nearly universal phenomenon (Figure 4). The author believes that the best current view of osteoporosis should be similar in many ways to our view of atherosclerosis. Thus, osteoporosis has multiple causes (or risk factors), occurs in all individuals with aging, but its onset may be delayed (Tables 1 and 2). High risk populations with low bone mass may be identified in the future, resulting in intervention to correct risk factors in order to prevent the occurrence of fracture. Since the etiology of osteoporosis is undoubtedly multifactorial, it must be stated that it would be incorrect to assume that any single factor (including vitamin D) is solely responsible for bone loss with aging.

One of the tasks in human skeletal research has been to describe the correct model for involutional bone loss. Studies that we have performed suggest that there is a normal distribution of bone mass in women who have recently entered the menopause and that the range of total skeletal mass is quite wide with differences of seven hundred grams of total body calcium.[3] Thus it seems likely that efforts to prevent osteoporosis must not only be directed at reduction of bone loss in adult life but also at increasing bone accretion during the phase of skeletal growth. There is also very little evidence to support the concept of osteoporosis arising in a group of women with rapid bone loss. While there may be some heterogeneity in the rates of bone loss in a postmenopausal population, they are of very low magnitude.

There are clear genetic and sex differences that must be considered in involutional bone loss; blacks have higher bone mass than whites, and men have higher bone mass than women. This is true even during the phase of skeletal growth. At the time of menopause, there is an accelerated loss of bone in women that can be prevented by administration of estrogen or calcium supplements.

Figure 4: A Conceptual Presentation of the Change in Bone Mass with Aging

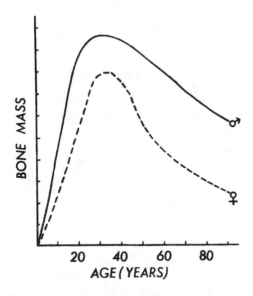

Bone mass is higher in men than in women throughout life. Following achievement of adult bone mass, there is a gradual loss that occurs in all individuals. This loss is accelerated around the time of menopause in females resulting in a still lower bone mass in the postmenopausal female population as compared to a male population.

Table 1: Nonendocrine Factors That May Contribute to Involutional Osteopenia

Dietary
    Low calcium content
    High phosphate content
    High protein content
    High acid ash content
    Vitamin D deficiency
    Vitamin C deficiency
    Ingestion phytates
    Lactase deficiency
Physical inactivity
Sex (women have less bone mass)
Race (whites have less bone mass than blacks)
Cigarette smoking
Local factors (mast cells)
Drugs (e.g., heparin)
Pregnancy and lactation

Table 2: Some Hormonal Factors That May be Associated with Involutional Osteopenia

(1) Estrogen withdrawal at menopause results in increased bone resorption (increased sensitivity of bone to PTH).
(2) Calcitonin levels are less in women than men and decrease with age.
(3) Parathyroid hormone secretion increases with aging.
(4) GH secretion decreases with aging.

Calcium malabsorption is present in postmenopausal osteoporosis.[17,30] Gallagher, et al.,[31] measured intestinal calcium absorption by a double isotope method in 94 normal subjects. They found a decrease of calcium absorption with age ($r = 0.22$, $P < 0.025$). Others have also noted a decrease in calcium absorption with age and impaired intestinal adaptation to a decrease in dietary calcium.[2,51]

Osteomalacia: Osteomalacia is a disorder of calcification of bone, characterized by an increase in the amount of unmineralized matrix (osteoid) and an abnormality in the calcification front. One of the major causes of osteomalacia is deficiency of vitamin D or failure to form the active vitamin D metabolites. The pediatric counterpart of osteomalacia is rickets. Thus, inadequate intake of vitamin D (from sunlight or diet), malabsorption, and altered hepatic or renal metabolism of vitamin D may lead to osteomalacia (Table 3).

Table 3: Some Conditions in the Elderly Resulting in Altered Metabolism of Vitamin D

I. Lack of D
   Malabsorption syndrome
   Dietary deficiency
   Inadequate sunlight
   Gastrectomy
   Laxative abuse
   Pancreatic insufficiency
II. Lack of $25OHD_3$
   Drugs (anticonvulsants, cholestyramine, glucocorticoids)
   Hepatic disease
   Biliary cirrhosis
III. Lack of $1,25(OH)_2D_3$
   Chronic renal failure
   Hypoparathyroidism
   Cadmium and strontium toxicity
   Postmenopausal osteoporosis

The clinical presentation of osteomalacia may be identical to osteoporosis (with fracture of the ribs, spine and long bones). However, rather nonspecific, generalized bone pain and muscle weakness may more often result; this presentation is often not correctly identified. A proximal myopathy may occur in half the patients with osteomalacia. It is of interest to speculate that fracture of the hip in patients with osteomalacia may result from the lack of protection normally afforded by surrounding musculature in a fall. Indeed, falling may result from the instability produced by the myopathy. The classic laboratory findings of osteomalacia include hypocalcemia, hypophosphatemia and elevated alkaline phosphatase. However, many cases with histologic osteomalacia may not have any of these chemical abnormalities.

AGING AND 25OHD$_3$

Table 4: Cause of Low Levels of 25OHD$_3$ in the Elderly

Deficient exposure to sunlight
Poor dietary intake of vitamin D
Poor absorption of vitamin D
Defective hydroxylation of vitamin D
Drug effects

Levels of Circulating 25OHD$_3$

Corless, et al.[15] in 1975, reported very low serum concentrations of 25OHD in patients in a British long stay geriatric hospital. This work was confirmed by other investigators as well.[64] Nordin, et al.[65] also in Great Britain, presented evidence that 25OHD levels decrease with aging. Lund,[56] in a study of 596 subjects, found lower levels of 25OHD in elderly Danes but noted that the mean values were similar to those reported in the United States and Sweden but were higher than those observed in England, Belgium and France. Seasonal variation in serum 25OHD levels presumably secondary to changes in sunlight, was noted to be present in the elderly. It is of interest that vitamin D levels were lower throughout the year in the elderly whether they were or were not given vitamin D supplements.

It should be cautioned that Davie, et al.,[18] reported several patients who had very low levels of 25OHD who did not have histologic evidence of osteomalacia. These authors suggested that care should be exercised in interpretation of serum 25OHD vitamin D levels vis-a-vis the presence of osteomalacia. Furthermore, it is of importance to note that vitamin D levels differ in different populations. For example, Lester et al.[52] were able to confirm decreased levels of 25OHD in the elderly, together with a diminished increment in 25OHD concentration in the fall in 36 normal people in England. However, Gallagher et al.[31] found normal levels for serum 25OHD in 20 elderly normal subjects in the United States. It has been established that the British have lower concentrations of 25OHD than do the Swiss, Germans or Americans.[34,38,66]

In the study reported by Corless et al.[15] it was observed that all patients with low 25OHD levels had dietary intakes less than 100 international units of vitamin D per day. Nayal et al.[64] studied dietary intake, sunlight exposure and 25OHD levels in 62 patients admitted to a geriatric unit. They found that the 25OHD levels were correlated with dietary intake of vitamin D but not with sunlight, suggesting that dietary deficiency may be a more important cause of low levels of 25OHD in the elderly.[31] On the other hand, it is clear that the provision of sunlight in a long term geriatric ward is capable of bringing 25OHD levels into the normal range. This was done with ultraviolet background radiation that was less than that necessary to produce erythema of the skin.[16]

MacLennan and Hamilton[58] measured $25OHD_3$ levels in long-stay geriatric patients and determined that in those with low levels, supplementation with 500 IU of vitamin D produced adequate serum levels in most patients within two months (from 8 to 21 ng/ml). Thus, it would seem possible to achieve adequate serum levels of $25OHD_2$ either with background ultraviolet radiation or with low dose vitamin D supplements. The possibility that there may be intestinal malabsorption of vitamin D in the elderly has also been raised. Barragry et al.[7] studied intestinal absorption of labeled cholecalciferol in 20 female geriatric patients. They found that absorption of vitamin D was lower than in the younger patients. The measured metabolites were also lower suggesting that this finding indeed resulted from reduced absorption rather than altered metabolism of the vitamin D.

Aging is associated with reduced liver function.[44] It is possible, therefore, that there is impairment in hydroxylation of vitamin D with aging. Rushton[81] injected 50,000 IU vitamin D subcutaneously in three groups: young controls, old people and old inpatients. He measured serum $25OHD_3$ levels. Responses were reduced in the elderly confirming the hypothesis that there is reduced hepatic hydroxylation of vitamin D in the elderly.

It is of interest that the revised 1980 recommended dietary allowance for vitamin D is only 200 IU in the elderly.[14] The lowered recommendation is probably based on concern for development of nephrolithiasis. However, if one were to accept either calcium absorption or serum levels of vitamin D in younger individuals as the determinant of adequate nutritional intake, it is unlikely that 200 IU daily would be sufficient. Further studies are needed to determine the amount of vitamin D that is necessary to restore calcium absorption and adaptation in the elderly to the level observed in youth.

Aging and $1,25(OH)_2D_3$

Renal function declines with aging.[19,65] Nordin et al.[65] demonstrated that there is a correlation between calcium absorption and creatinine clearance. It has been postulated that the impaired renal function results in decreased conversion of $25OHD_3$ to $1,25(OH)_2D_3$. Support for this concept of impaired conversion of $25OHD_3$ to $1,25(OH)_2D_3$ with aging comes from the finding that large doses of parent vitamin D are required to correct calcium malabsorption in the elderly. This is to be compared with much smaller doses of the active renal metabolite or its analogue.[60,65]

There are again differences in levels observed globally. Morita et al.[62] found that plasma $1,25(OH)_2D_3$ levels were in the normal range in most elderly subjects. Others, however, have found lower serum $1,25(OH)_2D_3$ levels in elderly individuals.[86] Furthermore, it has been noted that calcium absorption and serum $1,25(OH)_2D_3$ levels are positively correlated, as would be expected in normal subjects. The reduction in $1,25(OH)_2D_3$ levels presumably causes the decreased calcium absorption and adaptation to a low calcium diet that is seen in aging.

HORMONES THAT INTERACT WITH VITAMIN D

A number of changes in endocrine systems have been reported with aging. Since there is a body of evidence relating the endocrines to vitamin D metabolism these will be reviewed here, and the evidence for a relationship of these changes to the development of involutional osteopenia will be examined (Table 2).

Parathyroid Hormone

Parathyroid hormone levels (PTH) have been reported to increase with age.[62,72] Roof, et al.[80] analyzed this decline by decade to show that PTH decreased from childhood until age 49 years in women and then increased to a peak at age 60 to 69 years. Berlyne[8] suggested that involutional osteopenia may result from secondary hyperparathyroidism due to declining renal function. Higher PTH levels in osteoporotic subjects were also found in several studies from Japan[70,72] whereas normal values have also been reported.[25]

These findings are to be contrasted with the findings of Riggs et al.[77] and others, that PTH is normal or decreased in osteoporotic as compared to age matched nonosteoporotic individuals except for a small subset of patients.[28,31,78] The conflict in these findings are probably a reflection of different patient populations. Riggs et al.[77] agree that the subset (10% of their osteoporotic patients) has secondary hyperparathyroidism, perhaps because of inadequate conversion of $25OHD_3$ to $1,25(OH)_2D_3$.[79]

Calcitonin

Calcitonin levels are lower in women than men, and decrease with aging, leading to the speculation that calcitonin deficiency may play a role in involutional osteopenia.[21] Whether calcitonin plays a part in control of vitamin D metabolism in man remains controversial. One author reported enhanced conversion of 25OHD$_3$ to 1,25(OH)$_2$D$_3$ following administration of calcitonin.[32] Larkins et al.[49] reported that in the chick increased production of 1,25(OH)$_2$D$_3$ follows administration of synthetic salmon calcitonin but does not follow administration of human calcitonin.

Growth Hormone

Growth hormone levels decrease with age in women as contrasted with men.[82,95] A large body of animal studies and a study in humans suggest a relationship between growth hormone and vitamin D metabolism.[35,97] A relationship between declining levels of GH and reduced synthesis or activity of 1,25(OH)$_2$D$_3$ in osteoporosis remains to be demonstrated.

Gonadal Steroids

Estrogen withdrawal at the menopause is associated with accelerated bone loss. In animal experiments, estrogen increases the activity of the renal 25OHD-1-hydroxylase.[46] Estrogen replacement therapy has been shown to increase the reduced levels of 1,25(OH)$_2$D$_3$ in osteoporotic women.[28]

An indirect effect of estrogen deficiency on vitamin D metabolism is central to one model of postmenopausal bone loss.[39] Estrogen produces resistance of bone to PTH. Thus, menopause may be associated with increased bone resorption because of enhanced effectiveness of PTH. This results in an increase of serum calcium (with a subsequent decrease in PTH) and an increase of serum phosphate. Synthesis of 1,25(OH)$_2$D$_3$ would be diminished by both the decreased serum PTH and increased serum phosphate, and calcium absorption would secondarily be diminished.

Androgens have been implicated in vitamin D metabolism as well; studies in rats suggest that testosterone enhances the elevation of 25OHD produced by ultraviolet radiation.[67]

DRUGS AND ILLNESSES IN THE ELDERLY WHICH AFFECT VITAMIN D METABOLISM

There are a variety of drugs which may affect vitamin D metabolism. Since this is a relatively recent area of investigation, it is likely that even more drugs will be implicated in the future. Reduced serum levels of 25OHD$_3$ have been noted in patients receiving long term anticonvulsant therapy with diphenylhydantoin, mysoline, and phenobarbital.[10] It has been suggested that conversion of 25OHD$_3$ to more polar, less active metabolites might be responsible for their production of decreased levels of 25OHD$_3$.[22] Alternately, it has been suggested that low levels of $24,25(OH)_2D_3$ may result from the use of these drugs with consequent osteomalacia.[96]

Glucocorticosteroids reduce intestinal calcium absorption and increase bone resorption.[54,63] Their use may be associated with a reduced plasma level of 25OHD.[47] A variety of illnesses also interfere with vitamin D metabolism and may produce osteomalacia (or osteoporosis) in the elderly. These include intestinal malabsorption, gastric surgery, laxative abuse, pancreatic insufficiency and renal failure (Table 3).[13,54,63]

INVOLUTIONAL OSTEOPENIA

Role of 25OHD$_3$

In disease states where osteopenia achieves clinical significance in the form of fractures, caution must be exercised again not to extrapolate findings from one part of the world to another. Fracture of the hip had been considered to result from osteoporosis rather than osteomalacia. More recently, in Britain, Scotland and India it has become evident that there is a high prevalence of histologic osteomalacia (approximately 30%) in patients with hip fractures.[1,95] Measurement of 25OHD in these populations has demonstrated low levels consistent with levels expected in an osteomalacic population.[6] Indeed, 40% of patients had 25OHD levels below these very low levels (10 ng/ml). However, in Denmark, a high incidence of histologic osteomalacia has been reported and this population does not have reduced serum 25OHD$_3$ levels.[61,88]

No evidence has been found of reduced levels of 25OHD$_3$ in osteoporotic women who reside in the United States and Belgium.[11,31] However, a recent histologic study of hip fractures in the United States revealed a low grade osteomalacia that was detectable only by quantitative histologic measurement in 8 of 31 femoral heads.[87] This is a surprisingly high incidence of even minimal osteomalacia and suggests that vitamin D deficiency may play a role in fracture of the hip in the United States as well as abroad. This is to be contrasted with studies of spinal crush fractures where a low incidence of osteomalacia is found on iliac crest bone biopsy.[43]

In reviewing histologic studies, it is important to consider that vitamin D affects muscle metabolism and that its deficiency produces a myopathy. It has been postulated that the muscular weakness resulting from severe vitamin D deficiency could very well lead to a fall and therefore a higher incidence of detected osteomalacia in patients with fractures of the hip. Thus, patients with hip fractures may differ from patients with fractures of the spine because the former may be selected (with an osteomalacic component) as a result of falling because of muscle weakness.

Lawoyin et al.[50] administered 25OHD$_3$ to six women with postmenopausal osteoporosis. They noted an increase in calcium absorption and increased serum concentrations of $1,25(OH)_2D_3$ and $24,25(OH)_2D_3$. It has been suggested that administration of $25OHD_3$ may have an advantage over use of $1,25(OH)_2D_3$ because $24,25(OH)_2D_3$ may have an additional effect on bone mineralization.

Role of $1,25(OH)_2D_3$

Evidence for reduced levels of $1,25(OH)_2D_3$ in osteoporotic patients has been presented by Gallagher et al.[31] They found significantly lower $1,25(OH)_2D_3$ levels in North American osteoporotic women as compared to age and sex matched controls. This was especially apparent by examination of the ratio of $25OHD_3$ to $1,25(OH)_2D_3$ which suggested decreased production of $1,25(OH)_2D_3$ (although rapid metabolism of this hormone could not be excluded).

It has been believed that vitamin D deficiency would lead to osteomalacia rather than osteoporosis. It is however, still unsettled whether vitamin D acts primarily through regulation of serum concentrations of calcium and phosphorus. Nordin[65] has suggested that calcium malabsorption (or calcium deficiency) of any cause leads to osteoporosis rather than osteomalacia. He has postulated that osteomalacia may be associated with more severe vitamin D deficiency that is observed in osteoporosis resulting in reduced levels of calcium, phosphorus or both and impaired mineralization of bone. In this scheme, then, calcium malabsorption and osteoporosis occur from mild vitamin D deficiency (or a decrease in its metabolites) and severe deficiency of vitamin D (metabolites) may result in osteomalacia.

As a result of these studies a number of preliminary investigations have been reported in which $1,25(OH)_2D_3$ or its analogue $1\alpha OHD$ has been used to treat osteoporosis.[20,31,37,88] The majority of these reports are consistent with an overall effect of improved calcium absorption and calcium balance. However, it must be cautioned that these are preliminary results and that it is possible to produce hypercalcemia and increased bone resorption with these agents.[37]

SUMMARY

The field of vitamin D metabolism has experienced a knowledge explosion in the last decade. The advance from laboratory findings to clinical application has been extraordinarily rapid. Vitamin D deficiency occurs in elderly populations as a result of inadequate exposure to sunlight, dietary inadequacy and poor intestinal absorption. The lower $25OHD_3$ levels in elderly populations may also result from defective hepatic hydroxylation of vitamin D_3 and drug and illness effects on vitamin D metabolism. Decreased 25OHD levels are more prevalent in countries and populations with inadequate sunlight and unfortified foods.

Reduction in renal function with age is believed to result in decreased synthesis of $1,25(OH)_2D_3$, leading to the decreased intestinal absorption of calcium that is observed in the elderly. It is unlikely that the RDA for vitamin D of 200 IU in the elderly is adequate for correction of the calcium malabsorption.

The literature concerning the relationship of vitamin D metabolism to osteoporosis and osteomalacia in the elderly is confused as a result of differing global populations and selection of patients. It appears that fracture of the hip is frequently associated with histologic evidence of osteomalacia. Thus, demonstration of improvement in bone mass in "osteoporosis" when the latter is defined as fracture of the hip would be hazardous. The etiology of osteoporosis is multifactorial. In those patients in whom mild vitamin D deficiency, or altered metabolism of vitamin D, plays a major role, administration of vitamin D or its active metabolite should prove beneficial.

REFERENCES

(1) Aaron, J.E., Gallagher, J.C., Anderson, J., Stasiak, L., Longton, E.B., Nordin, B.E.C., and Nicholson, M. 1974. Frequency of osteomalacia and osteoporosis in fractures of the proximal femur. *Lancet* 1:229-233.

(2) Alevizaki, C.C., Ikkos, D.C., and Singhelakis, P. 1973. Progressive decrease of true intestinal calcium absorption with age in normal man. *J. Nucl. Med.* 14:760-762.

(3) Aloia, J.F., Cohn, S.H., Ross, P., Vaswani, A., Abesamis, C., Ellis, K., and Zanzi, I. 1978. Skeletal mass in postmenopausal women. *Am. J. Physiol.* 235:E82-E87.

(4) Avioli, L.V., McDonald, J.E., and Lee, S.W. 1965. The influence of age on the intestinal absorption of ^{47}Ca in women and its relation to ^{47}Ca absorption in postmenopausal osteoporosis. *J. Clin. Invest.* 44:1960-1967.

(5) Avioli, L.V., Lee, S.W., McDonald, J.E., Lund, J., and DeLuca, H.F. 1967. Metabolism of vitamin D_3-^3H in human subjects: Distribution blood, bile, feces and urine. *J. Clin. Invest.* 46:983-992.

(6) Baker, M.E., McDonnell, H., Peacock, M., and Nordin, B.E.C. 1979. Plasma 25-hydroxy vitamin D concentrations in patient with fractures of the femoral neck. *Brit. Med. J.* 1:589.

(7) Barragry, J.M., France, M.W., Corless, D., Gupta, S.P., Switala, S., Boucher, B.J., and Cohen, R.D. 1978. Intestinal cholecalciferol absorption in the elderly and in younger adults. *Clin. Sci. Molecular Med.* 55:213-220.

(8) Berlyne, G.M., Ben-Ari, J., Kushelvesky, A., Idelman, A., Galinsky, D., Hirsch, M., Shainkin, R., Yagil, R., and Zlotnik, M. 1975. The aetiology of senile osteoporosis: Secondary hyperparathyroidism due to renal failure. *Qtr. J. Med.* 44:505-521.

(9) Bhattacharyya, M.H., and DeLuca, H.F. 1973. The regulation of rat liver calciferol-25-hydroxylase. *J. Biol. Chem.* 248:2969-2973.

(10) Borgstedt, A.D., Bryson, M.F., Young, L.W., and Forbes, G.B. 1972. Long term administration of antiepileptic drugs and the development of rickets. *J. Pediat.* 81:9-15.
(11) Bouillon, R., Geusens, P., Dequeker, J., and DeMoor, P. 1979. Parathyroid function in primary osteoporosis. *Clin. Sci.* 57:167-171.
(12) Bullamore, J.R., Wilkinson, R., Gallagher, J.C., Nordin, B.E.C., and Marshall, D.H. 1970. Effect of age on calcium absorption. *Lancet* 2:535-537.
(13) Coburn, J.W., Hartenbower, D.L., and Birchman, A.S. 1976. Advances in vitamin D metabolism as they pertain to chronic renal disease. *Am. J. Clin. Nutr.* 29:1283-1299.
(14) Commentary—1980 Revised Recommended Dietary Allowances. 1979. *J. Amer. Diet. Assoc.* 75:623-625.
(15) Corless, D., Beer, M., Boucher, B.J., Gupta, S.P., and Cohen, R.D. 1975. Vitamin-D status in long-stay geriatric patients. *Lancet* 2:1404-1406.
(16) Corless, D., Gupta, S.P. 1978. Response of plasma-25-hydroxy-vitamin D to ultraviolet irradiation in long-stay geriatric patients. *Lancet* 2:649-651.
(17) Crilly, R., Horsman, A., Marshall, D.H., and Nordin, B.E.C. 1979. Prevalence, pathogenesis and treatment of postmenopausal osteoporosis. *Australian New Zealand J. Med.* 9:24-30.
(18) Davie, M., Lawson, D.E.M., and Jung, R.T. 1978. Low plasma-25-hydroxy-vitamin D without osteomalacia. *Lancet* 1:820.
(19) Davies, D.F., and Shock, N.W. 1950. Age change in glomerular filtration rate, effective renal plasma flow and tubular excretory capacity in adult males. *J. Clin. Invest.* 29:496-507.
(20) Davies, M., Mawer, E.B., and Adams, P.H. 1977. Vitamin D metabolism and the response to 1,25-dihydroxycholacalciferol in osteoporosis. *J. Clin. Endocrin. Metab.* 45:199-208.
(21) Deftos, L.J., Bury, A.E., Haebner, J.F., Singer, F.R., and Potts, J.T. 1971. Immunoassay for human calcitonin. *Metab. Clin. Exp.* 20:1129-1137.
(22) Dent, C.E., Richens, A., Rowe, D.J.E., and Stamp, T.C.B. 1970. Osteomalacia with long term anticonvulsant therapy in epilepsy. *Brit. Med. J.* 4:69-72.
(23) DeLuca, H.F. 1974. Vitamin D: The vitamin and the hormone. *Fed. Proc.* 33:2211.
(24) DeLuca, H.F., Blunt, K.W., and Rikkers, H. 1971. Biogenesis. In W.H. Sebrell, Jr. and R.S. Harris (eds.), *The Vitamins*, Vol. 3, pp. 213-230. Academic Press, New York.
(25) Dequeker, J., and Bouillon, R. 1977. Parathyroid hormone secretion and 25-hydroxyvitamin D levels in primary osteoporosis. *Calcified Tissue Res.* 22 (Suppl):495-496.
(26) Eddy, R.L. 1971. Metabolic bone disease after gastrectomy. *Am. J. Med.* 50:442-449.
(27) Fontaine, O., Pavlovitch, H., and Balsan, S. 1978. 25-hydroxycholecalciferol metabolism in hypohysectomized rats. *Endocrinology* 102:1822-1826.
(28) Franchimont, P., and Heynen, G. 1976 *Parathormone and Calcitonine Radioimmunoassay in Various Medical and Osteoarticular Disorders*, pp. 101-107. J.B. Lippincott, Philadelphia.
(29) Fraser, D.R., and Kodicek, E. 1970. Unique biosynthesis by kidney of a biologically active vitamin D metabolite. *Nature* 228:764-766.

(30) Gallagher, J.C., Aaron, J.E., Horsman, A., Marshall, D.H., Wilkinson, R., and Nordin, B.E.C. 1973. The crush fracture syndrome in post-menopausal women. *Clin. Endocrine Metab.* 2:293-315.
(31) Gallagher, J.C., Riggs, B.L., Eisman, J., Hamstra, A., Arnaud, S.B., and DeLuca, H.F. 1979. Intestinal calcium absorption and serum vitamin D metabolites in normal subjects and osteoporotic patients. *J. Clin. Invest.* 64:729-736.
(32) Galante, L., Coston, K., MacAuley, S.J., and MacIntyre, I. 1972. Effect of calcitonin on vitamin D metabolism. *Nature (London)* 238:271-273.
(33) Garabedian, M., Holick, M.F., DeLuca, H.F. and Boyle, I.T. 1972. Control of 25-hydroxycholecalciferol metabolism by the parathyroid glands. *Proc. Natl. Acad. Sci. USA* 69:1673-1676.
(34) Garcia-Psacual, B., Peytremann, A., Courvosier, B., and Lawson, D.E.M. 1976. A simplified competitive protein-binding assay for 25-hydroxycalciferol. *Clin. Chim. Acta* 68:99-105.
(35) Gertner, J.M., Horst, R.L., Broadus, A.E., Rasmussen, H., and Genel, M. 1979. Parathyroid function and vitamin D metabolism during human growth hormone replacement. *J. Clin. Endocrin. Metab.* 49:185-188.
(36) Ghazarian, J.G., Jefcoate, C.R., Knutson, J.C., Orme-Johnson, W.H., and DeLuca, H.F. 1974. Mitochondrial cytochrome P_{450}: A component of chick kidney 25-hydroxycholecalciferol-1α-hydroxylase. *J. Biol. Chem.* 249:3026-3033.
(37) Haas, H.G., Dambacher, M.A., Gucaga, J., Lauffenburger, T., Lammle, B., and Olah, J. 1979. 1,25(OH)$_2$ vitamin D$_3$ in osteoporosis—a pilot study. *Horm. Metab. Res.* 11:168-171.
(38) Haddad, J.G., and Chyu, K.J. 1971. Competitive protein-binding radioassay for 25-hydroxycholecalciferol. *J. Clin. Endocrin.* 33:992-995.
(39) Heaney, R.P. 1979. Calcium metabolic changes at menopause—their possible relationship to post-menopausal osteoporosis. In U.S. Barzel (ed.), *Osteoporosis II*, pp. 101-109. Grune and Stratton, New York.
(40) Holick, M.F., Baxter, L.A., Schraufrogel, P.K. Tavela, T.E., and DeLuca, H.F. 1976. Metabolism and biological activity of 24,25-dihydroxy vitamin D$_3$ in the chick. *J. Biol. Chem.* 251:397-402.
(41) Holick, M.F., Schnoes, H.K., DeLuca, H.F., Gray, R.W., Boyle, I.T., and Suda, T. 1972. Isolation and identification of 24,25-dihydroxycholecalciferol: A metabolite of vitamin D$_3$ made in the kidney. *Biochemistry* 11:4251-4255.
(42) Holick, M.F., Semmler, E.J., Schnoes, H.K., and DeLuca, H.F. 1973. 1α-hydroxy derivative of vitamin D$_3$: A highly potent analog of 1α-25-dihydroxy vitamin D$_3$. *Science* 180:190-191.
(43) Hulth, A.G., Nilsson, B.E., Westlin, N.E., and Wiklund, P.E. 1979. Bone biopsy in women with spinal osteoporosis. *Acta Med. Scand.* 206:205-206.
(44) Hyams, D.E. 1978. Liver and biliary system. In J.C. Brocklehurst (ed.), *Textbook of Geriatric Medicine and Gerontology*, pp. 385-417. Churchill Livingstone, Edinburgh.
(45) Ireland, P., and Fordtran, J.S. 1973. Effect of dietary calcium and age of jejunal calcium absorption in humans studied by intestinal perfusion. *J. Clin. Invest.* 52:2672-2681.
(46) Kenney, A.D. 1976. Vitamin D metabolism: Physiological regulation in egg-laying Japanese quail. *Am. J. Physiol.* 230:1609-1615.

(47) Klein, R.G., Arnand, S.B., Gallagher, J.C., DeLuca, H.F., and Riggs, B.L. 1977. Intestinal calcium absorption in exogenous hypercorisonism. Role of 25-hydroxyvitamin D and corticosteroid dose. *J. Clin. Invest.* 60:253-259.
(48) Lam, H.Y., Onisko, B.L., Schnoes, H.K., and DeLuca, H.F. 1974. Synthesis and biological activity of 3-deoxy-1α-hydroxyvitamin D_3. *Biochem. Biophys. Res. Commun.* 59:845.
(49) Larkins, R.G., MacAuley, S.J., Rapoport, A., Martin, T.J., Tulloch, B.R., Byfield, P.G.H., Matthews, E.W., and MacIntyre, I. 1974. Effects of nucleotides, hormones, ions and 1,25-dihydroxycholecalciferon on 1,25-dihydroxycholecalciferol production in isolated chick renal tubules. *Clin. Sci. Mol. Med.* 46:569-582.
(50) Lawoyin, S., Zerwekh, J.E., Glass, K., and Pak, C.Y.C. 1980. Ability of 25-hydroxyvitamin D_3 therapy to augment serum 1,25- and 24,25-dihydroxyvitamin D in postmenopausal osteoporosis. *J. Clin. Endocrinol. Metab.* 50:597-599.
(51) Lender, M., Verner, E., Stankiewicz, H., and Menczel, J. 1977. Intestinal absorption of Ca in elderly patients with osteoporosis, Paget's Disease and osteomalacia. Effects of calcitonin, oestrogen and vitamin D. *Gerontology* 23:31-36.
(52) Lester, E., Skinner, R.K., and Wills, M.R. 1977. Seasonal variation in serum-25-hydroxyvitamin-D in the elderly in Britain. *Lancet* 1:979-980.
(53) Lindholm, T.S., Sevastikoglou, A., and Lindgren, U. 1978. Short-term effects of varying doses of 1α-hydroxyvitamin D_3 on blood and urine chemistry and calcium absorption of osteoporotic patients. *Clin. Orth. Related Res.* 135:226-231.
(54) Long, R.G., Shiver, R.D., Willis, M.R., and Sherlock, S. 1976. Serum 25-hydroxy vitamin D in untreated parenchymal and cholestatic liver disease. *Lancet* 2:650-652.
(55) Lund, B., Kjaer, I., Friis, T., Hjorth, L., Reimann, I., Andersen, R.B., and Sorensen, O.H. 1975. Treatment of osteoporosis of ageing with 1α-hydroxycholecalciferol. *Lancet* 2:1168-1171.
(56) Lund, B. Sorenson, O.H., and Christensen, A.B. 1975. 25-hydroxycholecalciferol and fractures of the proximal femur. *Lancet* 1:300-302.
(57) Lund, B., and Sorenson, O.H. 1979. Measurement of 25-hydroxyvitamin D in serum and its relation to sunshine, age and vitamin D intake in the Danish population. *Scand. J. Clin. Lab. Invest.* 39:23-30.
(58) MacLennan, W.J., and Hamilton, J.C. 1977. Vitamin D supplements and 25-hydroxy vitamin D concentrations in the elderly. *Brit. Med. J.* 2:859-861.
(59) Marshall, D.H., Gallagher, J.C., Guha, P., Hanes, R., Oldfield, W., and Nordin, B.E.C. 1977. The effect of 1α-hydroxycholecalciferol and hormone therapy on the calcium balance of postmenopausal osteoporosis. *Calcif. Tissue Res.* 22:78-84.
(60) Marshall, D.H., and Nordin, B.E.C. 1977. The effect of 1 alpha-hydroxyvitamin D_3 with and without oestrongens on calcium balance in postmenopausal women. *Clin. Endocrinol.* 7(Suppl):159s-168s.
(61) Melsen, F., Mosekilde, L., and Beck-Nielsen, H. 1975. Quantitative histological investigation in metabolic diseases of bone. *Ugeskr Lager* 137:931-933.

(62) Morita, R., Yamamoto, I., Fukunaga, M., Dokoh, S., Konish, J., Kousaka, T., Nakajima, K., Toriz-ka, K., Aso, T., and Motohasi, T. 1979. Changes in sex hormones and calcium regulation hormones with reference to bone mass associated with aging. *Endocrinol. Japan* 1:15-22.
(63) Moss, A.J., Waterhouse, C., and Terry, R. 1965. Gluten-sensitive enteropathy with osteomalacia but without steatorrhea. *N. Eng. J. Med.* 272:825-830.
(64) Nayal, A.S., MacLennan, W.J., Hamilton, J.C., Rose, P., and Kong, M. 1978. 25-hydroxy-vitamin D, diet and sunlight exposure in patients admitted to a geriatric unit. *Gerontology* 24:117-122.
(65) Nordin, B.E.C., Wilkinson, R., Marshall, D.H., Gallagher, J.C., Williams, A.C., and Peacock, M. 1976. Calcium absorption in the elderly. *Calcified Tissue Res.* 21:442-451.
(66) Offerman, G., VonHerrath, D., Schaefer, B., and Lawson, D.E.M. 1974. Serum 25-hydroxycholecalciferol in uremia. *Nephron.* 13:269-277.
(67) Ohata, M., and Fujita, T. 1979. Vitamin D and osteoporosis. *Endocrinol. Japan* 1:7-13.
(68) Okamura, W.H., Mitra, M.N., Proscal, D.A., and Norman, A.W. 1975. Studies on vitamin D and its analogs. VIII. 3-deoxy-1α,25-dihydroxyvitamin D_3, a potent new analog of 1α,25-$(OH)_2 D_3$. *Biochem. Biophys. Res. Commun.* 65:24-30.
(69) Okamura, W.H., Mitra, M.N., Wing, R.M., and Norman, A.W. 1974. Chemical and biological activity of 3-deoxy-1α-hydroxyvitamin D_3, the active form of vitamin D_3. *Biochem. Biophys. Res. Commun.* 60:179-185.
(70) Okano, K., Nakai, R., and Harasawa, M. 1979. Endocrine factors in senile osteoporosis. *Endocrinol. Japan* 1:23-30.
(71) Olson, E.B., Jr., Knutson, J.C., Bhattacharyya, M.H., and DeLuca, H.F. 1976. The effect of hepatectomy on the synthesis of 25-hydroxyvitamin D_3. *J. Clin. Invest.* 57:1213-1220.
(72) Orimo, H., and Shiraki, M. 1979. Role of calcium regulating hormones in the pathogenesis of senile osteoporosis. *Endocrinol. Japan* 1:1-6.
(73) Parfitt, A.M., Miller, M.J., Frame, B., Villaneva, A.R., Raos, D.S., Oliver, I., and Thompson, D.I. 1978. Metabolic bone disease after intestinal bypass for treatment of obesity. *Ann. Intern. Med.* 89:193-199.
(74) Ponchon, G., and DeLuca, H.F. 1969. The role of the liver in the metabolism of vitamin D. *J. Clin. Invest.* 48:1273-1279.
(75) Prost, C., Hanniche, M., Bordier, P., Miravet, L., DeSeze, S., and Rambaud, J.C., 1975. Osteomalacia in chronic pancreatitis. *Nouv. Presse. Med.* 4:1561-1566.
(76) Rasmussen, H., Bordier, P., Marie, P., Miravet, L., and Rackwaert, A. 1977. Action of vitamin D metabolites in humans. *Trans. Assoc. Amer. Phy.* 90:380-389.
(77) Riggs, B.L. 1979. Postmenopausal and senile osteoporosis: Current concepts of etiology and treatment. *Endocrinol. Japan* 1:31-41.
(78) Riggs, B.L., Arnaud, C.D., Jowsey, J., Goldsmith, R.S., and Kelly, P.J. 1973. Parathyroid function in primary osteoporosis. *J. Clin. Invest.* 52:181-184.
(79) Riggs, B.L., Gallagher, J.C., DeLuca, H.F., Edis, A.J., Lambert, P.W., and Aranaud, C.D. 1978. A syndrome of osteoporosis, increased serum immunoreactive parathyroid hormone, and inappropriately low serum 1,25-dihydroxyvitamin D. *Mayo Clin. Proc.* 53:701-706.

(80) Roof, B.S., Piel, C.F., Hansen, J., and Fundenberg, H.H. 1976. Serum parathyroid hormone levels and serum calcium levels from birth to senescence. *Mech. Aging Dev.* 5:289-304.
(81) Rushton, C. 1978. Vitamin D hydroxylation in youth and old age. *Age and Aging* 7:91-95.
(82) Sandberg, H., Yoshimine, N., Maeda, S., Symons, D., and Zavodnick, J. 1973. Effects of an oral glucose load on serum immunoreactive insulin, free fatty acid, growth hormone and blood sugar levels in young and elderly subjects. *J. Am. Geriatr.* 21:433-439.
(83) Saville, P.D. 1973. The syndrome of spinal osteoporosis. *Clin. Endocr. Metab.* 2:177-185.
(84) Schachter, D., Finkelstein, J.D., and Kowarski, S. 1964. Metabolism of vitamin D. I. Preparation of radioactive vitamin D and its intestinal absorption in the rat. *J. Clin. Invest.* 43:787-796.
(85) Sebrell, W.H., Jr., and Harris, R.S. 1954, *The Vitamins*, 1st ed. Vol. II. pp. 131-266. Academic Press, New York.
(86) Sherk, H.H., Cruz, M.D., and Stamburgh, J. 1977. Vitamin D prophylaxis and the lowered incidence of fractures in anticonvulsant rickets and osteomalacia. *Clin. Orth. Rel. Res.* 129:251-257.
(87) Sokoloff, L. 1978. Occult osteomalacia in American (U.S.A.) patients with fracture of the hip. *Am. J. Surg. Path.* 2:21-30.
(88) Sorenson, O.H., Lund, B., Friis, T.H., Hjorth, L., Reinmann, I., Kjaer, I., and Anderson, R.B. 1977. Effect of 1α-hydroxycholecalciferol in senile osteoporosis and in bone loss following prednisone treatment. *Israel J. Med. Sci.* 12:253-258.
(89) Spanos, E., Pike, J.W., Haussler, M.R., Coston, K.W., Evans, I.M.A., Goldner, A.M., McCain, T.A., and MacIntyre, I. 1976. Circulating 1α-25-dihydroxyvitamin D in the chicken: Enhancement by injection of prolactin and during egg laying. *Life Sci.* 19:1751-1756.
(90) Spencer, E.M., and Tobiassen, O. 1977. The effects of hypophysectomy on 25-hydroxyvitamin D_3 metabolism in the rat. In A.W. Norman, K. Schaefer, J.W. Coburn, H.F. DeLuca, D. Fraser, H.G. Grigoleit, and D.V. Herrath (eds.), *Vitamin D Biochemical, Chemical and Clinical Aspects Related to Calcium Metabolism*, pp. 197-199. de Gruyter, New York.
(91) Tanaka, Y., and DeLuca, H.F. 1973. The control of 25-hydroxyvitamin D metabolism by inorganic phosphorus. *Arch. Biochem. Biophys.* 154:566-576.
(92) Tanaka, Y., and DeLuca, H.F. 1974. Role of 1,25-dihydroxyvitamin D_3 in maintaining serum phosphorus and curing rickets. *Proc. Natl. Acad. Sci. USA* 71:1040-1044.
(93) Tanaka, Y., Frank, H., and DeLuca, H.F. 1973. Biological activity of 1,25-dihydroxyvitamin D_3 in the rat. *Endocrinology* 92:417-422.
(94) Vaishnava, H., and Rissi, S.N.A. 1974. Frequency of osteomalacia and osteoporosis in fractures of proximal femur. *Lancet* 1:676-677.
(95) Vidalon, C., Khurana, R.C., Chae, S., Gegick, C.G., Stephan, T., Nolan, S., and Danowski, T.S. 1973. Age-regulated changes in growth hormone in nondiabetic women. *J. Am. Geriatr. Soc.* 21:253-255.
(96) Weisman, Y., Fattal, A., Eisenberg, Z., Harel, S., Spirer, Z., and Harell, A. 1979. Decreased serum 24,25-dihydroxy vitamin D concentrations in children receiving chronic anticonvulsant therapy. *Brit. Med. J.* 2:521-523.
(97) Yeh, J.K., and Aloia, J.F. 1979. The influence of growth hormone on vitamin D metabolism. *Biochem. Med.* 21:311-322.

–10–
Vitamin E and Aging

Linda H. Chen

INTRODUCTION

Many theories have been proposed to explain the biological changes associated with the phenomenon of senescence.[18] By reviewing these theories, possible factors contributing to biological aging can be listed. They include: genetic time table, rate of living, increased abnormal chromosomes, deterioration of deoxyribonucleic acid and ribonucleic acid, deterioration of protein synthesis, age-dependent changes in enzymes systems, increased crosslinkage of collagen, breakdown in immunologic processes, changes in membrane diffusion, slowed adaptation of stress, accumulation of free radicals, cellular mutation, accumulated results of ionizing radiation, waste product accumulation, nutritional improprieties and loss of cells.[48]

Recently, Leibovitz and Siegel[50] have reviewed aspects of free radical reactions with regard to the aging process. Free radicals induce lipid peroxidation which results in damage to cell membrane and other cell components and the accumulation of aging pigments. Enzymatic defense systems including superoxide dismutase and glutathione peroxidase and nonenzymatic defense systems including vitamin E, vitamin C, selenium, sulfur-containing amino acids are of importance with respect to free radical and lipid peroxide quenching. The above systems in this regard may play a role against such deteriorative reactions in the aging process.

The current view on the relationship between vitamin E and aging has been discussed in two review articles.[3,4] There is some support for the theory that lipid peroxide formed in the tissues can interact with protein to produce so-called "lipofuscin pigment" that may be related to aging processes in various tissues. Since vitamin E interacts with free radicals, it has been proposed that vitamin E may prevent lipid peroxidation and slow down the aging process. However, this theory is still at an early stage and controversial.

CHEMISTRY, PHYSIOLOGY, FOOD SOURCES AND DAILY ALLOWANCE OF VITAMIN E

Vitamin E activity is exhibited by eight isomers of related chemical structure, which are derivatives of chroman-6-ol. The first series α-, β-, γ- and δ-tocopherols are derived from tocol which contains 16 carbon saturated isoprenoid side chains. The second series of four isomers, the tocotrienols, contains a triply unsaturated side chain. Within each series the compounds differ only in the number and position of methyl groups in the ring structure. α-Tocopherol is the most active biologically of all the isomers. It accounts for about 80% of the total activity of the vitamin.

Vitamin E is fat soluble. It requires the presence of fat and bile salts for absorption. The vitamin is carried with the chylomicrons into the lymphatic circulation and to the liver, and is associated with β-lipoprotein in the blood. Relatively constant concentrations of vitamin E are maintained in each tissue, generally related to the log of dietary intake,[6] with one exception that adipose tissue accumulates this vitamin progressively.

Current concepts regarding the biological role of vitamin E may be grouped broadly into two categories: (a) Vitamin E acts solely as an antioxidant, preventing the peroxidation of polyunsaturated fatty acids, and thus serving as a biological free-radical scavenger,[57] and (b) vitamin E has nonantioxidant metabolic roles since the antioxidant function of vitamin E, by itself, cannot account for the wide range of manifestations associated with vitamin E deficiency symptoms of animals. The nonantioxidant roles of vitamin E have not yet been delineated clearly, and may include roles in heme synthesis,[7] nucleic acid and protein metabolism,[11] mitochondrial metabolism,[17] and maintaining function and ultrastructure of cellular membranes.[61] The richest sources of vitamin E are vegetable oils, including corn, soybean, cottonseed and sunflower oils, and hydrogenated products from these oils. These contribute about 70% of the average daily intake for humans. Whole cereal grains, nuts, legumes, dark-green vegetables, liver, eggs and dairy products also provide moderate amounts.

Food processing and storage destroy some of the tocopherol content of most foods; however, tocopherols are fairly stable to heat, and are not destroyed during normal cooking. There is considerable loss of this vitamin in fried foods, and also in the heating of oils. Losses also occur in frozen storage, where the temperatures do not completely prevent oxidative destruction of tocopherols.

The Foods and Nutrition Board, National Academy of Sciences, recommends a daily intake of 3 to 4 International Units (IU) for infants, 5 to 7 IU for children, 8 IU for women, 10 IU during pregnancy, 11 IU during lactation, and 10 IU for men. One mg of dl-α-tocopheryl acetate provides 1 IU of the vitamin activity whereas 1 mg of naturally-occurring d-α-tocopherol provides 1.49 IU.

Vitamin E requirement is influenced by a number of factors in the diet such as the presence of sulfur-containing amino acids, selenium and unsaturated oil. In general, increased levels of dietary polyunsaturated acids demand an increased intake of vitamin E.

Typical American diets provide 4.4 to 12.7 mg α-tocopherol, with an average of 9 mg (13.4 IU).[8] In addition, the γ-tocopherol content was estimated to be about 2.5 times that of α-tocopherol. The biological activity of this amount of γ-tocopherol was estimated to be about 25% of α-tocopherol. Thus, the typical diet would supply adequate amounts to meet body needs.

VITAMIN E DEFICIENCIES IN MAN AND ANIMALS

Vitamin E deficiency in adult men and women is rare, since vitamin E is commonly available in foods, is stored throughout the body in tissues, and is turned over slowly in the body. Deficiencies are found in patients with malabsorption of fats and oils due to diseases such as cystic fibrosis, liver cirrhosis, obstructive jaundice, pancreatic insufficiency, sprue or steatorrhoea.[9,28,42]

Both premature infants and full-term newborn infants may have relatively low levels of vitamin E in blood and tissues because there is little transfer of vitamin E across the placenta to the fetus. Hemolytic anemia in premature infants has often been observed.[7] Infants and children with Kwashiorkor have been shown to develop vitamin E deficiency.

Vitamin E deficiencies in animals can be demonstrated when animals are fed with diets free of vitamin E. A wide range of vitamin E deficiency symptoms is observed with considerable variation from one species to another. They include: embryonic degeneration in the female rat, hen, turkey; sterility in the male rat, guinea pig, hamster, dog and cock; liver necrosis in the rat and pig; erythrocyte hemolysis in the rat and chick; encephalomalacia and exudative diathesis in the chick; nutritional muscular dystrophy in the rabbit, guinea pig, monkey, duck, mouse, chicken, mink; and white muscle disease of lamb and calf.[62,69] Some, but not all, of the symptoms of vitamin E deficiency can be prevented by nonspecific antioxidants or selenium.

VITAMIN E AND AGING

Free Radical Reactions in Biological Systems

Free radicals are produced at random in enzymatic and nonenzymatic reactions involving molecular oxygen throughout biological systems, during irradiation damage, and are found as environmental pollutants. They are highly reactive due to the presence of unpaired electrons, and produce subsequent free radical chain reactions which go on continuously in biological systems, thus contributing to the degradation of these systems.[41] The free radical reactions have been considered as some of the basic deteriorative reactions in cellular mechanisms of aging processes.[15,31,33,63]

The biochemical changes which may be produced by free radical reactions include: accumulation of metabolically inactive lipofuscin age pigments;[64,73] polymerization and crosslinking of enzymes and proteins;[22,73] oxidative alteration in collagen,[46] elastin,[47] chromosomes,[70] and mucopolysaccharides;[56] and alteration in membrane structure and subcellular organelles due to lipid peroxidation.[38,73] These changes may result in a decline in cellular integrity. There is considerable evidence to substantiate some of these changes, but insufficient information to determine whether they are of primary or secondary importance to organismic aging.

Aging, Vitamin E Deficiency and Lipofuscin Pigments

Among several changes occurring in tissues of aging man and animals, the formation of lipofuscin age pigments is very interesting since it can also be induced by vitamin E deficiency.

Pigment Accumulation as a Function of Age: Different tissues age differently, but pigmentation can be found in many tissues of man and animals such as the heart, blood vessel, brain, muscle, liver, kidney, adrenal gland, pituitary gland and reproductive organs.[19,60,64,66,68,75] The concentration of these insoluble yellowish brown substances increases with age. The pigment granules are commonly observed in clusters of round or oblong bodies varying from less than 1 to 3 microns in diameter. The pigment shows yellow to brown or blue fluorescence under ultraviolet light. Some of the pigment is intracellular, some extracellular. The intracellular pigment is found to scatter throughout the cytoplasm in younger specimens, but forms aggregates in the older. Terms like lipofuscin (most common), ceroid, chromolipoid, yellow pigment, lipopigment, senile pigment, and "wear and tear" pigment[44] are used to describe the pigment. It is formed from the interaction of autooxidized lipid, especially unsaturated lipid, with protein and other biochemical factors. Pigment accumulation in cells could impede the diffusion and transport of essential metabolites and macromolecules.[29]

Pigment Accumulation as a Function of Vitamin E Deficiency: In studies of vitamin E-deficient rat tissues, deposit of pigment was observed in the uteri by Desai et al,[21] Elftman et al,[24] Radice and Herraiz,[65] Sulkin and Srivanij,[71] Martin and Moore,[55] and Horowitz and Hartrott.[39] Similar pigment deposit was found by Sulkin and Srivanij,[71] and Wünscher and Küstner[82] in nerve cells; Radice and Herraiz,[65] and Martin and Moore[55] in sex glands; Weglicki et al[80] in adrenal glands; Miyagishi[60] in cerebral cortex; and Reddy et

al[67] in adipose tissue, bone marrow, heart and spleen. The pigment was also found in tissues of vitamin E-deficient man[49] and other animals.[23,58] Tappel et al[74] reported that diet supplemented with vitamin E and other antioxidants significantly inhibited the accumulation of pigment in testis and heart of mice. Studies have suggested that it is impossible to distinguish the pigment from lipofuscin pigment derived through peroxidation and polymerization of unsaturated lipids.[64]

The above findings that lipofuscin pigment is present in tissues of animals markedly deficient in vitamin E and the absence of such pigment in young animals with vitamin E supplementation demonstrate that vitamin E protects against lipofuscin pigment accumulation in young animals. In addition, Sulkin and Srivanij[71] showed that the accumulation of pigment in the cytoplasm of nerve cells of rats was accelerated by vitamin E deficiency.

It thus appears that the causes of lipofuscin pigment formation described above, namely, in vitamin E deficiency and in aging process may be similar. Chio et al[14] postulated that pigment formation was related to autooxidation of cellular membranes which is secondary to antioxidant deficiency. Autooxidation of lipid is a free radical chain reaction which can be stopped by vitamin E or other antioxidants including free radical inhibitors. When there is insufficient antioxidant available, the free radical chain reaction will go on to produce more oxidized lipid. Since lipofuscin pigment contains peroxidized lipid, aging may be the result of not having enough antioxidants to prevent lipid peroxidation. It has been suggested that the addition of one or more antioxidants or free radical inhibitors to nutritionally adequate and acceptable natural diets, and the reduction of the intake of substances which increase *in vivo* free radical reactions (such as polyunsaturated lipids and copper) may increase the life span of man.[35,72]

Vitamin E and Aging

There is little evidence in man or animals at present that vitamin E retards aging processes or prevents age-related biological changes.

Many reports on the plasma vitamin E levels of control populations from many different countries have appeared in the literature. Plasma vitamin E levels in general populations seem to increase with increasing age as reported by Chen et al,[13] Desai,[20] Vatassery et al,[77] Leitner et al[51] and Lewis et al.[53] This is probably due to the lipid-soluble property of vitamin E which allows vitamin E to be stored in the body and accumulated with age. It could also be due to the increase in blood β-lipoprotein level with increasing age.[52] In studies by Wei and Draper[81] and Barnes and Chen,[6] plasma levels of vitamin E decreased with increasing age in healthy elderly people above age of 60. In other studies,[43,77] plasma levels of vitamin E did not decrease with increasing age in the elderly. Underwood *et* al[76] measured liver level of vitamin E in human subjects at various ages who died accidentally and found that there was not any tendency for tissue levels to decrease with increasing age.

Weglicki et al[79] studied the vitamin E contents of tissues of male and female rats. They found that, with increasing age, there was an increase in vitamin E contents in the liver, heart, adrenal glands but not in the brain. Although the concentration of vitamin E in the brain decreased with increasing age, the difference was not significantly different. The study of Hirai and Yoshikawa[37] showed a similar result with respect to the brain.

Studies showing a lack of effect of vitamin E in modifying age-related change have been reported also. Tappel[74] reported that parameters used to measure aging such as kidney function, treadmill ability and calcium uptake by muscle microsomes were not improved in aging rats fed with high levels of vitamin E and other antioxidants. Grinna[30] reported that the need for dietary tocopherol appears to decrease with increasing animal age. The older the rat the more difficult it was to produce signs of deficiency or to deplete the liver of its store of vitamin E. Whether a similar situation can be assumed for elderly human individuals is not known, though the results of Underwood et al[76] would suggest that tissue levels of vitamin E are adequate in elderly people. Grinna[30] also reported that age-associated functional and compositional changes were not modified to any extent by alteration of the dietary level of vitamin E. Increased levels of vitamin E did not retard age-associated changes at the subcellular level, including: changes in microsomal membrane composition or function, phospholipid loss, and changes in enzyme activity.

Vitamin E, Synthetic Antioxidants and Longevity

If free radical injury constitutes an environmental source of aging, the ability of vitamin E and other antioxidants to quench free radical reactions should, theoretically, lead to enhanced longevity, but to date very few investigations of this nature have been reported. Harman[34] reported that the average life spans of mice and rats were shortened by increasing the amount and/or degree of unsaturation of the dietary lipid. Later he reported that 2-mercaptoethylamine (a drug which protects against radiation) and butylated hydroxytoluene (BHT) or ethoxyquin (commercial food additives) could prolong the life spans of certain strains of mice when added to their diets.[32] Comfort[15] supported the effect of ethoxyquin in prolonging life span of mice. Emanuel[25] used 2-ethyl-6-methyl-3-hydroxypyridine·HCl in the diet to increase the life span of mice. Addition of tocopherol-p-chlorophenoxyacetate to the diet of drosophila[59] and α-tocopherol to that of nematode[26] also increased the life span. Contradicting the above studies was a report by Tappel et al[74] in which effects of three levels of the dietary components vitamin E, vitamin C, BHT, selenium and methionine on various age-related factors in mice were studied. These dietary antioxidants did not prolong the life span of mice.

Kent[45] has carefully reviewed the above studies and suggested alternative explanations for antioxidants' benefits, if dietary antioxidants do not retard the aging process: (a) Antioxidants may either spoil the appetite or hinder the assimilation, which leads to caloric restriction. Calorie restriction may be the real reason for increased longevity. (b) Antioxidants may reduce the toxicity of normal laboratory diets. (c) Antioxidants may induce enzyme activity that stimulates reactions leading to increased life expectancy. (d) Antioxidants may have an antitumor effect that delays the onset of cancer. (e) Antioxidants may inhibit pathophysiologic vascular changes that delay the onset of cardiovascular disease. (f) Antioxidants may enhance the immunologic system to bolster the body's resistance to disease. Kent concluded that further research is needed to determine the precise effect of dietary antioxidants on biological systems, and whether these compounds can be used to protect man against degenerative disease or the aging process.

VITAMIN E CONTROVERSY

Ingestion of massive doses of vitamin E has been a fad for years. Various claims have been made that vitamin E can be used in large doses as a drug to treat many human ailments such as ischemic heart disease, infertility, spontaneous abortion, muscular dystrophy, cancer and many other disorders. However, as reported by the Institute of Food Technologist's Expert Panel on Food Safety and Nutrition and the Committee on Public Information,[3] there are few conclusive data to substantiate such claims. Farrell[27] has also extensively reviewed the clinical studies relevant to therapeutic use of vitamin E and found that most of the published studies involving very large doses of vitamin E for the above disorders, both positive and negative, were not adequately controlled and would not withstand careful scientific scrutiny. Vitamin E has also been proposed for increased athletic ability, physical vigor, strength and endurance. However, wherever these claims have been subjected to rigorous testing, they have not been substantiated.[3] Much of the misinformation probably has rooted from the diverse syndromes encountered with vitamin E deficiency in different species of animals.

There are at least three special situations in which supplemental vitamin E has proven to be beneficial.[3] They are: hemolytic anemia in premature infants, vitamin E deficiencies caused by malabsorption of lipids, and intermittent claudication.

MEGADOSES OF VITAMIN E

Toxicity resulting from excessive doses of vitamin E supplements has not been unequivocally established. Information currently available suggests that daily doses of less than 300 mg are safe for most people.[7] However, it is uncertain that chronic ingestion of vitamin E in large doses is entirely safe. Concern about this problem has been expressed repeatedly.[7,10,40]

In human studies, large doses of vitamin E produced nausea, intestinal distress[36,78] and increased creatinuria.[36] March et al[54] demonstrated hypervitaminosis E in chicks with the following symptoms: depression in growth, decreased bone calcification, reduced hematocrit, and a decline in respiration of skeletal muscle mitochondria. Increased prothrombin times were also observed. The effect on bone calcification and prothrombin time suggest an interference with the normal metabolism of vitamins D and K by hypervitaminosis E. Corrick[16] also reported that vitamin E excess depressed growth in rats. Alfin-Slater et al[1] reported that liver total lipid and cholesterol were increased, and tissue fatty acid patterns were altered by high vitamin E dosage in rats. It has been suggested that an unbalanced ratio between vitamin E and vitamin K may lead to impairment of the blood coagulation mechanisms.[2]

CONCLUDING REMARKS

Although there is some support that the causes of lipofuscin pigment formation in aging process may be similar to those of the pigment formation in vitamin E deficiency, there is little evidence at the present time that excess dosage of vitamin E retards aging process or prevents age-related biological changes. The controversy regarding the effect of vitamin E on senescense will continue until further evidence can be documented.

REFERENCES

(1) Alfin-Slater, R.B., Aftergood, L. and Kishineff, S. 1972. Investigations on hypervitaminosis E in rats. *IX. Intl. Cong. Nutr. Abstr.* p. 191.
(2) Anonymous. 1975. Hypervitaminosis E and coagulation. *Nutr. Rev.* 33:269-270.
(3) Anonymous. 1977. Vitamin E. *Nutr. Rev.* 35:57-62.
(4) Anonymous. 1977. Effect of dietary α-tocopherol in aging rats. *Nutr. Rev.* 35:50-52.
(5) Barber, A.A. and Bernheim, F. 1967. Lipid peroxidation: its measurement, occurrence and significance in animal tissue. *Adv. Gerontol. Res.* 2:355-403.
(6) Barnes, K. and Chen, L.H. 1981. Vitamin E status of the elderly in Central Kentucky. *J. Nutr. for Elderly* (in press).
(7) Bieri, J.G. 1972. Kinetics of tissue α-tocopherol depletion and repletion. *Ann. N.Y. Acad. Sci.* 203:181-191.
(8) Bieri, J.G. 1975. Vitamin E. *Nutr. Rev.* 33:161-167.

(9) Bieri, J.G. and Evarts, R.P. 1973. Tocopherols and fatty acids in American diets. *J. Am. Diet. Assoc.* 62:147-151.
(10) Binder, J.J., Herting, D.C., Hurst, V., Finch, S.C. and Spiro, H.M. 1965. Tocopherol deficiency in man. *N. Eng. J. Med.* 273:1289-1297.
(11) Briggs, M.H. 1974. Vitamin E in clinical medicine. *Lancet* 1:220.
(12) Catinani, G.L. 1980. Vitamin E: Role in nucleic acid and protein metabolism. In L.J. Machlin (Ed.), *Vitamin E: A Comprehensive Treatise*, pp. 318-331. Marcel Dekker, NY.
(13) Chen, L.H., Hsu, S.J., Huang, P.C. and Chen, J.S. 1977. Vitamin E status of Chinese population in Taiwan. *Am. J. Clin. Nutr.* 30:728-735.
(14) Chio, K.S., Reiss, U., Fletcher, B. and Tappel, A.L. 1969. Peroxidation of subcellular organelles: Formation of lipofuscin-like fluorescent pigments. *Science* 166:1535-1536.
(15) Comfort, A., Youhotsky-Gore, I. and Pathmanathan, K. 1971. Effect of ethoxyquin on the longevity of C3H mice. *Nature* 229:254-255.
(16) Corrick, J.A. 1969. Growth and reproduction of albino rats as affected by various excessive levels of dietary vitamin E. *Diss. Abstr.* B 29:2249B.
(17) Corwin, L.M. 1980. The role of vitamin E in mitochondrial metabolism. In L.J. Machlin (Ed.), *Vitamin E: A Comprehensive Treatise*, pp. 318-331. Marcel Dekker, NY.
(18) Curtis, H. 1966. *Biological Mechanism of Aging.* Charles C. Thomas Co., Springfield.
(19) Csallany, A.S., Ayaz, K.L. and Su, L.C. 1977. Effect of dietary vitamin E and aging on tissue lipofuscin pigment concentration in mice. *J. Nutr.* 107:1792-1799.
(20) Desai, I.D. 1968. Plasma tocopherol level in normal adults. *Can. J. Physiol. Pharmacol.* 46:819-822.
(21) Desai, I.D., Fletcher, B.L. and Tappel, A.L. 1975. Fluorescent pigments from uterus of vitamin E-deficient rats. *Lipids* 10:307-309.
(22) Desai, I.D. and Tappel, A.L. 1963. Damage to proteins by peroxidized lipids. *J. Lipid Res.* 4:204-207.
(23) Einarson, L. 1962. Cellular structure in the CNS of vitamin E-deficient monkeys. In N.W. Shock (Ed.), *Biological Aspects of Aging*, p. 131. Columbia University Press, NY.
(24) Elftman, H., Kaunitz, H. and Slanetz, C.A. 1949. Histochemistry of uterine pigment in vitamin E-deficient rats. *Ann. N.Y. Acad. Sci.* 52:72-79.
(25) Emanuel, N.M. 1976. Free radicals and the action of inhibitors of radical process under pathological states and aging in living organisms and in man. *Quant. Rev. Biophysics* 9:283-309.
(26) Epstein, J. and Gershan, D. 1972. Studies on aging in nematodes. IV. The effect of antioxidants on cellular damage and life span. *Mech. Age. Dev.* 1:257-264.
(27) Farrell, P.M. 1980. Deficiency states, pharmacological effects and nutrient requirements. In L.J. Machlin (Ed.), *Vitamin E: A Comprehensive Treatise*, pp. 520-620. Marcel Dekker, NY.
(28) Farrell, P.M., Bieri, J.G., Fratantoni, J.F., Wood, R.E. and DiSaut'Agnese, P.A. 1977. The occurrence and effects of human vitamin E deficiency. *J. Clin. Invest.* 60:233-241.
(29) Gordon, P. 1974. Free radicals and the aging process. In M. Rockstein, M.L. Sussman and J. Chesky (Eds.), *Theoretical Aspects of Aging*, pp. 61-81. Academic Press, NY.

(30) Grinna, L.S. 1976. Effect of dietary α-tocopherol on liver microsomes and mitochondria of aging rats. *J. Nutr.* 106:918-929.
(31) Harman, D. 1962. Aging: A theory based on free radical and radiation chemistry. *J. Gerontol.* 11:298-300.
(32) Harman, D. 1968. Free radical theory of aging: Effect of free radical reaction inhibitors on the mortality rate of male LAF_1 mice. *J. Gerontol.* 23:476-482.
(33) Harman, D. 1969. Prolongation of life: Role of free radical reactions in aging. *J. Amer. Geriat. Soc.* 17:721-735.
(34) Harman, D. 1971. Free radical theory of aging: Effect of the amount and degree of unsaturation of dietary fat on mortality rate. *J. Gerontol.* 26:451-457.
(35) Harman, D. 1978. Free radical theory of aging: Nutritional implications. *Age* 1:145-152.
(36) Hillman, R.W. 1957. Tocopherol excess in man. Creatinuria associated with prolonged ingestion. *Am. J. Clin. Nutr.* 5:597-600.
(37) Hirai, S. and Yoshikawa, M. 1972. Vitamin E and aging of the nervous system. In N. Shimazono and Y. Takagi (Eds.), *International Symposium on Vitamin E*, pp. 228-237. Kyoritsu Shuppan Co., Tokyo.
(38) Hoffsten, P.E., Hunter, F.E., Gerbicki, J.M. and Winstein, J. 1962. Formation of lipid peroxide under conditions which lead to swelling and lysis of rat liver mitochondria. *Biochem. Biophys. Res. Commun.* 7:276-280.
(39) Horowitz, I. and Hartrott, W.W. 1971. Ceroid in the products of conception of normal and vitamin E-deficient rats. *J. Nutr.* 101:959-965.
(40) Horwitt, M.K. 1976. Vitamin E: A reexamination. *Am. J. Clin. Nutr.* 29:569-578.
(41) Isenberg, I. 1964. Free radicals in tissue. *Physiol. Rev.* 44:487-517.
(42) Kelleher, J. and Losowsky, M.S. 1970. The absorption of α-tocopherol in man. *Brit. J. Nutr.* 24:1033-1047.
(43) Kelleher, J. and Losowsky, M.S. 1977. Vitamin E in the elderly. In C. deDuve and O. Hayaishi (Eds.), *Tocopherol, Oxygen and Bimembranes*, pp. 311-325. Elsevier, North Holland Biochemical Press, Amsterdam.
(44) Kent, S. 1976. Solving the riddle of lipofuscin's origin may uncover clues to the aging process. *Geriatrics* 31(5):128-137.
(45) Kent, S. 1977. Do free radicals and dietary antioxidants wage intracellular war? *Geriatrics* 32:127-136.
(46) LaBella, F.S. and Paul, G. 1965. Structure of collagen from human tendon as influenced by age and sex. *J. Gerontol.* 20:54-59.
(47) LaBella, F.S., Vivian, S. and Thornhill, D.P. 1966. Amino acid composition of human aortic elastin as influenced by age. *J. Gerontol.* 21:550-555.
(48) Lawton, A.H. 1976. Introductory remarks: Trace elements in aging. In J.M. Hsu, R.L. Davis and R.W. Neithamer (Eds.), *Biomedical Role of Trace Elements in Aging*, pp. 1-6. Eckerd College, St. Petersburg, FL.
(49) Lee, S.P. and Nicholson, G.I. 1976. Ceroid enteropathy and vitamin E deficiency. *New Zealand Med. J.* 76:318-320.
(50) Leibovitz, B.E. and Siegel, B.V. 1980. Aspects of free radical reactions in biological systems: Aging. *J. Gerontol.* 35:45-56.
(51) Leitner, Z.A., Moore, T. and Sharman, I.M. 1960. Vitamin A and vitamin E in human blood. 2. Levels of vitamin E in the blood of British men and women. *Brit. J. Nutr.* 14:281-287.

(52) Lewis, B., Chait, A., Wootton, I.D.P., Oakley, C.M., Krikler, D.M., Sigurdsson, G., February, A., Maurer, B. and Birkhead, J. 1974. Frequency of risk factors for ischaemic heart disease in a healthy British population. *Lancet* 1:141-146.

(53) Lewis, J.S., Pian, A.K., Baer, M.T., Acosta, P.B. and Emerson, G.A. 1973. Effect of long-term ingestion of polyunsaturated fat, age, plasma cholesterol, diabetes mellitus and supplemented tocopherol upon plasma tocopherol. *Am. J. Clin. Nutr.* 26:136-143.

(54) March, B.E., Wong, E., Seier, L., Sim, L. and Biely, J. 1973. Hypervitaminosis E in the chick. *J. Nutr.* 103:371-377.

(55) Martin, A.J. and Moore, T. 1939. Some effects of prolonged vitamin E deficiency in rats. *J. Hyg. Lond.* 39:643-650.

(56) Matsumura, G., Harp, A. and Pigmon, W. 1966. Depolymerization of hyaluronic acid by autooxidants and radiations. *Radiation Res.* 28:735-752.

(57) McCay, P.B. and King, M.M. 1980. Vitamin E: Its role as a biological free radical scavenger and its relationship to the microsomal mixed-function oxidase system. In L.J. Machlin (Ed.), *Vitamin E: A Comprehensive Treatise*, pp. 289-317. Marcel Dekker, NY.

(58) Menschik, Z., Munk, M.K., Rogalski, T., Rymaszewski, O. and Szczensnik, T.J. 1949. Vitamin E studies in mice with special reference to the distribution and metabolism of lipids. *Ann. N.Y. Acad. Sci.* 52:94-103.

(59) Miquel, J. and Johnson, J.E., Jr. 1975. Effects of various antioxidants and radiation protectants on the life span and lipofuscin of drosophila and C57BL/6L mice. *The Gerontologist* 15:25-29.

(60) Miyagishi, T. 1966. Electron microscopic studies on the cerebral cortex of senile and vitamin E-deficient rats—with special reference to the morphopathogenesis of lipo-pigments. *Psychiatr. Neurol. Japan* 68:885-871.

(61) Molenaar, I., Hulstaert, C.E. and Hardonk, M.J. 1980. Vitamin E: Role in function and ultrastructure of cellular membranes. In L.J. Machlin (Ed.), *Vitamin E: A Comprehensive Treatise*, pp. 372-390. Marcel Dekker, NY.

(62) Nelson, J.S. 1980. Pathology of vitamin E deficiency. In L.J. Machlin (Ed.), *Vitamin E: A Comprehensive Treatise*, pp. 397-428. Marcel Dekker, NY.

(63) Packer, L., Deamer, D.W. and Heath, R.L. 1967. Regulation and determination on structure in membranes. *Adv. Gerontol. Res.* 2:77-120.

(64) Porta, E.A. and Hartroft, W.S. 1969. Lipid pigments in relation to aging and dietary factors (lipofuscin). In M. Wolman (Ed.), *Pigments in Pathology*, pp. 192-236. Academic Press, NY.

(65) Radice, J.C. and Herraiz, M.L. 1949. Fluorescent pigments in the uteri of vitamin E-deficient rats. *Ann. N.Y. Acad. Sci.* 52:126-128.

(66) Reagen, J.W. 1950. Geroid pigment in the human ovary. *Am. J. Obstat. Gynec.* 59:433-436.

(67) Reddy, K., Fletcher, B.L., Tappel, A. and Tappel, A.L. 1973. Measurement and spectral characteristics of fluorescent pigments in tissues of rats as a function of dietary polyunsaturated fats and vitamin E. *J. Nutr.* 103:908-915.

(68) Reichel, W. 1972. Lipid pigment formation as a function of age, disease and vitamin E deficiency. In N. Shimazono and Y. Takagi (Eds.), *International Symposium on Vitamin E*, pp. 207-227. Kyoritsu Shuppan Co., Tokyo.

(69) Scott, M.L. 1969. Studies on vitamin E and related factors in nutrition and metabolism. In H.G. DeLuca and J.W. Suttie (Eds.), *Fat-Soluble Vitamins*, pp. 357-369. The University of Wisconsin Press, Madison.
(70) Sinex, F.M. 1962. Chemical changes in irreplaceable macromolecules. In N.W. Shock (Ed.), *Biological Aspect of Aging*, pp 307-311. Columbia Univ. Press, NY.
(71) Sulkin, N.W. and Srivanij, D. 1960. The experimental production of senile pigments in the nerve cells of young rats. *J. Gerontol.* 15:2-9.
(72) Tappel, A.L. 1968. Will antioxidant nutrients slow aging process. *Geriatrics* 23(10):97-105.
(73) Tappel, A.L. 1973. Lipid peroxidation damage to cell components. *Fed. Proc.* 32:1870-1874.
(74) Tappel, A.L., Fletcher, B. and Deamer, D. 1973. Effect of antioxidants and nutrients on lipid peroxidation fluorescent products and aging parameters in the mouse. *J. Gerontol.* 28:415-425.
(75) Trombly, R., Tappel, A.L., Coniglio, J.G., Jr., Grogan, W.M. and Rhamy, R.K. 1975. Fluorescent products and polyunsaturated fatty acids of human testis. *Lipids* 10:591-596.
(76) Underwood, B.A., Seigel, H., Dolinski, M. and Weisell, R.C. 1970. Liver store of α-tocopherol in a normal population dying suddenly and rapidly from unnatural causes in New York City. *Am. J. Clin. Nutr.* 23:1314-1321.
(77) Vatassery, G.T., Alter, M. and Stadlan, E.M. 1971. Serum tocopherol levels and vibratory threshold with age. *J. Gerontol.* 26:481-484.
(78) Vogelsany, A.B., Shute, E.V. and Shute, W.E. 1947. Vitamin E in heart disease. IV. General remarks. *Med. Record* 160:279-284.
(79) Weglicki, W.B., Luna, Z. and Nair, P.P. 1969. Sex and tissue-specific references in concentrations of α-tocopherol in mature and senescent rats. *Nature* 221:185-186.
(80) Weglicki, W.B., Reichel, B. and Nair, P.P. 1968. Accumulation of lipofuscin-like pigment in the rat adrenal gland as a function of vitamin E deficiency. *J. Gerontol.* 23:469-475.
(81) Wei, C.K. and Draper, H.H. 1975. Vitamin E status of Alaskan Eskimos. *Am. J. Clin. Nutr.* 28:808-813.
(82) Wünscher, W. and Küstner, R. 1967. Untersuchungen über die mengenmassige verteilung von Lipofuszin and Vitamin E-mangel-pigment in Nervenzellen von Ratten verschiedener Alterstufen. *Gerontologia* 13:153-164.

–11–
The Current Status of Zinc, Copper, Selenium and Chromium in Aging

Jeng M. Hsu and H. Steve Hsieh

INTRODUCTION

Gerontologists have initiated many theories to explain the process of human aging but none have been definitely proven or generally accepted. Based on the observation that the major steps of genetic information transfer are all mediated by metal ions, Eichhorn[35] speculated that metal ion concentrations may be a factor in aging. He felt that it might be possible to ameliorate the aging process by controlling the metal ion environment.

The metabolic role of trace elements has recently become a great concern to nutritionists, dieticians, and physicians. Many of the trace elements are specifically involved in enzyme activity and in a variety of diseases.[143] The changes in quantity of a single trace element or the alterations in trace element ratio seem to be of significance in the process of mammalian physiologic aging.[101]

The purpose of this review is to update the current knowledge and to present a general review of four essential trace elements: zinc, copper, selenium, and chromium. Emphasis will be given with respect to their involvement with the aging process.

THE STATUS OF ZINC IN HUMAN NUTRITION

Introduction

The essentiality of zinc for human health and optimal growth has been well established. This element had been relatively ignored in the field of biological science until 1934 when Todd et al.[205] showed the importance of zinc in the growth of rats. In 1961, deficiency of zinc in Iranian males was first suspected by Prasad et al.[160] and was established following their later detailed studies in Egypt.[161-163] Since then, much research and experimentation has been carried out concerning the function and clinical application of zinc metabolism in the well-being of man as well as animal.

Symptoms of Zinc Deficiency

Zinc deficiency has been demonstrated in mice, rats, pigs, guinea pigs, birds, sheep, cattle, goats, monkeys, and man.[209] Among the most frequently reported symptoms of zinc deficiency are loss of appetite, retardation or cessation of growth, lesions of the skin and its appendages, and impairment of reproductive development and function. Alopecia and skin lesions in the paws are commonly observed in the investigations of zinc deficiency in rats and mice. In zinc-deficient pigs, parakeratosis can appear around the eyes and mouth, on the scrotum, and on the lower parts of the legs. In growing chicks, turkeys and Japanese quail, poor feathering and skeletal abnormalities are a regular and conspicuous feature of zinc deficiency.

In the human, dwarfism, hypogonadism, poor hair growth, and roughness of the skin are prominent features of zinc deficiency.[161-164] Similar syndromes have been described in women.[170]

Zinc Toxicity

Inhalation of zinc oxide fumes in high concentrations, as may occur in industrial workers, produces an acute illness characterized by fever, chills, leukocytosis, cough, headache, gastroenteritis, and pulmonary distress.[150] Acute intoxication following ingestion of 12 g of metallic zinc in a 16-year-old male has been reported.[135] This was characterized by drowsiness, unusual lethargy, lightheadedness, slight staggering of gait and difficulty in writing, which was followed by an elevated zinc concentration in whole blood. Chelation therapy promoted dramatic clinical improvement and a fall in blood zinc levels.

Food poisoning has also been observed in large numbers of people who have consumed zinc-contaminated food and drink.[14] The symptoms and signs of toxicity include nausea, vomiting, stomach cramps, diarrhea, and fever. More recently, a case involving zinc toxicity in a patient with renal failure following hemodialysis has been described by Gallery et al.[41] The cause of toxicity was attributed to the fact that the water for hemodialysis was stored in a galvanized tank. The patient suffered from nausea, vomiting, fever, and severe anemia.

Numerous studies have been published concerning the toxicity of zinc in animals which have been extensively reviewed by Underwood[209] and Van Reen.[214]

Dietary Source

Since recommended dietary allowances (RDA)[37] have been established for zinc, many dieticians and nutritionists are anxious to determine the zinc content of the diets of individuals or groups. To date, four major sources of references are available for the estimation of zinc content in American diets. Gormican,[48] using emission spectroscopy, determined the zinc concentration in 128 foods commonly used in hospital menus. Murphy et al.[134] provided a critical and exhaustive review of published and unpublished data on zinc content of 212 foods. Freeland and Cousins[38] reported the zinc content of 174 foods as determined by atomic absorption spectrophotometry. More recently, Haeflein and Rasmussen[61] have published an extensive compilation of data on the zinc content of selected food items appearing on 24-hour dietary recalls of 19 volunteers. Results of all these surveys indicate that meats, seafoods, and dairy products are the food groups that provide the richest source of zinc. Interestingly, however, milk was identified as a poor source of zinc.[38] The foods lowest in zinc include fruits and vegetables.[61] Such dietary items as coffee, tea, and carbonate beverages are characteristically low in zinc.[134]

Pathways Followed by Dietary Zinc

Zinc absorption occurs mainly in the second portion of duodenum.[128,129] Oral zinc is also absorbed by other locations in the small and large intestine. Data concerning the facultative transport are meagre. It is known that intestine-zinc absorption can be influenced by many dietary components and factors such as the amount of food intake and the phytic acid, fiber, calcium or phosphorus content of food. New findings[36,63,99] indicate that the metal-binding ligands in the rat intestine may be involved in transporting copper and zinc across the intestinal epithelium. More recently, Hurley[88] discovered a low molecular weight zinc-binding ligand (ZBL) in human milk which was not found in cow's milk and suggested that ZBL may enhance the absorption of zinc.

Following the intestinal absorption of zinc or the low molecular weight zinc complex, approximately two-thirds of it is loosely bound to albumin,[45,216] presumably to one of the histidine moieties of this molecule.[56] The remainder is more firmly bound to the globulins.[220] The zinc-albumin complex has been called the major macromolecular ligand of zinc.[75] The micromolecular zinc ligands which are available for transport to all tissues including liver, brain, and red cells are primarily zinc-histidine or zinc-cystine complex.

In liver, zinc is incorporated into macromolecular zinc ligands; namely, α_2-macroglobulin[151] and other zinc metalloenzymes. Zinc also has been shown to form metallothionine in the liver, kidney, and intestine. This cytoplasmic metalloprotein is characterized by a high content of cysteine which accounts for its unusually high binding properties to zinc, cadmium, and mercury.[92] Although this metalloprotein was first isolated two decades ago and numerous studies have been subsequently conducted, its precise role in the biological system has not been established. This protein may play a role in the absorption, storage, and detoxication of zinc.[166,217]

Regardless of whether zinc enters the body orally or parenterally, it is secreted mainly into the small intestine, chiefly via the pancreatic juice and is excreted in the feces. Only small amounts are secreted into the bile, the cecum, and the colon. The capacity of normal kidneys to excrete zinc is limited presumably because of the binding of zinc to serum albumin. Urinary excretion of zinc in normal subjects and those with various diseases has been summarized by Roman.[169] Significant quantities of zinc can be lost in sweat, especially in a hot climate.[162] Additional routes of zinc elimination include hair and dermal detritus, menses, semen, prostatic fluid, and milk.

Metabolic Function of Zinc

For comprehensive discussions of metabolic functions of zinc, the reader is referred to several excellent reviews.[143,167,211]

Since zinc is an integral component of various metalloenzymes and can activate a wide variety of enzymes, it seems logical to suspect that the syndrome of zinc deficiency is due to a decrease in the activity of essential zinc-containing enzymes. This hypothesis could prove to be a correct one in the future. The most perplexing problem in the mind of many zinc investigators is the rapid onset of symptoms when an experimental animal receives a zinc-deficient diet. Equally puzzling is the rapid recovery from these zinc-deficient symptoms after zinc repletion. A review of a number of investigators indicates that zinc may be intimately involved in protein, RNA, and DNA synthesis. This possibility is further strengthened by the recent discovery of three additional zinc-containing enzymes: DNA polymerase (EC2.7.7.7),[191] DNA-dependent RNA polymerase (EC 2.7.7.6),[188] and reverse transcriptase (EC 2.7.7.—)[3]

Effect of Age on Zinc Metabolism

Within the last few years, many contributions have been made toward trace element nutrition of the elderly; however, little is known with regard to the effect of age on zinc metabolism. The reader is referred to *The Biomedical Role of Trace Elements in Aging.*[87]

A recent survey[54] on 44 elderly participants in Indianapolis with a mean of 69 years of age indicates that dietary intake showed the mean consumption of zinc was 10.1 mg. This value, calculated on the basis of the dietary records and recalls, revealed that 59% of the subjects consumed less than two-thirds of the allowance for this mineral.[37] In her second study including 31 men and 34 women, with a mean age of 75 years, Greger[53] again found the zinc content of the diets below recommended levels. In addition, 5% of the subjects had a hair zinc level less than 75 µg/g which is indicative of zinc deficiency. Whether the low intake of dietary zinc affects their general health or is due to changes of zinc requirement associated with aging is not clear.

Conflicting data exist with regards to the effect of age on plasma zinc concentrations. Somewhat higher levels have been observed in newborns and children than in adults,[10,110] but another report showed the opposite.[67] Herring et al.[76] were unable to detect significant difference in plasma zinc concentration with advancing age. Perhaps it should be emphasized that the subjects studied were aged between 10 and 50 years and the techniques employed for the determination of plasma zinc concentration can be the bases for causing some doubt about the accuracy of their findings. Davis et al.[29] observed no significant change in plasma zinc concentration attributable to ages 20 to 60 years. Their conclusion appears questionable since plasma zinc concentration was measured on specimens from both healthy subjects and those with various diseases. Still, a significant linear decrease in plasma zinc concentration with increasing age of both sexes was reported by Lindeman et al.[103] Their results were based of 204 males and 54 females, aged 20 to 84 years. Although the mechanism of a reduced plasma zinc concentration with advancing age is not known, it is possible that the reduction is associated with the decrease in plasma albumin in the aged.[21]

The influence of age on the absorption of zinc has been studied in a rat model using radioisotope retention measurements by means of a whole body counter.[197] Results indicate that radiozinc-65 sulfate was absorbed and retained more by young (1.5 month) than mature (12 month) rats. This observation is consistent with the decreasing permeability of the intestinal wall with age.[197] Alternatively, the finding may be an indication of different zinc requirements in different age groups. Additional data are needed to confirm these apparent age-related trends in human subjects.

The concentration of zinc in hair appears to reflect zinc nutrition.[12] If the hair has been growing for a reasonable time, hair zinc becomes a useful clinical index of zinc status.[98] Strain et al.[198] have presented a convincing report indicating that zinc deficiency in humans results in a lowered level of this element in the hair. This is similar to the experience of Reinhold et al.[165] who found a good correlation with respect to the hair zinc content in Near Eastern subjects living in villages and having a poor nutritional state and those living in cities and having a good nutritional state. Recent studies of Hambidge and Walravens[71] also illustrate the value of the determination of hair zinc concentration in the diagnosis of zinc deficiency in infants and preadolescent children.

In a relatively large sample of human population, Petering et al.[154] found that the content of zinc in the hair of males increased from 105 ppm at age 2 to 180 ppm at age 12 and thereafter declined slowly to 125 ppm at age 80. Both slopes were statistically significant. The average content of zinc in the hair of females was very similar to that in the hair of males of comparable age. It is not clear that the decrease in zinc level in the hair of subjects from 12 to 80 years of age is a reflection of nutritional change with increasing age or an indication of environmental effects on trace metal metabolism.

Alcoholism and Other Factors of Zinc Metabolism

Alcoholism: Vallee et al.[210,213] postulated that zinc deficiency may play a central role in alcoholic liver cirrhosis. These investigators observed that patients with cirrhosis had a reduced zinc content in serum and liver. Paradoxically, there was an increase of urinary zinc excretion. Since then, these findings have been confirmed by many workers.[194] A low plasma zinc concentration is a nonspecific diagnostic value in distinguishing the various types of liver disease.[65] Excessive loss of zinc in urine will deplete zinc retention in body tissue which, in turn, will develop conditioned zinc deficiency. It has been demonstrated by in vitro experiments that a variable amount of serum zinc was not bound by serum proteins, but rather was freely diffusible and probably bound to amino acids. Thus, hyperzincuria could be explained on the basis of an increase in the diffusable fraction in the blood. To test this hypothesis, Sullivan and Burch[194] recently compared the urinary excretion of amino acid concentrations between cirrhotic patients and control subjects. They reported that the only significant difference in urinary amino acid excretion was decreased taurine, asparagine, glutamine, histidine and 3-methylhistidine. Although there may be a relationship between certain urinary amino acids and zinc, the mechanism of hyperzincuria in cirrhosis remains poorly understood. Precise correlations of plasma zinc, urinary zinc, plasma protein levels, free amino acid concentrations, and changes in zinc metalloenzymes will be necessary for a better understanding of zinc metabolism in liver disease.

Acrodermatitis Enteropathica (AE): This is a heriditary disease appearing in early infancy. It is characterized by acral distribution of skin lesions and associated gastrointestinal abnormalities. The beneficial effects of zinc sulfate in the treatment of a patient with AE and an associated lactose intolerance was first reported by Moynahan and Barns.[133] Based on a careful calculation of the micronutrient content of the diet and a complete recovery of skin lesions and bowel symptoms after zinc administration, these authors suggested that a zinc deficiency exists in AE. Their original observation was reported in 8 other infants under the care of Moynahan with equally good results.[133] Later, other physicians confirmed this dramatic therapeutic efficacy of zinc sulfate in AE.[78,130,139,159,204]

The mechanism of zinc deficiency may be linked to malabsorption, as indicated by the study of Lombeck et al.,[106] showing a significant reduction of ^{65}Zn absorption by whole body measurements in 3 patients with AE. Though the exact nature and sites of molecular defect have not been determined, it appears that uptake of zinc by the intestinal epithelial cells is diminished. The block in intestinal absorption is only partial; therefore, oral zinc supplementation is effective in correcting the deficient state. New speculation indicates that an abnormality[104] of Paneth cells in the small intestine of patients with AE may in some way be related to the primary defect of this disease.[154]

Zinc Deficiency During Parenteral Nutrition: The low levels of zinc present in most parenteral commercial solutions coupled with the tendency toward increased urinary losses suggest that patients receiving total parenteral nutrition (TPN) may develop zinc deficiency. The importance of zinc, copper, selenium, and chromium during illness which requires TPN has been recently emphasized by Greene.[52] A number of reports have illustrated the relationship between zinc levels and treatment with TPN.[4,51,94] Kay et al.[94] reported the results of a comprehensive study concerning the effect of intravenous alimentation on plasma zinc levels of 37 patients. In addition to the decreased plasma zinc levels, they also observed acute zinc deficiency syndrome such as diarrhea, mental depression, oral dermatitis, and alopecia. These observations suggest that additional zinc should be provided to patients receiving TPN despite the possible risk of zinc toxication.

Wound Healing: The biochemical and physiological observations indicate the important role of zinc in a system undergoing rapid cellular division such as wound healing. Thus, it was not surprising that zinc therapy was found to enhance wound healing. In 1967, Pories et al.[156] reported that oral zinc sulfate accelerates the healing rate of the wound following surgical excision of a pilonidal sinus. Since that time, numerous studies have been attempted to examine these provocative claims. Both single- and double-blind studies have been performed in man and animal with conflicting results.[75]

Halbook and Lanner[64] recently demonstrated that in patients with venous leg ulcers and a decreased serum zinc concentration, wound healing was accelerated by zinc therapy. Yet, administration of zinc was without effect on patients with normal serum zinc concentrations. Similar results have been found in patients with ischemic leg ulcers.[62] Such results suggest that zinc has a role in wound healing only when the patient is in zinc-deficient status. The question is then what is zinc-deficient status? What criteria should be used to define the zinc-deficient status in man? If a clear understanding of the role of zinc in wound healing is needed, our knowledge concerning the metabolic role of zinc and the physiologic and pathologic factors which regulate the dynamic nature of these processes will have to be carefully evaluated.

Pregnancy and Contraceptives: There is some conflict in the literature with regard to the effect of pregnancy and oral contraceptive steroids on zinc metabolism as indicated by changes in plasma or serum zinc concentrations. Much of this has been reviewed.[192] A significant decrease in plasma zinc concentration has been reported during pregnancy[90] and during the third trimester of pregnancy.[65] A comprehensive study involving a greater number of subjects also demonstrated a decreased plasma zinc concentration during pregnancy while taking the oral contraceptive.[67] This reducing effect of oral contraceptives was also observed by Briggs et al.[13] and Schenker et al.[181] In a subsequent paper, Schenker et al.[180] presented the data indicating a significant increase of erythrocyte carbonic anhydrase during the use of oral contraceptives as well as during pregnancy. These findings led to the belief that oral contraceptives may cause a redistribution of zinc between plasma and red cells where this metal is incorporated into carbonic anhydrase.

In contrast, O'Leary and Spellacy[145] and McKenzie[119] noted in a preliminary report that plasma zinc levels were not significantly altered in pregant women and nonpregnant women after administration of the oral contraceptives, respectively. No human studies have been reported on the effect of oral contraceptives taken by different age groups in relation to zinc metabolism.

Requirement for Zinc

The single new addition to the 1974 RDA table is the allowance for zinc, 15 mg per day for adult men and women.[37] This allowance is based on data from metabolic studies and the average zinc intake of mixed diets. The recommended level for pregnant women is 20 mg and for lactating mothers, 25 mg per day. An allowance of 10 mg per day is recommended for children to age 10 years. No exact data are available to determine the zinc requirement of the infant; the allowance is tentatively set at 3 mg per day for the first 6 months of life and 5 mg per day for those less than 1 year of age. No recommendation was given regarding the requirement of daily zinc intake for the aged.

THE STATUS OF COPPER IN HUMAN NUTRITION

Introduction

Copper is essential to human life and health and, like many heavy metals, it is also potentially toxic. This cation plays a key physiological role as the prosthetic element of more than a dozen specific copper proteins such as cytochrome oxidase and tyrosinase. During the past years, copper deficiency has been reported in infants and in adults. On the other hand, copper toxicity induced Wilson's disease has been known for many years. In this illness, the liver and the lenticular nucleus of the brain contain abnormally large amounts of copper. Recently, it has been shown that serum copper concentration is increased with age in male population, suggesting its possible linkage to the acceleration of the aging process.[72]

Symptoms of Copper Deficiency

A wide variety of symptoms have been associated with experimentally produced copper deficiency in animals. These include anemia,[17] poor growth,[131] lesions in the central nervous system,[7] achromotrichia,[193] reproductive failure[95] and congenital abnormalities.[144] Swayback is an ataxic disorder of lambs which is produced when copper deficiency develops in the central nervous system.[85] Changes in the growth and a reduction in the quantity and quality of wool produced by copper-deficient sheep was recognized early by Australian investigators.[8] In cattle, intermittent diarrhea occurs in the severely copper-deficient areas of Australia.[9] O'Dell et al.[144] revealed histological evidence of an impairment in the elastic tissue of the aortas of copper-deficient chicks which resulted in ruptures of the major blood vessels and ultimate death of the fowl. Similar findings were reported with copper-deficient pigs.[16] In addition to these external symptoms, copper deficiency reduced the concentration of copper in plasma,[109] liver, kidney, femur,[1] and brain.[221] A fall in serum ceruloplasmin, a specific copper protein, has been observed in copper-deficient rats.[131] The most commonly observed decreases in copper containing enzymes are cytochrome oxidase in the heart and liver,[55] amine oxidase in the bone[174] and aorta.[73]

Copper deficiency is rare condition in man because of its abundance in the liver. The first reported cases of mixed iron and copper deficiencies in milk-fed infants with anemia were described by Josephs.[91] Since then, copper deficiency has been observed in premature infants,[2] in malnourished infants,[49,93] and in adults.[33] The clinical symptoms of copper deficiency include recurrent diarrhea, anemia, neutropenia, and demineralization of bone.[24] The earliest detectable manifestation of copper depletion was the decrease of serum copper and ceruloplasmin.[80] The earliest clinical evidence of copper deficiency was persistent neutropenia. In some of the more severe and prolonged cases, an arrest of maturation of the granulocyte series in the bone marrow was detected.[24]

Copper Toxicity

Acute copper poisoning in man has been reported when a large dose of copper sulfate was ingested, or when acidic food or drink, previously in prolonged contact with the metal, was consumed.[84,148,189] The toxic symptoms including metallic taste, nausea, vomiting, epigastric burning and diarrhea have been well documented.[84] The vomiting and diarrhea induced by ingesting milligram quantities of copper generally prevent the patient from its serious systemic toxic effects, such as hemolysis, hepatic necrosis, gastrointestinal bleeding, proteinuria, convulsion, coma, and even death.[23] Acute copper poisoning can also occur as an industrial hazard for workers engaged in copper smelters and refiners.[108]

Chronic copper poisoning occurred in a 15-month-old infant

by drinking water with an unusually high copper concentration (800 μg).[177] The symptoms such as gingivitis,[206] lichen planus,[39] and eczematous dermatitis[5] have been shown to be related with the copper alloys in some dental and other prostheses.

Dietary Source

The copper content of selected common foods and beverages has been reported by several investigators.[113,183] Although many values were approximately agreeable, some differed widely. These discrepancies are primarily related to the different methodology and the altered practice in feeding poultry and cattle and in the use of fertilizers for grain crops and vegetables. The richest sources of copper are oysters, liver, mushrooms, nuts and chocolate, but their copper content, like that of other foods, varies with the copper content of the soil or water from which they originate. The poorest sources of copper are the dairy products. The contribution of drinking water to the total copper intake varies with the type of piping and the hardness of water.

Pathways Followed by Dietary Copper

Copper is absorbed mainly from the duodenum in man.[200] Immediately after absorption, copper, loosely bound to albumin[6] and perhaps to amino acids,[140] is transported by plasma. More than 90% of absorbed copper is soon deposited in the liver where it is incorporated into a variety of metalloproteins and enzymes. The liver is the organ responsible for the storage of copper, for its incorporation into ceruloplasmin, a blue copper protein which accounts for most of the copper in serum, and for its excretion into bile. Biliary copper is not available for reabsorption because it is bound to a protein.[47] In contrast to bile, urine contains minute amounts of copper. Studies on man using radioactive copper-64 or copper-67 indicated that about half of dietary copper is not absorbed but excreted directly in the feces.[195,199] The mechanisms regulating the passage of copper through the body are still unknown.

Metabolic Function of Copper

The literature related to the biochemical function of copper in man and animal has been extensively reviewed.[142] Only a brief discusion will be given here in connection with the metalloenzymes and metalloproteins, which are known to have physiological significance.

Tyrosinase and Melanin Formation: Tyrosinase (EC 1.10.3.1) catalyzes two types of reactions, hydroxylation of tyrosine to 3,4-dihydroxyphenylalanine (dopa) and oxidation of dopa to the quinone, leading to the formation of melanin. Failure of pigment production in hair and wool due to copper deficiency has been observed in numerous mammalian species. Lack of pigmentation in feathers of copper-deficient turkey poults has also been reported. The precise mechanism involving copper in the pigmentation process is unknown. It seems most probable that depressed tyrosinase activity is responsible for the lack of melanin formation.

Amine Oxidase (EC 1.14.2.1): Copper deficiency results in a failure of crosslink formation of connective tissues.[22] An accumulation of a soluble elastinlike protein, an uncrosslinked precursor of elastin, has been observed in aortic tissue from copper-deficient chicks[144] and pigs.[190] Aortic tissues from bovine and chick contain at least three types of amine oxidases, two of which catalyze the oxidative deamination of benzylamine and one of which acts on peptidyllysine. Both the soluble benzylamine oxidases and the lysol oxidase activities in aorta are decreased by dietary copper deficiency.[144] The role of copper in the formation of aortic elastin has been described by Hill et al.[77] The primary biochemical lesion is a reduction in amine oxidase activity of the aorta. This reduction results in a decreased capacity for oxidatively deaminating the ϵ-amino group of the lysine residues in elastin. Thus, less lysine is being converted to desmosine, the crosslinkage groups of elastin, and few crosslinkages are formed, thereby reducing aortic elasticity.

Ceruloplasmin (EC 1.12.3.1): Ceruloplasmin, the major serum component containing copper, is a metalloprotein and is tightly bound to an α_2-globulin. A postulated role for ceruloplasmin is the promotion of iron utilization by stimulation of iron oxidation and transferrin saturation. According to Osaki et al.,[146] iron is absorbed from the intestine as ferrous iron and must be oxidized to ferric iron before it binds to transferrin, the plasma iron transport protein. Ceruloplasmin is presumably responsible for the oxidation of ferrous iron.

Cytochrome Oxidase (EC 1.9.3.1): Cytochrome oxidase is composed of cytochromes a and a_3, and the enzyme contains 1 g atom of copper per mol of heme A. It is bound to mitochondria membranes and is liproprotein in nature. This enzyme is the terminal oxidase in the mitchondrial electron transport system, catalyzing the transfer of electrons and cytochrome c, another essential pigment required in cellular respiration, to oxygen. Although the activities of cytochrome oxidase in the liver and brain are reduced in the copper-deficient rats[40] and swine,[55] it is difficult to associate the enzyme with a specific pathology.

Other Copper Metalloproteins: Referral should be made to less well-defined copper metalloproteins. These include superoxide dismutase, uricase, dopamine-β-hydroxylase and ferroxidase II. For those who wish to know more about these copper metalloenzymes, the reader is directed to the recent view of O'Dell.[142]

Effect of Age on Copper Metabolism

Serum Copper and Ceruloplasmin Levels: Herring et al.[76]

surveyed a population from 10 to 50 years of age and found a positive correlation between serum concentration and age. These findings were later confirmed by Harman[72] who showed a linear increase with a mean serum concentration of 124 μg/100 ml at age 20 years and 145 μg/100 ml at age 60 years. The increase in serum copper with age may increase the rate of lipid peroxidation which, in turn, may result in an acceleration of the processes of atherosclerosis and of arteriolocapillary fibrosis. Recently studies by Yunice et al.[221] indicated that aging in male population increased serum copper concentration but maintained the normal values of serum ceruloplasmin concentration. The failure of serum ceruloplasmin to increase suggests that there might be an increased amount of copper present as the free ion or bound to amino acids. Very few studies have shown age-related effects on copper accumulation in human tissues.

Liver Copper Levels: Copper concentrations in animal liver have been studied in great detail. With most species, including man, liver copper concentrations are higher in the newborn than in the adult.[209] It has been observed that normal newborn liver contains much larger concentrations of copper than other tissues.[25] Porter[158] found that the largest proportion of this increased copper in newborn bovine liver was localized in the mitochondrial fraction. In the newborn human liver, this fraction contained more than 30 times as much copper as the mitochondrial fraction from adult human liver. The localization of neonatal mitochondrocuprein to the inner mitochondrial membrane, together with its extraordinarily high copper content, suggests that this copper protein may serve as a storage for copper in the formation of hepatic cytochrome oxidase during the neonatal period.

Effects of Diseases and Other Conditions on Copper Metabolism

A critical evaluation on the relationship of various diseases and serum copper concentrations is beyond the scope of this review. Instead, a brief description of selected diseases will be presented.

Serum copper concentrations are increased in a number of chronic illnesses such as coronary and cerebral atherosclerosis, hypertension, diabetes mellitus, chronic pulmonary disease, and in various hematologic disorders.[76] Hypocupremia is associated with nephrosis which is accompanied by an increased urinary excretion of copper. Hypercupremia is evident in most chronic and acute infections, in leukemia, and in Hodgkins's disease. In hyperthyroidism, an increase in plasma copper is accompanied by a decrease in erythrocyte copper.

Wilson's Disease: Wilson's disease[179] is a rare, recessive inherited disease in which copper accumulates principally in the liver, brain, and kidneys. Paradoxically, the concentration of copper and ceruloplasmin in plasma is much reduced and the urinary copper is markedly elevated. The mechanism of copper deposition is probably the result of the deficiency or absence of a specific copper-containing enzyme in the liver, leaving the metal more readily available to diffuse into other tissues. Copper accumulation can be reversed by administration of a chelating agent such as D-penicillamine. Although penicillamine can significantly lower the serum and liver copper levels, it has no effect on the concentration of copper in hair. These findings suggest that excess copper is removed from the body.

Menkes Disease: Menkes disease is a sex-linked disorder which results in cerebral and cerebellar degeneration, scorbutic-like changes in bone, and a peculiar steel-like hair.[121] Low levels of serum copper and ceruloplasmin were found in all patients studied and a defect in the intestinal absorption of copper has been demonstrated.[26] Recent studies by Lott et al.[107] indicate that the copper absorption block is only partial in the steely hair syndrome and is overcome by high doses of oral copper supplement.

Pregnancy, Parturition and Oral Contraceptives: The observation that women had an elevation in plasma copper concentration during pregnancy has been known for many years. Nielsen[141] found that the average serum copper of 31 pregnant women during the third month was 270 µg/100 ml in comparison with a normal nonpregnant level of 120 µg/100 ml. Similar results were obtained by other investigators.[111] The possible involvement of sex hormones in copper metabolsim comes from the original observation of Lahey et al.[100] indicating that serum copper concentrations were significantly higher in females than in males. Later, Russ and Raymunt[175] reported marked increases in both serum copper and ceruloplasmin in 8 patients following 3 to 4 weeks treatment of 1 mg daily of ethinylestradiol. Gault et al.[44] found serum ceruloplasmin concentration was increased in 4 subjects taking estrogen. Other workers, including Cartwright and Wintrobe,[18,19] Halsted et al.[66] and Schenker et al.,[181] all indicated that the intake of oral contraceptive preparations is responsible for the increase of serum copper and ceruloplasmin. However, it is equally important to mention that preliminary studies by Briggs and Briggs[11] showed no effect upon serum ceruloplasmin levels in 5 women receiving a daily dose of progesterones for 21 days.

Human Requirements for Copper

Since factors such as sex, age, other nutrients, and certain drugs could affect copper metabolism, the basic minimum requirements for copper cannot be given with certainty. The RDA for copper is based on balance studies, but as the amount of information available is limited, they are not included in the table. The requirement of infants and children has been estimated at 0.05 to 0.1 mg/kg of body weight per day; an intake of 0.08 mg/kg appears to be sufficient. The adult human requirement for copper intake is not known but has been estimated at 2 mg per day.

THE STATUS OF SELENIUM IN HUMAN NUTRITION

Introduction

Selenium is a toxicant but is also an essential nutrient. Incidences of selenium toxicity or selenium deficiency occur naturally in animals in the United States. It is unique that selenium at dietary levels of 0.1 to 0.5 ppm is beneficial while at 5 ppm, a level comparable to the nutritional requirement of many other trace elements, it could become toxic. As an essential nutrient, selenium is best known as an antioxidant, a role reminiscent of vitamin E. Although the essentiality of selenium in animals has long been established, the role of selenium in human nutrition is just being unveiled.

Symptoms of Selenium Deficiency

Selenium deficiency in animals can affect capillaries, muscles and reproductive systems. The deficient diseases include: (a) microcirculatory disorders, such as exudative diathesis in chicks,[152] mulberry heart and hepatosis dietetica in pigs,[34] and liver necrosis in rats;[185] (b) myopathies, such as white muscle disease in sheep and calves[136] and gizzard myopathy in turkeys;[187] and (c) reproductive disorders such as infertility in sheep[4] and rats,[118] reduced hatchability in quail,[89] and retained placenta in cattle.[207] Others such as loss of hair or feathers[118,202] and pancreatic degeneration[203] also have been demonstrated.

Disorders associated with selenium deficiency in humans have not been observed until recently. Van Rij et al.[215] reported that in New Zealand, where soil selenium levels are low, patients receiving total parenteral nutrition showed reduction in plasma selenium and in activity of erythrocyte glutathione peroxidase. Weak muscle was also observed, which responded to selenium treatment. In China, a selenium responsive disease, Keshan disease, was recently reported.[96,97] Keshan disease affects children or young women, causing cardiomyopathy and death. Oral doses of sodium selenite reduced the severity and prevented the death of the affected patients. Thus, Keshan disease becomes the first selnium-responsive disease occurring naturally in humans.

Selenium Toxicity

In the chronic form of selenium toxicity, i.e., alkali disease, the affected animals showed emaciation, loss of hair, sloughing of the hoofs and atrophy of the heart.[171] Failure of reproduction was also observed. Acute selenosis, or blind staggers, caused blindness, paralysis, salivation, respiration disturbance and death.[171] Cataracts due to chronic selenosis have been experimentally produced.[147]

Although selenium poisoning occurs in industrial workers, there is no conclusive evidence that dietary selenium induces selenosis in humans even in areas where soil selenium is high.[137] It has been suggested that high selenium may be associated with high incidence of dental caries,[60] but evidence for such association is not unequivocal.

Metabolic Function of Selenium

The role of selenium as an antioxidant has long been recognized. However, a biochemical basis for this role was not available until recently when Rotruck et al.[173] showed that selenium is an intrinsic part of the enzyme glutathione peroxidase. This also helps to distinguish between the role of selenium and that of vitamin E. Selenium as glutathione peroxidase functions to reduce lipid peroxides in the cytosol, while vitamin E is an antioxidant of the membrane components. Activity of glutathione peroxidase in erythrocytes correlated well with selenium status and the degree of hemolysis.[79]

Whagner et al.[218] isolated a selenium-containing protein which bore resemblance to cytochrome c in molecular weight, spectra and amino acid compositions. Such selenoprotein was absent in the heart and muscle of selenium-deficient lambs. Thus, in addition to its antioxidant role, selenium may have a role in oxidation-reduction reactions.

Selenium can also protect animals from damages by heavy metals such as mercury and cadmium. Ganther et al.[42,43] showed that dietary selenite could reduce the chronic poisoning effect of methyl mercury or cadmium in rats or Japanese quail. This protective effect of selenite probably depends on its conversion to reactive selenide, which in turn binds the heavy metals.[86]

Metabolism of Selenium

Selenium is mainly absorbed in the duodenum. In humans, as high an 70 to 80% of the dietary selenium is absorbed.[196] The form of selenium apparently affects its absorption: 50% absorption for selenite, 75% for selenomethionine, and 66% for selenium in fish.[168] The absorbed selenium is transported in the blood plasma,[115] first in the albumin, then in the globulins.[117] When selenite is absorbed, it is first converted by erythrocytes to a yet undefined form, which is then rapidly taken up by plasma proteins.[157]

Ingested selenium is excreted mainly in the urine and feces, and in the expired air when subacute doses of selenium is taken. Women consuming a diet containing 24.2 μg/selenium/day excreted 13.1 μg/day and 10 μg/day in the urine and feces, respectively.[196] A similar result of urinary excretion was found by Thompson.[201] Trimethyl selenide is the major component found in the urine,[149] while dimethyl selenide is found in the exhaled air.[116]

Effect of Age on Metabolism

The level of serum selenium increases with age.[105] It is approximately 50 ng/ml in newborns, and following a drop to 34 ng/ml in the early infancy, it increases steadily to 58 ng/ml in the second half of the first year, to 82 ng/ml for 1 to 5 year olds and to 92 ng/ml for school children, compared to 102 ng/ml for adults. The activity of selenium enzyme, glutathione peroxidase, was found to be higher in adults than in children.[120,153] On the other hand, Robinson et al.[168] found that patients over 65 years old had lower blood selenium levels than the young and medium aged. Along this line, it is interesting to note that patients of Legionnaire's disease, which mainly affects older people, have lower serum selenium levels compared to similarly paired samples of serum.[20]

Requirement for Selenium

Based on balance studies with normal women in New Zealand, Steward et al[196] suggested that the minimum dietary requirement for selenium is probably no more than 20 µg/day. On the other hand, the safe maximum daily intake of selenium was suggested by Sakurai and Tshuchya[176] to be 500 µg. According to RDA, 50 to 200 µg/day is considered to be adequate and safe for adults, with caution not to exceed 200 µg/day in long-term consumption.[138]

THE STATUS OF CHROMIUM IN HUMAN NUTRITION

Introduction

Evidence is accumulating that chromium is an essential micronutrient for man. Marginal deficiency of chromium in the human during pregnancy and in elderly subjects may exist in the United States because of inadequate intake due to its loss in processing of foodstuff, e.g., flour and sugar. This trace element probably acts as a cofactor for insulin. Many of these and earlier findings have been extensively reviewed.[31,70,124]

Symptoms of Dietary Chromium Deficiency

Deficiency of chromium has been induced in rats,[186] mice,[184] and squirrel monkeys[27] by feeding diets low in available chromium. The first known sign of mild chromium deficiency in rats is an impairment of intravenous glucose tolerance. Glucose removal rates, the rate of disappearance of excess glucose from the blood in percent per minute, declined within a few weeks to approximately half their original values.[127] This impairment can be restored to normal by oral doses of 20 to 50 µg of any of a variety of chromium compounds.[186] Furthermore, as chromium markedly enhanced the incorporation of glucose into fat by insulin, this trace metal was considered necessary for the action of insulin at the cellular level.[126] With environmental restriction of chromium contamination, rats had impaired growth, fasting hyperglycemia and glucosuria, as well as elevated serum cholesterol levels.[182] Squirrel monkeys fed a commercial laboratory chow low in available chromium also developed impaired glucose and tolbutamide tolerance. This impairment can be prevented by adding chromium to the drinking water (10 μg/l).[27]

In genetically diabetic mice in which hyperglycemia and hyperinsulinemia coexist, glucose tolerance factor (GTF) supplement but not inorganic chromium was effective in reducing the elevated blood glucose levels to normal. This observation theorized that the diabetic mice may have an impairment of conversion from inorganic chromium to GTF.[30]

Metabolism of Chromium

Inorganic trivalent chromium is poorly absorbed from the gastrointestinal tract while the absorption of chromate is slightly better. Animal studies suggest that chromium is absorbed in the upper small intestine.[32] After absorption, there are at least three forms of chromium circulating in the blood stream. The trivalent form of chromium is specifically bound and transported by transferin.[83] The hexavalent form of chromium is absorbed preferentially by erythrocytes where it is bound within the cells.[50] Once inside the cells, the chromate is reduced to the trivalent chromium resulting in a stable tagging of the red cells. The third form is presumably GTF-bound chromium, but definite studies on this complex binding have yet to be done. Studies by Mertz et al.[125] have indicated that inorganic chromium does not cross the placenta to any significant extent while chromium in the form of GTF is efficiently transferred to the fetuses. These observations suggest that a naturally occurring complex of chromium is transported across the placenta whereas the ion and simple complex are not.

Orally absorbed chromium is excreted mainly by the kidneys. The dialyzable portion of chromium in blood is filtered at the glomerules but up to 63% is reabsorbed in the tubules.[125]

Glucose Tolerance Factor

The GTF is a biologically active form of chromium.[124] Its mode of action was described as facilitating the formation of a ternary complex between chromium, insulin, and insulin receptors of cell membranes. It occurs preformed in certain foods such as preparation of brewer's yeast and can be utilized directly by animal and man. Its exact structure has not been identified yet and is probably a tetraaquo, di-nicotinato chromium complex. Analysis of highly purified GTF preparations from brewer's yeast indicate the presence of glycine, glutamic acid, and a sulfur-containing amino acid. GTF can be synthesized in the intestine and liver from niacin and amino acids in man.

Effect of Age on Chromium Metabolism

The indirect evidence summarized by several investigators[31,124] strongly suggests that chromium deficiency does occur in man. The first evidence of the existence of chromium deficiency in human tissues in the United States came from an analytical survey.[182] A marked decline of chromium concentration with age was reported in samples of kidney, liver, aorta, heart, and spleen. Cause and effect have not been demonstrated but it is believed that the kind of foods ingested in the United States may predispose people to the observed low tissue concentrations.

Maturity-onset diabetics[46] and middle-aged subjects[81] residing in the United States showed improvement of impaired glucose tolerance by simple addition of 150 μg of inorganic trivalent chromium to their daily diets. Children in Jordan[57,58] suffering from protein-caloric malnutrition restored the intravenous glucose tolerance to normal by receiving a chromium supplementation (250 μg/day) to their formula.

The observation in which 4 out of 10 elderly subjects who received chromium supplement displayed an improved glucose tolerance[102] was the indication that dietary intake of chromium may be inadequate in an aging population. It further indicates that chromium is required for normal carbohydrate metabolism. The effectiveness of restoring the impaired glucose tolerance by inorganic chromium was again demonstrated in 4 institutionalized elderly subjects, 2 young volunteers, and 1 patient with hemochromatosis.[31] In a subsequent study,[31] brewer's yeast extract known to contain chromium in the form of GTF was used to treat 12 volunteers over the age of 65 with an impaired glucose tolerance. After 1 and/or 2 months on the oral supplement, 6 of the 12 subjects showed improved glucose tolerance within normal values. These combined results suggest that impaired GTF is a common finding in elderly subjects and that GTF or chromium supplementation of the diet is likely to be beneficial. Perhaps it is appropriate to mention that McCay[114] recommended a daily dietary supplement of brewer's yeast for the elderly and suggested that it would reduce their insulin requirement. It appears likely that GTF or chromium content of the yeast is responsible for McCay's recommendation.

Urinary excretion of chromium has been measured to determine the status of chromium in man. The daily excretion of chromium is in the range of 3 to 50 μg per 24 hr.[28,69,219] Recently, urinary excretion of chromium and chromium content in hair samples were determined in a series of Turkish subjects of different age groups and of different nutritional background.[59] Their findings indicate urinary excretion of chromium in well-nourished subjects was lowest during infancy and showed stepwise increases with age to adult levels. No correlation was found between hair chromium content and age or nutritional state.

New studies[208] have demonstrated that active GTF extracts contain volatile chromium and that the biological activity is directly correlated with the volatile chromium content. Thus, it becomes important to determine what forms of chromium were excreted in the urine of elderly subjects and diabetics. The results of such a study[15] on urinary analysis reveal that the total chromium concentration decreases with increasing age of the subjects, e.g., from 32.9 down to 8.0 μg/ml. The healthy subjects have the highest concentration of volatile chromium in their urine, while the elderly and the diabetics have lower concentrations of volatile chromium in their urine. In addition, the decrease of urinary chromium with age was greater in the volatile form than in the nonvolatile form. Although the nature of the volatile chromium fraction in urine has not been investigated, its presence in the urine,[172] foodstuff, and plants[112,219] has been reported previously. Additional work is needed to establish whether the volatile chromium in urine is in GTF itself or some form of GTF.

Requirement for Chromium

In view of the greatly varying availability of chromium in different foods,[122] a meaningful recommendation of chromium intake cannot be given as yet. It has been tentatively suggested that a daily intake of 10 to 30 μg of GTF-chromium per day is enough to meet daily human requirements.[123]

CONCLUSION

An attempt has been made to illustrate the importance of trace elements in human health and disease. Emphasis, whenever possible, is given to their involvement with the aging process. A selection of four essential elements, zinc, copper, selenium and chromium, which have given ample evidence of their participation in human metabolism and certain diseases, is covered in this review. The literature cited with few exceptions, is that published up to December 1979.

During the last 10 years, we have witnessed remarkable advances in the field of trace elements. Basic research has demonstrated the role of zinc in protein and nucleic acid metabolism, the role of copper in collagen and elastin synthesis, the role of selenium as a component of glutathione peroxidase, and the role of chromium in glucose metabolism. The recent discovery that AE is most likely due to impairment of zinc absorption provides an exciting opportunity for both clinicians and nutritionists to have detailed studies on zinc metabolism. Our current knowledge that chromium may play an important role in late onset diabetes will encourage greatly future research in clinical medicine. It is hoped that additional evidence will enhance our understanding of how these biochemical reactions can be modulated for improving health and increasing longevity.

REFERENCES

(1) Alfaro, B., and Heaton, F.W. 1974. The subcellular distribution of copper, zinc, and iron in liver and kidney: changes during copper deficiency in the rat. *Br. J. Nutr.* 32:435-445.
(2) al-Rashid, R.A., and Spangler, J. 1971. Neonatal copper deficiency. *New Engl. J. Med.* 285:841-843.
(3) Auld, D.S., Kawaguchi, H., Livingston, D.M., and Vallee, B.L. 1974. RNA-dependent DNA polymerase (reverse transcriptase) from avian mycloblastosis virus: a zinc metalloenzyme. *Proc. Natl. Acad. Sci. USA* 71:2091-2095.
(4) Arakawa, T., Tamura, T., Igarashi, Y., Suzuki, H., and Sandstead, H. 1976. Zinc deficiency in two infants during total parenteral alimentation for diarrhea. *Am. J. Clin. Nutr.* 29:197-204.
(5) Barranco, V.P. 1972. Eczematous dermatitis caused by internal exposure to copper. *Arch. Derm.* 106:386-387.
(6) Bearn, A.G., and Kunkel, H.G. 1955. Metabolic studies in Wilson's disease using ^{64}Cu. *J. Lab. Clin. Med.* 45:623-631.
(7) Bennetts, H.W. 1932. Enzootic ataxia of lambs in Western Australia. *Aust. Vet. J.* 8:137.
(8) Bennetts, H.W., and Black, A.B. 1942. Enzootic ataxia and copper deficiency of sheep in Western Australia. Commonwealth Australia Council Sci. Ind. Red. Bull. No. 147, p. 4.
(9) Bennetts, H.W., and Hall, H.T.B. 1939. Falling disease of cattle in the southwest of Western Australia. *Aust. Vet. J.* 15:152-159.
(10) Berfenstam, R. 1952. Studies on blood zinc. *Acta Paediat. Stockholm* 41:7.
(11) Briggs, M.H., and Briggs, M. 1972. Preliminary studies on metabolic effects of a continuous-dose, progestogen-only, oral contraceptive. *S. Afr. J. Med. Sci.* 3:105-115.
(12) Briggs, M.H., Briggs, M., and Austin, J. 1971. Effects of steroid pharmaceuticals on plasma zinc. *Nature, London* 232:480-481.
(13) Briggs, M.H., Briggs, M., and Wakatama, A. 1971. Trace elements in human hair. *Experientia* 28:406-407.
(14) Brown, M.A., Thom, J.V., Orth, G.L., Cova, P., and Juarez, J. 1964. Food poisoning involving zinc contamination. *Arch. Environ. Health* 8:657-660.
(15) Canfield, W.K., and Doisy, R.J. 1976. Chromium and diabetes in the aged. In Hsu, Davis and Neithamer (eds.), *The Biomedical Role of Trace Elements in Aging*, pp. 119-128. Eckerd College, St. Petersburg, FL.
(16) Carnes, W.H., Shields, G.S., Cartwright, G.E., and Wintrobe, M.M. 1961. Vascular lesions in copper defieient swine (abstr). *Fed. Proc. Fed. Am. Soc. Exp. Biol.* 20:118.
(17) Cartwright, G.E., Gubler, C.J., Bush, J.A., and Wintrobe, M.M. 1956. Studies on copper metabolism. XVII. Further observations on the anemia of copper deficiency in swine. *Blood* 11:143-153.
(18) Cartwright, G.E., and Wintrobe, M.M. 1964. The question of copper deficiency in man. *Am. J. Clin. Nutr.* 15:94-110.
(19) Cartwright, G.E., and Wintrobe, M.M. 1964. Copper metabolism in normal subjects. *Am. J. Clin. Nutr.* 14:224-232.
(20) Chen, J.R., and Anderson, J.M. 1979. Legionnaire's disease: concentrations of selenium and other elements. *Science* 206:1426-1427.

(21) Chesrow, E.J., Bronsky, D., Orfei, E., Dyniewicz, H., Dubin, A., and Musci, J. 1958. Serum proteins in the aged: means and stability of mucoprotein levels and electrophoretic partitions. *Geriatrics* 13:20-24.
(22) Chou, W.S., Savage, J.E., and O'Dell, B.L. 1969. Role of copper in biosynthesis of intramolecular cross-links in chick tendon collagen. *J. Biol. Chem.* 244:5785-5789.
(23) Chuttani, H.K., Gupta, P.S., Gulati, S., and Gupta, D.N. 1965. Acute copper sulfate poisoning. *Am. J. Med.* 39:849-854.
(24) Cordano, A., Baetl, J.M., and Graham, G.G. 1964. Copper deficiency in infancy. *Pediatrics* 34:324-336.
(25) Cunningham, I.J. 1931. Some biochemical and physiological aspects of copper in animal nutrition. *Biochem. J.* 25:1267-1294.
(26) Dank, D.M., Campbell, P.E., Stevens, B.J., Mayne, V., and Cartwright, E. 1972. Menkes' kinky hair syndrome—an inherited defect in copper absorption with widespread effects. *Pediatrics* 50:188-201.
(27) Davidson, I.W.F., and Blackwell, W.L. 1968. Changes in carbohydrate metabolism of squirrel monkeys with chromium dietary supplementation. *Proc. Soc. Exp. Biol. Med.* 127:66-70.
(28) Davidson, I.W.F., and Secrest, W.L. 1972. Determination of chromium in biological materials by atomic absorption spectrometry using a graphite furnace atomizer. *Analyt. Chem.* 44:1808-1813.
(29) Davis, I.J., Musa, M., and Dormandy, T.L. 1968. Measurements of plasma zinc. I. In health and diseases. *J. Clin. Path.* 21:359-365.
(30) Doisy, R.J., Jastremski, M.S., and Greenstein, F.L. 1973. Metabolic effects of glucose tolerance factor and trivalent chromium in normal and genetically diabetic mice (abstr). Int. Congr. Ser. No. 280, p. 155 *Excerpta Medica*, Amsterdam.
(31) Doisy, R.J., Streeten, D.H.P., Frieberg, J.M., and Schneider, A.J. 1976. Chromium metabolism in man and biochemical effects. In Prasad (ed.), *Trace Elements in Human Health and Disease*, Vol. 2, pp. 79-101. Academic Press, New York.
(32) Donaldson, R.M., and Barreras, R.F. 1966. Intestinal absorption of trace quantities of chromium. *J. Lab. Clin. Med.* 68:484-493.
(33) Dunlap, W.M., James, G.W. III, and Hume, D.M. 1974. Anemia and neutropenia caused by copper deficiency. *Ann. Intern. Med.* 80:470-476.
(34) Eggert, R.C., Patterson, E., Akers, W.T., and Stockstad, E.L.R. 1957. The role of vitamin E and selenium in the nutrition of the pig. *J. Animal Sci.* 16:1037.
(35) Eichhorn, G.L. 1973. Cited in meeting brief from Chicago. *Chem. Engr. News.* 50:12.
(36) Evans, G.W. 1976. Zinc absorption and transportation. In Pradad (ed.), *Trace Elements in Human Health and Disease*, Vol. 2, pp. 181-186. Academic Press, New York.
(37) Food and Nutrition Board: *Recommended Dietary Allowance*, 8th rev. ed. National Academy of Science, Washington, DC.
(38) Freeland, J.H., and Cousins, R.J. 1976. Zinc content of selected food. *J. Am. Diet. Assn.* 68:526-529.
(39) Frykholm, K.O., Frithiof, L., Fernstron, A.I.B., Moberger, G., Blohm, S., and Bjourn, E. 1969. Allergy to copper derived from dental alloys as a possible cause of oral lesions of lichen planus. *Acta derm. vener.*, Stockholm 49:268-281.

(40) Gallagher, C.H., and Reeve, V.E. 1971. Copper deficiency in the rat. Effect of synthesis of phospholipids. *Aust. J. Exp. Biol. Med. Sci.* 49:21-31.
(41) Gallery, E.D.M., Blomfield, J., and Dixon, S.R. 1972. Acute zinc toxicity in haemodialysis. *Br. Med. J.* 4:331-333.
(42) Ganther, H.E., Goudie, C., Sunde, M.L., Lopecky, M.J., Wagner, P., Oh, S.H., and Hoekstra, W.G. 1972. Selenium: relation to decreased toxicity of methylmercury added to diets containing tuna. *Science* 175:1122-1124.
(43) Ganther, H.E., Wagner, P.A., Sunde, M.L., and Hoekstra, W.G. 1973. Protective effects of selenium against heavy metal toxicities. In Hemphill (ed.), *Trace Substance in Environmental Health*, Vol. 6, pp. 247-259. University of Missouri, Columbia, MO.
(44) Gault, M.E., Stein, J., and Aronoff, A. 1966. Serum ceruloplasmin in heptobiliary and other disorders: significance of abnormal values. *Gastroenterology* 50:8-18.
(45) Giroux, E.L., and Henkin, R.I. 1972. Competition for zinc among serum albumin and amino acids. *Biochem. Biophys. Acta* 273:64-72.
(46) Glinsmann, W.H., and Mertz, W. 1966. Effect of trivalent chromium on glucose tolerance. *Metabolism* 15:510-520.
(47) Gollan, J.L., and Deller, D.J. 1973. Studies on the nature and excretion of biliary copper in man. *Clin. Sci.* 44:9-15.
(48) Gormican, A. 1970. Inorganic elements in foods used in hospital menu. *J. Am. Diet Assn.* 56:397.
(49) Graham, G., and Cordano, A. 1969. Copper depletion and deficiency in the malnourished infants. *Johns Hopkins Med. J.* 124:139-150.
(50) Gray, S.J., and Sterling, K. 1950. The tagging of red cells and plasma protein with radioactive chromium. *J. Clin Invest.* 29:1604-1613.
(51) Greene, H.L. 1975. Vitamins and trace elements. In Ghadimi (ed.), *Total Parenteral Nutrition: Promises and Premises*, pp. 351-371. Wiley & Sons, Chichester.
(52) Greene, H.L. 1977. Trace metals in parenteral nutrition. *Prog. Clin. Biol. Res.* 14:87-97.
(53) Greger, J.L. 1977. Dietary intake and nutritional status in regards to zinc of institutionalized aged. *J. Geront.* 32:549-553.
(54) Greger, J.L., and Sciscoe, B.S. 1977. Zinc nutriture of elderly participants in an urban feeding program. *J. Am. Diet. Assn.* 70:37-41.
(55) Gubler, C.J., Cartwright, G.E., and Wintrobe, M.M. 1957. Studies on copper metabolism. XX. Enzyme activities and iron metabolism in copper and iron deficiencies. *J. Biol. Chem.* 224:533-546.
(56) Gurd, F.R.N., and Goorman, D.S. 1952. Preparation and properties of serum and plasma protein. XXXII. The interaction of human serum albumin with zinc ions. *J. Am. Chem. Soc.* 74:670-675.
(57) Gurson, C.T., and Saner, G. 1971. Effect of chromium on glucose utilization in marasmic protein-caloric malnutrition. *Am. J. Clin. Nutr.* 24:1313-1319.
(58) Gurson, C.T., and Saner, G. 1973. Effect of chromium supplementation on growth in marasmic protein-caloric malnutrition. *Am. J. Clin. Nutr.* 26:988-991.
(59) Gurson, C.T., Saner, G., Mertz, W., Wolf, W.R., and Sokucu, S. 1975. Nutritional significance of chromium in different age groups and in populations differing in nutritional backgrounds. *Nutr. Reports Intl.* 12:9-17.

(60) Hadjimarkos, D.M. 1968. Effect of trace elements on dental caries. *Adv. Oral Biol.* 31:253-292.
(61) Haeflein, K.A., and Rasmussen, A.L. 1977. Zinc content of selected foods. *J. Am. Diet. Assn.* 70:610-616.
(62) Haeger, K., and Lanner, E. 1974. Oral zinc sulfate and ischaemic leg ulcer. *J. Vasc. Dis.* 3:77-81.
(63) Hahn, C., and Evans, G.W. 1973. Identification of a low molecular weight ^{65}Zn complex in rat intestine. *Proc. Soc. Exp. Biol. Med.* 144:793-795.
(64) Halbrook, T., and Lanner, E. 1972. Serum zinc and healing of venous leg ulcers. *Lancet* 2:780-782.
(65) Halsted, J.A., Hackely, B.M., Rudzki, C., and Smith, J.C. Jr. 1968. Plasma zinc concentrations in liver diseases. *Gastroenterology* 54: 1098-1105.
(66) Halsted, J.A., Hackley, B.M., and Smith, J.C. Jr. 1968. Plasma zinc and copper in pregnancy and after oral contraceptives. *Lancet* 2:278.
(67) Halsted, J.A., and Smith, J.C. Jr. 1970. Plasma zinc in health and disease. *Lancet* 1:322-324.
(68) Halsted, J.A., Smith, J.C. Jr., Hackley, B.M., and McBean, L. 1969. Plasma zinc and copper levels. *Am. J. Obstet. Gynec.* 105:645-646.
(69) Hambidge, K.M. 1971. Chromium nutrition in the mother and the growing child. In Mertz and Cornatzer (eds.), *Newer Trace Elements in Nutrition*, pp. 171-193. Dekker, New York.
(70) Hambidge, K.M. 1974. Chromium nutrition in man. *Am. J. Clin. Nutr.* 27:505-514.
(71) Hambidge, K.M., and Walravens, P.A. 1976. Zinc deficiency in infants and preadolescent children. In Prasad (ed.), *Trace Elements in Human Health and Disease*, pp. 21-31. Academic Press, New York.
(72) Harman, D. 1965. The free radical theory of aging: effect of age on serum copper levels. *J. Geront.* 20:151-153.
(73) Harris, E.D., and O'Dell, B.L. 1974. Copper and amine oxidases in connective tissue metabolism. *Adv. Exp. Med. Biol.* 48:267-284.
(74) Hartley, W.J. and Grant, A.B. 1961. A review of selenium responsive diseases of New Zealand livestock. *Fed. Proc.* 20:679-688.
(75) Henkin, R.I. 1974. Zinc in wound healing. *New Engl. J. Med.* 291:675-676.
(76) Herring, B.W., Leavell, B.S., Paixao, L.M., and Yoe, J.H. 1960. Trace metals in human plasma and red blood cells. I. Observations of normal subject. *Am. J. Clin. Nutr.* 8:846-854.
(77) Hill, C.H., Starcher, B., and Kim, C. 1967. Role of copper in the formation of elastin. *Fed. Proc.* 26:129-133.
(78) Hirsh, F.S., Michel, B., and Strain, W.H. 1976. Gluconate zinc in acrodermatitis enteropathica. *Arch. Derm.* 112:475-478.
(79) Hoekstra, W.G. 1974. Biochemical role of selenium. In Hoekstra, Suttie, Ganther and Mertz (eds.), *Trace Elements Metabolism in Animals*, 2, pp. 61-77. University Park Press, Baltimore.
(80) Holtzman, N.A., Charache, P., Cordano, A., and Graham, G.G. 1970. Distribution of serum copper in copper deficiency. *Johns Hopkins Med. J.* 126:34-42.
(81) Hopkins, L.L. Jr., and Price, M.G. 1968. Effectiveness of chromium(III) in improving the impaired glucose tolerance of middle-aged Americans. In *Western Hemisphere Nutrition Congr.*, 2, p. 40.

(82) Hopkins, L.L. Jr., Ransome-Kuti, O., and Majaj, A.D. 1968. Improvement of impaired carbohydrate metabolism by chromium(III) in malnourished infants. *Am. J. Clin. Nutr.* 21:203-211.
(83) Hopkins, L.L. Jr., and Schwarz, K. 1964. Chromium(III) binding of serum proteins, specifically siderophilin. *Biochim. Biophys. Acta* 90:484-491.
(84) Hopper, S.H., and Adams, H.S. 1958. Copper poisoning from vending machine. *Publ. Hlth. Rep.* 73:910-914.
(85) Howell, J. McC. 1970. The pathology of swayback. In Mills (ed.), *Trace Element Metabolism in Animals*, pp. 103-105. Livingstone, Edinburgh.
(86) Hsieh, H.S., and Ganther, H.E. 1977. Biosynthesis of dimethyl selenide from sodium selenite in rat liver and kidney cell-free systems. *Biochem. Biophys. Acta* 497:205-217.
(87) Hsu, J.M., Davis, R.L., and Neithamer, R.W. 1976 (eds.), *The Biomedical Role of Trace Elements in Aging*, pp. 15-237, Eckerd College, St. Petersburg, FL.
(88) Hurley, L.S. 1977. Zinc deficiency in prenatal and neonatal development. *Proc. Clin. Biol. Res.* 14:47-58.
(89) Jensen, L.S. 1968. Selenium deficiency and impaired reproduction in Japanese quail. *Proc. Soc. Exp. Biol. Med.* 128:970-972.
(90) Johnson, N.C. 1961. Study of copper and zinc metabolism during pregnancy. *Proc. Soc. Exp. Biol. Med.* 108:518-519.
(91) Josephs, H.W. 1931. Treatment of anemia in infants with iron and copper. *Bull Johns Hopkins Hosp.* 49:246-258.
(92) Kagi, J.H.R., and Valle, B.L. 1960. Metallotheonine: Cadmium and zinc containing protein from equine renal cortex. *J. Biol. Chem.* 235:3460-3465.
(93) Karpet, J.T., and Peden, V.H. 1972. Copper deficiency in long-term parenteral nutrition. *J. Pediat.* 80:32-36.
(94) Kay, R., Tasman-Jones, C., Pybus, J., Whiting, R., and Black, H. 1976. A syndrome of acute zinc deficiency during total parenteral nutrition in man. *Ann. Surg.* 183:331-340.
(95) Keil, H.L., and Nelson, V.E. 1931. The role of copper in haemoglobulin regeneration and reproduction. *J. Biol. Chem.* 93:49-57.
(96) Keshan Disease Research Group of Chinese Academy of Medical Science. 1979. Observations on effect of sodium selenite in prevention of Keshan disease. *Chinese Med. J.* 92:471-476.
(97) Keshan Disease Research Group of the Chinese Academy of Medical Science. 1979 Epidemiological studies on the etiologic relationship of selenium and Keshan disease. *Chinese Med. J.* 92:477-482.
(98) Klevay, L.M. 1970. Hair as a biopsy material. I. Assessment of zinc nutriture. *Am. J. Clin Nutr.* 23:284-289.
(99) Kowarsky, S., Blair-Stanck, C.S., and Schachter, D. 1974. Active transport of zinc and identification of binding protein in rat jejunal mucosa. *Am. J. Physiol.* 226:401-407.
(100) Lahey, M.E., Gubler, C.J., Cartwright, G.E., and Wintrobe, M.M. 1953. Studies on copper metabolism. VI. Blood copper in normal human subjects. *J. Clin. Invest.* 32:322-328.
(101) Lawton, A.H. 1976. Trace elements in aging. In Hsu, Davis and Neithamer (eds.), *The Biomedical Role of Trace Elements in Aging*, pp. 1-6. Eckerd College, St. Petersburg, FL.

(102) Levine, R.A., Streetan, D.P.H., and Doisy, R. 1968. Effects of oral chromium supplementation on the glucose tolerance of elderly human subjects. *Metabolism* 17:114-125.
(103) Lindeman, R.D., Clark, M.L., and Colemore, J.P. 1971. Influence on age and sex on plasma and red cell zinc concentrations. *J. Geront.* 26:358-363.
(104) Lombeck, I., Bassewitz, D.B. von. Becker, K., Tinschmann, P., and Kastner, H. 1974. Ultrastructural finding in acrodermatitis enteropathica. *Pediat. Res.* 8:82.
(105) Lombeck, I., Kasperek, K., Harbisch, H.D., Feinendegen, L.E., and Bremer, H.J. 1977. The selenium state of healthy children. *Eur. J. Pediat.* 125:81-88.
(106) Lombeck, I., Schnippering, H.G., Ritzl, F., Feinendegen, K.E., and Bremer, H.J. 1975. Absorption of zinc in acrodermatitis enteropathica. *Lancet* i:855.
(107) Lott, I.T., Dipaolo, R., Schwartz, D., Janoski, S., and Kanfer, J.N. 1975. Copper metabolism in the steely hair syndrome. *New Engl. J. Med.* 292:197-199.
(108) Lyle, W.H., Payton, J.E., and Hui, M. 1976. Haemodialysis and copper fever. *Lancet* ii:1324-1325.
(109) MacPherson, A., Brown, N.A., and Hemingway, R.C. 1964. The relationship between the concentration of copper in the blood and livers of sheep. *Vet. Rec.* 76:643.
(110) Mehanand, D., and Houck, J.C. 1968. Fluorometric determination of zinc in biological fluids. *Clin. Chem.* 14:6-11.
(111) Markowitz, H., Gubler, C.J., Mahoney, J.P., Cartwright, G.E., and Wintrobe, M.M. 1955. Studies on copper metabolism. XIV. Copper ceruloplasmin and oxidase activity in sera of normal subjects: pregnant women. *J. Clin Invest.* 34:1498-1508.
(112) Maxia. V., Meloni, S., Rollier, M.A., Brandone, A., Patwardhan, V.N., Wahlien, C.I., and El-Shami, S. 1972 Selenium and chromium assay in Egyptian foods and blood of Egyptian children by activation analysis. Nuclear activation techniques in the life sciences. *JAEA-SM* 157/67, pp. 527-550 (Int. Atomic Engery Agency, Wien).
(113) McCance, R.A., and Widdowson, E.M. 1947. *The Chemical Composition of Foods.* Chemical Publishing, Brooklyn, NY.
(114) McCay, C.M. 1952. Chemical aspects of aging and the effect of diet upon aging. In Lansing (ed.), *Cowdry's Problems of Aging,* 3rd ed, pp. 139-202. Williams and Williams, Baltimore, MD.
(115) McConnell, K.P. 1941. Distribution and excretion studies in the rat after a single subtoxic subcutaneous injection of sodium selenate containing radioselenium. *J. Biol. Chem.* 141:427-437.
(116) McConnell, K.P., and Portman, O.W. 1952. Excretion of dimethyl selenide by the rat. *J. Biol. Chem.* 195:277-282.
(117) McConnell, K.P., Wabnitz, C.H., and Roth, D.M. 1960. Time-distribution studies of selenium-75 in dog serum proteins. *Tex. Rep. Biol. Med.* 18:438-445.
(118) McCoy, K.E.M., and Weswig, P.H. 1969. Some selenium responses in the rat not related to vitamin E. *J. Nutr.* 98:383-389.
(119) McKenzie, J.M. 1974. Influence of oral contraceptives on serum zinc and copper concentrations (abstr). *Fed. Proc.* 33:2734.

(120) McKenzie, R.L., Rea, H.M., Thompson, C.D., and Robinson, M.F. 1978. Selenium concentration and glutathione peroxidase activity in blood of New Zealand infants and children. *Am. J. Clin. Nutr.* 31:1413-1418.
(121) Menkes, J.H., Alter, M., Steigleder, G.K., Weakley, D.R., and Sung, J.H. 1962. A sex-linked recessive disorder with retardation of growth, peculiar hair and focal cerebral and cerebellar degeneration. *Pediatrics* 29:764-779.
(122) Mertz, W. 1969. Chromium occurrence and function in biological systems. *Physiol. Rev.* 49:163-239.
(123) Mertz, W. 1971. Human requirements: basic and optimal. *Ann. N.Y. Acad. Sci.* 199: 191-199.
(124) Mertz, W. 1976. Effects and metabolism of glucose tolerance factors. In *Present Knowledge in Nutrition*, 4th ed, pp. 365-375. The Nutrition Foundation, Washington, DC.
(125) Mertz, W., Roginski, E.E., Feldman, F.J., and Thurman, D.E. 1969. Dependence of chromium transfer into rat embryo on the chemical form. *J. Nutr.* 99:363-367.
(126) Mertz, W., Roginski, E.E., and Schwarz, K. 1961. Effect of trivalent chromium complex on glucose uptake by epididymal fat tissue of rats. *J. Biol. Chem.* 236:318-322.
(127) Mertz, W., and Schwarz, K. 1959. Relation of glucose tolerance factor to impaired glucose tolerance in rats on stock diets. *Am. J. Physiol.* 196:614-618.
(128) Methfessel, A.H., and Spencer, H. 1973. Zinc metabolism in the rat. I. Intestinal absorption of zinc. *J. Appl. Physiol.* 34:58-62.
(129) Methfessel, A.H., and Spencer, H. 1973. Zinc metabolism in the rat. II. Secretion of zinc into intestine. *J. Appl. Physiol.* 34:63-67.
(130) Michaelsson, G. 1974. Zinc therapy in acrodermatitis enteropathica. *Acta. Derm. Vener. Stockholm* 54:377-381.
(131) Milne, D.M., and Weswig, P.H. 1968. Dietary copper on blood and liver copper containing fractions in rats. *J. Nutr.* 95:429-433.
(132) Moynahan, E.J. 1974. Acrodermatitis enteropathica: a lethal inherited zinc deficiency disorder. *Lancet* ii:399-400.
(133) Moynahan, E.J., and Barnes, P.M. 1973. Zinc deficiency and a synthetic diet for lactose intolerance. *Lancet* i:676-677.
(134) Murphy, E.W., Willis, B.W., and Watt, B.K. 1975. Provisional tables on the zinc content of foods. *J. Am. Diet. Assn.* 66:345-355.
(135) Murphy, J.V. 1970. Intoxication following ingestion of elemental zinc. *J. Am. Med. Assn.* 212:2119-2120.
(136) Muth, O.H., Oldfield, J.E., Remmert, L.F., and Schubert, J.R. 1958. Effects of selenium and vitamin E on white muscle disease. *Science* 128:1090-1091.
(137) National Research Council. 1976. *Selenium*, pp. 116-118. National Academy of Science, Washington, DC.
(138) National Research Council. 1980. *Recommended Dietary Allowances*, pp. 162-164. National Academy of Science, Washington, DC.
(139) Nelder, K.H., and Hambidge, K.M. 1975. Zinc therapy of acrodermatitis enteropathica. *New Engl. J. Med.* 292:879-882.
(140) Neuman, P.Z., and Sass-Kartsak, A. 1967. The state of copper in human serum: evidence for an amino acid found fraction. *J. Clin. Invest.* 46:646-658.

(141) Nielsen, A.L. 1944. On serum copper. III. Normal values. *Acta. Med. Scand.* 118:87-92.

(142) O'Dell, B.L. 1976. Biochemistry and physiology of copper in vertebrates. In Prasad (ed.), *Trace Elements in Human Health and Disease*, Vol. 1, pp. 391-410. Academic Press, New York.

(143) O'Dell, B.L., and Campbell, B.J. 1971. Trace elements: metabolism and metabolic function. *Compreh. Biochem.* 21:179-217.

(144) O'Dell, B.L., Hardwich, B.C., Reynolds, G., and Savage, G.E. 1961. Connective tissue defect in the chicks resulting from copper deficiency. *Proc. Soc. Exp. Biol. Med.* 108:402-405.

(145) O'Leary, J.A., and Spellacy, W.N. 1969. Zinc and copper levels in pregnant women and those taking oral contraceptives. *Amer. J. Obstet. Gynecol.* 103:131-132.

(146) Osaki, S., Johnson, D.A. and Frieden, E. 1966. The possible significance of the ferrous-oxidase of ceruloplasmin in normal human serum. *J. Biol. Chem.* 241:2746-2751.

(147) Ostadalova, I., Babicky, A., and Obenberger, J. 1977. Cataract induced by administration of a single dose of sodium selenite to suckling rats. *Experientia* 34:222-223.

(148) Paine, C.H. 1968. Food poisoning due to copper. *Lancet* ii:520.

(149) Palmer, I.S., Fischer, D.D., Halverson, A.W., and Olson, O.E. 1969. Identification of a major selenium excretory product in rat urine. *Biochem. Biophys. Acta* 177:336-342.

(150) Papp, J.P. 1968. Metal fume fever. *Postgrad. Med.* 43:160-163.

(151) Parisi, A.F., and Vallee, B.L. 1970. Isolation of a zinc, α_2-macroglobulin from human serum. *Biochem. N.Y.* 9:2421-2426.

(152) Patterson, E.L., Milstrey, R., and Stokstad, E.L.R. 1957. Effect of selenium in preventing exudative diathesis in chicks. *Proc. Soc. Exp. Biol. Med.* 95:617-620.

(153) Perona, G., Guidi, G.C., Piga, A., Cellerino, R., Milani, G., Colautti, P., Moschini, G., and Stievano, B.M. 1979. Neonatal erythrocyte glutathione peroxidase deficiency as a consequence of selenium imbalance during pregnancy. *Br. J. Haematol.* 42:567-574.

(154) Petering, H.G., Yeager, D.W., and Witherup, S.O. 1971. Trace metal content of hair. I. Zinc and copper content of human hair in relation to age and sex. *Arch. Environmental Health* 23:202-207.

(155) Polanco, I., Nistal, M., Guerrero, J., and Vasquez, C. 1976. Acrodermatitis enteropathica zinc and ultrastructural lesions in Paneth cells. *Lancet* i:430.

(156) Pories, W.H., Henzel, J.H., Rob, C.G., and Strain, W.H. 1967. Acceleration wound healing in man with zinc sulfate given by mouth. *Lancet* i:121-124.

(157) Porter, E.K., Karle, J.A., and Shrift, A. 1979. Uptake of selenium-75 by human lymphocytes in vitro. *J. Nutr.* 109:1901-1908.

(158) Porter, H. 1970. Neonatal hepatic mitochondrocuprein: the nature, submitochondrial localization and possible function of the copper accumulating physiologically in the liver of new-born animals. In Mills (ed.), *Trace Element Metabolism in Animals*, pp. 237-244. Livingstone, London.

(159) Portnoy, B., and Molokhia, M. 1974. Zinc in acrodermatitis enteropathica. *Lancet* ii:663-664.

(160) Prasad, A.S., Halsted, J.A., and Nadimi, M. 1961. Syndrome of iron deficiency anemia, hepatosplenomegaly, hypogonadism, dwarfism and geophagia. *Am. J. Med.* 31:532-546.
(161) Prasad, A.S., Miale, A. Jr., Farid, Z., Sandstead, H.H., and Darby, W.J. 1963. Biochemical studies on iron deficiency anemia, dwarfism and hypogonadism. *Arch. Intern. Med.* 111:407-428.
(162) Prasad, A.S., Miale, A., Farid, Z., Schulert, A., and Sandstead, H.H. 1963. Zinc metabolism in patients with the syndrome of iron deficiency anemia, hypogonadism and dwarfism. *J. Lab. Clin. Med.* 61:537-549.
(163) Prasad, A.S., Sandstead, H.H., Schulert, A.R., and El Rooby, A.S. 1963. Urinary excretion of zinc in patient with the syndrome of anemia, hepatosplenomegaly, dwarfism and hypogonadism. *J. Lab. Clin. Med.* 62:591-599.
(164) Prasad, A.S., Schulert, A.R., Sandstead, H.H., Miale, A. Jr., and Farid, Z. 1963. Zinc, iron and nitrogen content of sweat in normal and deficient subjects. *J. Lab. Clin. Med.* 62, 84-89.
(165) Reinhold, J.G., Kfoury, G.A., Ghalambor, M.A., and Bennett, J.C. 1966. Zinc and copper concentrations in hair of Iranian villagers. *Am. J. Clin. Nutr.* 18:294-300.
(166) Richards, M.P., and Counsin, R.J. 1975. Influence of parenteral zinc and antimomycin D on tissue zinc uptake and the synthesis of a zinc-binding protein. *Bioinorgan. Chem.* 4:215-224.
(167) Riordan, J.F., and Vallee, B.L. 1976. Structure and function of zinc metalloenzymes. In Prasad (ed.), *Trace Elements in Human Health and Disease*, Vol. 1, pp. 227-251. Academic Press, New York.
(168) Robinson, M.F., Godfrey, P.J., Thompson, C.D., Rea, H.M., and vanRij, A.M. 1979. Blood selenium and glutathione peroxidase activity in normal subjects and in surgical patients with and without cancer in New Zealand. *Am. J. Clin. Nutr.* 32:1477-1485.
(169) Roman, W. 1969. Zinc in porphyria. *Am. J. Clin. Nutr.* 22:1290-1303.
(170) Ronaghy, H.A., Barakat, R., Prasad, A.S., Reinhold, J.G., Haghshenas, M., Abadee, P., and Halsted, J.A. 1970. *Symposium on Food, Science and Nutritional Diseases in the Middle East*. Shiraz, Iran.
(171) Rosenfield, I., and Beath, O.A. 1964. *Selenium: Geobotany, Biochemistry, Toxicity and Nutrition*, pp. 299-332. Academic Press, New York.
(172) Ross, R.T., Gonzales, J.G., and Segar, S.A. 1973. The direct determination of chromium in urine by selective volatilization with atom reservoir atomic absorption. *Anal. Chim. Acta* 63:205-209.
(173) Rotruck, J.T., Pope, A.L., Ganther, M.E., Swanson, A.B., Hafeman, D.G., and Hoekstra, W.G. 1973. Selenium: biochemical role as a component of glutathione peroxidase. *Science* 179:588-590.
(174) Rucker, R.G., Parker, H.E., and Rogler, J.C. 1969. Effect of copper deficiency on chick bone collagen and selected bone enzymes. *J. Nutr.* 98:57-63.
(175) Russ, E.M., and Raymunt, J. 1956. Influence of estrogen on total serum copper and ceruloplasmin. *Proc. Soc. Exp. Biol. Med.* 92:465-466.
(176) Sakurai, H., and Tsuchya, K. 1975. A tentative recommendation for the maximum daily intake of selenium. *Environ. Physiol. Biochem.* 5:107-118.
(177) Salmon, M.A., and Wright, T. 1971. Chronic copper poisoning presenting as pink disease. *Arch. Dis. Childh.* 46:108-110.

(178) Sakar, B., and Kruck, T.P.A. 1973. Copper-amino acid complex in human serum. In Peisach, Aisen and Blumberg (eds.), *The Biochemistry of Copper*, pp. 183-196. Academic Press, New York.
(179) Schelinberg, I.H., and Sternlieb, I. 1976. Copper toxicity and Wilson's disease. In Prasad (ed.), *Trace Elements in Human Health and Disease*, Vol. 1, pp. 415-431. Academic Press, New York.
(180) Schenker, J.G., Ben-Yoseph, Y., and Shapira, E. 1972. Erthrocyte carbonic anhydrase B levels during pregnancy and use of oral contraceptives. *Obstet. Gynec.* 39:237-240.
(181) Schenker, J.G., Jungreis, E., and Polishuk, W.Z. 1971. Oral contraceptives and serum copper concentration. *Obstet. Gynec.* 37:233-243.
(182) Schroeder, H.A. 1968. The role of chromium in mammalian nutrition. *Am. J. Clin. Nutr.* 21:230-244.
(183) Schroeder, H.A., Nason, A.P., Tipton, I.H., and Balassa, J.J. 1966. Essential trace metals in man. *J. Chron. Dis.* 19:1007-1034.
(184) Schroeder, H.A., Winton, W.H. Jr., and Balassa, J.J. 1963. Effect of chromium, cadmium and other trace metals on the growth and survival of mice. *J. Nutr.* 80:39-47.
(185) Schwarz, K., and Foltz, C.M. 1957. Selenium as an integral part of factor 3 against dietary necrotic liver degeneration. *J. Am. Chem. Soc.* 79:3292-3293.
(186) Schwarz, K., and Mertz, W. 1959. ChromiumIII and the glucose tolerance factor. *Arch. Biochem. Biophys.* 85:292-295.
(187) Scott, M.L., Olson, G., Krook, L., and Brown, W.R. 1967. Selenium-responsive myopathies of myocardium and smooth muscle in young poultry. *J. Nutr.* 91:573-583.
(188) Scrutton, M.C., Wu, C.W., and Goldthwait, D.A. 1971. The presence and possible role of zinc in RNA polymerase obtained from Escherichia coli. *Proc. Natl. Acad. Sci. USA* 68:2497-2501.
(189) Semple, A.B., Parry, W.H., and Phillips, D.E. 1960. Acute copper poisoning. An outbreak traced to contaminated water from a corroded geyser. *Lancet* ii:700-701.
(190) Shields, G.S., Coulson, W.F., Kimball, D.A., Carnes, W.H., Cartwright, G.E., and Wintrobe, M.M. 1962. Studies on copper metabolism. XXXII. Cardiovascular lesions in copper deficient swine. *Am. J. Path.* 41:603-621.
(191) Slater, J.P., Mildvan, A.S., and Loeb, L.A. 1971. Zinc in DNA polymerases. *Biochem. Biophys. Res. Commun.* 44:37-43.
(192) Smith, J.C. Jr., and Brown, E.D. 1976. Effect of oral contraceptive agent on trace element metabolism—a review. In Prasad (ed.), *Trace Elements in Human Health and Disease*, Vol. 2, pp. 315-341. Academic Press, New York.
(193) Smith, S.E., and Ellis, G.H. 1947. Copper deficiency in rabbit achromotrichia, alopecia, dermatosis. *Arch. Biochem. Biophys.* 15:81-88.
(194) Sullivan, J.F., and Burch, R.E. 1976. Potential role of zinc in liver disease. In Prasad (ed.), *Trace Elements in Human Health and Disease*, Vol. 1, pp. 67-82. Academic Press, New York.
(195) Sternlieb, I. 1967. Gastrointestinal copper absorption in man. *Gastroenterology* 52:1038-1041.
(196) Stewart, R.D., Griffiths, N.M., Thompson, C.D., and Robinson, M.F. 1978. Quantitative selenium metabolism in normal New Zealand women. *Br. J. Nutr.* 40:45-54.

(197) Strain, W.H., Pories, W.J., Michael, E., Peer, R.M., and Zaresky, S.A. 1976. Influence of age on absorption and retention of trace elements. In Hsu, Davis and Neithamer (eds.), *The Biomedical Role of Trace Element in Aging*, pp. 161-173. Eckerd College, St. Petersburg, FL.

(198) Strain, W.H., Steadman, L.T., Lankau, C.A., Berliner, W.P., and Pories, W.J. 1966. Analysis of zinc levels in hair for the diagnosis of zinc deficiency in man. *J. Lab. Clin. Med.* 68:244-249.

(199) Strickland, G.T., Beckner, W.M., and Leu, M.L. 1972 Absorption of copper in homozygotes and heterozygotes for Wilson's disease and control: isotope tracer studies with ^{67}Cu and ^{64}Cu. *Clin. Sci.* 43:617-625.

(200) Thompsett, S.L. 1940. Factors influencing the absorption of iron and copper from the elementary tract. *Biochem. J.* 34:962-969.

(201) Thompson, C.D. 1972. Urinary excretion of selenium in some New Zealand women. *Proc. Univ. Otago, Med. Sch.* 50:31-33.

(202) Thompson, J.N., and Scott, M.L. 1969. Role of selenium in the nutrition of the chick. *J. Nutr.* 97:335-342.

(203) Thompson, J.N., and Scott, M.L. 1970. Impaired lipid and vitamin E absorption related to atrophy of the pancreas in selenium deficient chicks. *J. Nutr.* 100:797-809.

(204) Thyresson, N. 1974. Acrodermatitis enteropathica reported of a case healed with zinc therapy. *Acta Derm. Vener. Stockh.* 54:383-385.

(205) Todd, W.R., Elvehjem, C.A., and Hart, E.B. 1934. Zinc in the nutrition of the rat. *Am. J. Physiol.* 107:146-156.

(206) Trachtenberg, D.I. 1972. Allergic response to copper—its possible gingival implication. *J. Periodont.* 43:705-707.

(207) Trinder, N., Woodhouse, C.D., and Kenton, C.P. 1969. The effect of vitamin E and selenium on the incidence of retained placenta in dairy cows. *Vet. Res.* 85:550-553.

(208) Tuman, R.W. 1975. Biological Effects of Glucose Tolerance Factor (GTF) and Inorganic ChromiumIII in Normal and Genetically Diabetic Mice. Doctoral Dissertation, State Univ. of NY, Upstate Medical Center, Syracuse, NY.

(209) Underwood, E.J. 1971, *Trace Elements in Human and Animal Nutrition*, 3rd ed. Academic Press, New York.

(210) Vallee, B.L., and Hoch, F.L. 1957. Zinc in horse liver alcohol dehydrogenase. *J. Biol. Chem.* 225:185-196.

(211) Vallee, B.L., and Wacker, W. 1970. Metalloproteins. In Neurath (ed.), *The Proteins. Composiiton, Structure and Function*, 2nd ed., Vol. 5. Academic Press, New York.

(212) Vallee, B.L., Wacker, W.E.C., Bartholomay, A.F., and Hoch, F.L. 1975. Zinc metabolism in hepatic dysfunction. II. Correlation of metabolic patterns with biochemical findings. *New Engl. J. Med.* 257:1055-1065.

(213) Vallee, B.L., Wacker, W.E.C., Bartholomay, A.F., and Robin, E.D. 1956. Zinc metabolism in hepatic dysfunction. I. Serum zinc concentrations in Laennec's cirrhosis and their validation by sequential analysis. *New Engl. J. Med.* 255:402-408.

(214) Van Reen, R. 1953. Effects of excessive dietary zinc in the rat and interrelationship with copper. *Arch. Biochem. Biophys.* 46:337-344.

(215) Van Rij, A.M., Thompson, C.D., McKenzie, J.M., and Robinson, M.F. 1979. Selenium deficiency in total parental nutrition. *Am. J. Clin. Nutr.* 32:2076-2085.

(216) Vikbladh, I. 1951. Studies on zinc in blood. II. *Scand. J. Clin. Lab. Invest.* 3:suppl. 2, 1-74.
(217) Webb, M. 1972. Protection by zinc against cadmium toxicity. *Biochem. Pharmacol.* 21:2767-2771.
(218) Whagner, P.D., Pedersen, N.D., and Weswig, P.H. 1973. Selenium proteins in ovine tissues. II. Spectra properties of a 10,000 molecular weight selenium protein. *Biochem. Biophys. Res. Commun.* 53:1031-1035.
(219) Wolf, W., Mertz, W., and Marironi, R. 1974. Determination of chromium in refined and unrefined sugars by oxygen plasma ashing flameless atomic absorption. *J. Agric. Food Chem.* 22:1037-1043.
(220) Wolff, H. 1956. Untersuchungen Zur Pathophysiologie des Zinkstoffwechsels. *Klin. Wschr.* 34:409-418.
(221) Yunice, A.A., Lindeman, R.D., Czerwinski, A.W., and Clark, M. 1974. Influence of age and sex on serum copper and ceruloplasmin levels. *J. Geront.* 29:277-281.
(222) Zimmerman, A.W., Mathhien, J.M., Quarles, R.H., Brady, R.O., and Hsu, J.M. 1976. Hypomyelination in copper deficient rats. *Arch. Neurol.* 37:111-119.

–12–
Magnesium, Phosphorus and Calcium Needs for Bone Health

Anthony A. Albanese

The elements magnesium, phosphorus and calcium are most abundant in all forms of animal and plant life. Alone or in combination with other nutrients they participate in the activation of innumerable enzyme reactions and almost every known nutritional need.

MAGNESIUM

Magnesium is closely related to both calcium and phosphorus in its location and its functions in the body. About 70% of the magnesium in the body is in the bones. Muscle tissue contains more magnesium than calcium. Blood contains more calcium than magnesium.[12] Magnesium acts as a catalyzer, for some of the chemical reactions within the body. It also becomes a part of some of the complex molecules that are formed as the body uses food for growth, maintenance and repair. It plays an important role as a coenzyme in the building of tissue proteins. There is some relation between magnesium and the hormone cortisone as they affect the amount of phosphate in the blood.

Animals on a diet that is deficient in magnesium become extremely nervous and give an exaggerated response to even small noises or disturbances. Such unnatural sensitiveness disappears when they are given enough magnesium. In extreme deficiencies, the blood vessels expand, the heart beats faster and damage in the midbrain causes such irritability that the animals die in convulsions.

A deficiency of magnesium in human nutrition is not common, but it may occur more frequently than it is diagnosed. It disturbs the calcification of bone. An excess of magnesium causes a deposition of calcium in the soft tissues. An adult requires about 0.3 gram of magnesium a day. It is present in foods from both animal and plant sources, meats, milk, cereals, vegetables and fruit which vary greatly in the amount of magnesium they contain. Nuts, legumes and cereal grains have more magnesium than other foods but fresh fruits contain less than other foods. A diet that is adequate in other essential nutrients, especially protein of high quality, is likely to supply enough magnesium.

Pharmacologists speak of the antagonism between calcium and magnesium, and when magnesium salts are injected intravenously calcium is eliminated rapidly and there appears to be a very definite antagonism. Magnesium in the diet does not appear to show this antagonism, probably because it is poorly absorbed.

About 70% of the magnesium which is present in the body is in the bones. Blood has been shown to contain from 1 to 3 mg per 100 ml and the soft tissues contain more magnesium than calcium. Magnesium in the body appears to follow the metabolic pathways of phosphorus rather than of calcium, whereas in absorption from the intestine, it is more closely related to calcium absorption.

While a magnesium deficiency is unlikely to occur in the human diet, experimental production of a magnesium deficiency in animals has shown the production of vasodilation and a hyperirritability with convulsions and death. There was a lowered magnesium content of the blood while the calcium remained normal. In phosphorus deficiency there is a lowered phosphorus and also a lowered magnesium content of the blood. There is an increased content of calcium and phosphorus in the ash of bones that are low in magnesium. Magnesium is necessary for the efficient utilization of amino acids in the formation of protein. The fact that magnesium is necessary for normal functions of the body has led to some consideration of the requirement, and while a definite level has not been established, a magnesium intake of 0.27 g per day appears to be completely adequate for humans.

Symptomatic human deficiency observed in patients always develops in a setting of predisposing and complicating disease states. The latter include severe malabsorption of various etiologies, chronic alcoholism with malnutrition, prolonged magnesium-free parenteral feeding—usually in association with prolonged losses of gastrointestinal secretions, acute or chronic renal disease involving tubular dysfunction, large lactation losses, childhood malnutrition, postnatal tetany syndromes, familial disorders of renal or intestinal conservation, and in hyperparathyroidism—especially in the immediate postparathyroidectomy period. In such clinical circumstances, associated complex and uncontrolled variables—such as multiple nutritional inadequacies, metabolic abnormalities, manifestations of basic disease, infection and modifications in oral and parenteral nutrients, and medications occurring in close proximity to the magnesium therapy—often made it difficult and potentially misleading to ascribe certain clinical manifestations specifically to magnesium deficiency. Recent studies clarified the clinical and biochemical parameters of this deficiency.

The signs and symptoms noted in experimental depletion cover a wide spectrum, including personality change, spontaneous generalized muscle spasm, tremor, fasciculations, and Trousseau and Chvostek signs. These have been described separately or in various combinations in clinical cases of hypomagnesemia as have myoclonic jerks, athetoid movements, convulsions and coma. Convulsions with or without coma seem to occur much more frequently in acutely deficient infants than in adults. Magnesium deficiency in man leads to neuromuscular dysfunction as manifested by hyperexcitability, with tremor and convulsions, and is sometimes accompanied by behavioral disturbances. Nevertheless, since magnesium occurs widely in foods, particularly those of vegetable origin, a dietary deficiency of magnesium seems to be rare.

PHOSPHORUS

Phosphorus has more functions than any other element in the body; at least 9 major vital functions have been determined for this element: metabolism of proteins, carbohydrates and fats; vitamin and enzyme activity; acid-base regulation; maintenance of neurological functions, skeletal growth and tooth development. About 20% of the phosphorus located in tissues other than bone and teeth is distributed in every cell of the body and is vitally involved with their functions.[6] Phosphorus is linked to most of the vitamins in the enzyme systems of the body and is closely allied with carbohydrate functioning. The interrelationship of calcium with phosphorus will be discussed in connection with the formation of bone.

Whole blood contains from 35 to 45 mg of phosphorus per 100 ml of blood. Of this amount, the inorganic phosphorus composes from 3-5 mg per 100 ml. While the inorganic phosphorus is that portion readily available for chemical reaction and is affected most readily by intake of phosphorus, it is in constant exchange with the organic phosphorus in the blood and there is a constant replacement of the inorganic phosphorus if it is eliminated by disease or nutritional conditions.

Actually, the level of inorganic phosphorus is influenced by factors affecting assimilation, absorption, excretion and mobilization of the phosphorus. Because of this, normal levels in the blood do not necessarily guarantee a normal nutritional status of phosphorus. On the other hand, an abnormal low inorganic phosphorus value in the blood is a rather positive indication of disturbed phosphorus nutrition.

As with calcium, absorption is favored by an acid medium and for this reason, lactose (milk sugar) favors absorption by establishing an acid medium in the intestinal tract. Excesses of iron, aluminum and magnesium interfere with phosphorus absorption through the formation of insoluble phosphates that are not available to the animal. An abnormal calcium-phosphorus ratio in food interferes with the absorption of these elements and may cause deficiencies.

Interest has centered around the role of fiber in the absorption of calcium and phosphorus from the intestine, and it has been readily demonstrated that calcium and phosphorus in foods high in fiber are poorly available through digestion. Other experiments, however, have shown that fiber per se has little effect on the absorption of calcium and phosphorus, but rather forms a natural barrier around the calcium and phosphorus. Since fiber is poorly digested, the calcium and phosphorus are not available for absorption.

Foods which have a laxative effect tend to reduce mineral absorption. This may be true with items such as molasses which have a laxative effect because of their high mineral content, and it also may be associated with fiber content of certain foods.

Phosphorus is required for the formation of bone mineral in the ratio of 1 g of phosphorus per 2 g of calcium retained, and for the organic phosphorus of tissue cells which is approximately 1 g of phosphorus per 17 g of nitrogen retained. The daily requirement of phosphorus in the growing subject is in considerable excess of this because of allowances for incomplete intestinal absorption and urinary excretion. Similarly the adult requirement also allows for obligatory fecal and urinary loss of phosphate. The efficiency of phosphate absorption varies with the source of phosphorus and the ratio of calcium to phosphorus in the diet.

A major source of phosphorus in cereals and grains is phytic acid, inositol hexaphosphoric acid. The availability of phytic acid phosphorus has been questioned both with regard to absorption of phosphorus and also with its effect on calcium absorption since calcium phytate is insoluble and excreted in the stools. In a recent study of normal adult subjects, only 43 to 46% of dietary phosphorus was absorbed when rice was the principal diet constituent and 47 to 64% when diets of rice and milk or rice and wheat were fed. Although whole wheat also contains phosphorus as phytate much of the phytic acid phosphorus in wheat is converted to orthophosphate by the hydrolytic action of phytase during the leavening and baking process so that malabsorption of calcium and phosphate due to whole wheat ingestion ordinarily is not a problem. However, if the wheat source is the unleavened bread used in the Near East and the Asian subcontinent, and the calcium intake is low and exposure of the skin to sunshine limited due to climate conditions or cultural clothing patterns, the high phytic acid intake could be a contributing factor in the occurrence of osteomalacia.

Chronic renal hypophosphatemia can result from nutritional vitamin D deficiency, from blocks in metabolism of vitamin D due to genetic abnormalities, liver disease, or pharmacologic agents such as the anticonvulsants, diphenylhydantoin and phenobarbital. The major consequence is failure of bone mineralization with rickets in the growing bone and osteomalacia in adult bone. In addition to the disabilities resulting from osteomalacia, severe hypophosphatemia may also cause pronounced muscle weakness possibly due to reduction of organic phosphates including ATP in muscle cells.

The reduction of absorbability of phosphate by ingestion of large amounts of aluminum hydroxide or calcium as calcium carbonate, lactate or gluconate, can in patients with normal kidney function lead to phosphate depletion with severe hypophosphatemia. This has been of clinical importance particularly as the result of the excessive use of antacids based on aluminum hydroxide. Patients addicted to antacid medication may develop severe muscle weakness as the result of profound hypophosphatemia. They may in addition have bone pains and disability due to hypophosphatemic osteomalacia. Severe hypophosphatemia has also been observed in patients receiving "intravenous hyperalimentation" with solutions low in phosphate for periods longer than a few days. In such patients alterations of red blood cell ATP and 2,3-diphosphoglycerate have been found. Deficiency of ATP may be associated with membrane abnormalities and decreased life span of the red blood cells. The reduced 2,3-DPG is accompanied by an increased affinity of hemoglobin for oxygen and decreased delivery of oxygen to tissues at ordinary oxygen tensions. It has been suggested that this may be of importance in the muscle weakness, malaise and anorexia seen in the phosphate depletion syndrome.

CALCIUM

The high incidence of fractures in the elderly, especially women, makes an understanding of calcium nutrition, clinically very important.[2] Calcium comprises 1.5-2.0% of an adult's body weight. At all ages it is the most abundant mineral in the body. In bones, it is combined with phosphorus, which is 0.8-1.1% of the body weight and magnesium which is 0.05% of body weight. A 154-pound individual would have 1050-1540 g of calcium, 560-840 g of phosphorus and 35 g of magnesium in his body. About 99% of the calcium and 80-90% of the phosphorus and 60% of magnesium are found in the bones and teeth. The rest appear in soft tissue and body fluids and are highly important to many essential biochemical functions. Unlike calcium, deficiencies of phosphorus and magnesium are unlikely because of their high content in most animal and plant foods. Dairy products are the principal dietary source of calcium.

Calcium is essential for the clotting of blood, action of certain enzymes and control of the flow of fluids through cell walls. The right amount of calcium in the blood is needed for the rhythmic contraction and relaxation of heart musculature. Nerve irritability increases when calcium level is below normal. Calcium phosphate salts give rigidity and hardness to the bone and teeth. Both calcium and phosphorus are essential for the work of muscles and normal response of nerves to stimulation.

The calcium content of the body increases faster in relation to size during the first year of life than any other time. All the gains in calcium content during growth and maintenance in adulthood depend on adequate supply of calcium in the diet and ability of the body to utilize it. Phosphorus and calcium are of equal importance in bone formation. Phosphorus is involved in ossification or calcification just as much as calcium. When bone is formed, phosphorus is deposited with calcium as crystals containing about twice as much calcium as phosphorus. When bone loses calcium, it also loses phosphorus. Changes occur on both the inside and outside as a bone grows larger. New bone is formed around the outside of the shaft of long bone. Bone cell calcium from the inside of the shaft is absorbed into the blood at the same time and utilized in the body where needed. Adding bone substance on the outside and removing from the inside gives the skeleton added strength without unnecessary weight.

SOURCES OF CALCIUM

Dairy products are the major source of calcium. Two cups of milk or an equivalent amount of cheese or other dairy products (excluding butter) make an important contribution to the daily calcium intake. Foods other than dairy products which furnish a sizeable amount of calcium include: bakery products made with enriched white flour, dark green leafy vegetables such as collards, mustard greens or turnip greens; and salmon and sardines including the tiny bones.

Milk and milk products are the best nutritional sources of calcium, not only because of the significant amount of the mineral present, but also because of the Ca:P ratio (1:1.2) that is conducive to skeletal growth and the presence of nutrients such as vitamin D (fortified) and lactose which favor calcium absorption. One cup of fluid whole milk provides 291 mg of calcium. Thus, two cups of fluid whole milk contribute about three-fourths of the RDA of 800 mg for adults and most children. Various technological processes used in the production of the many milk products may increase the concentration of milk calcium. Specific dairy products, excluding butter, collectively contributed almost 75% of the available calcium in the U.S. The Department of Agriculture (USDA) estimates that 943 mg calcium and 1569 mg phosphorus were available per capita each day in 1977. Many other foods provide smaller quantities. Results of USDA surveys indicate that calcium intake in the diets of girls and women is less than 50% of the RDA values.

BIOAVAILABILITY

The calcium actually available to the body is the amount absorbed from the intestinal tract. To be absorbed, calcium must initially be dissolved or soluble in solution. Absorption occurs throughout the length of the small intestine, being greater in the duodenum and proximal jejunum (upper portion of the intestine) than in the ileum (lower portion). The mechanisms involved in intestinal calcium transport are not completely understood, but entail free or simple passive diffusion, facilitated diffusion, and active transport. The intestinal absorption of dietary calcium is incomplete, ranging from 10 to 50% of intake depending on several nondietary and dietary factors.

The relative bioavailabilities of calcium in non-fat dry milk, (NFDM), yogurt and rennet-precipitated casein (RPC) are illustrated in Figure 1. NFDM represents a source of calcium as it is normally found in milk, consisting of colloidal calcium phosphate and ionic calcium. Yogurt represents a fermented or acid-precipitated product in which the calcium is primarily ionic. RPC is a product similar to cheddar cheese curd and contains calcium as calcium paracaseinate. The effects of milk fat were eliminated by making yogurt and RPC from skim milk. Both products were freeze-dried then mixed into the rat diet. These results indicated that calcium availability in dairy products is affected by the nature of the calcium complex and products that contain colloidal calcium phosphate or calcium caseinate are better sources of calcium than those that contain only ionic calcium.[17]

Endogenous factors affecting the rate of calcium absorption include:

(1) nutritional status or calcium needs of the body,
(2) general health,
(3) age,
(4) emotional state,
(5) physical activity and immobilization, and
(6) medication.

An insufficient intake of calcium or an increased requirement for the mineral results in an increased absorption of calcium. For example, active intestinal calcium absorption is enhanced during growth, pregnancy, and lactation, which are physiological states that increase the body's need for calcium beyond maintenance requirements. Conversely, in the presence of abundant calcium intakes and/or an increased availability of calcium in the gut, the percentage of calcium absorbed decreases, although the absolute quantity of calcium absorbed is greater. It is well known that elderly individuals may not adapt to low calcium intakes or increased calcium demands as readily as younger persons.

General health of an individual influences calcium absorption, with absorption decreasing during illness. Calcium absorption tends

Figure 1

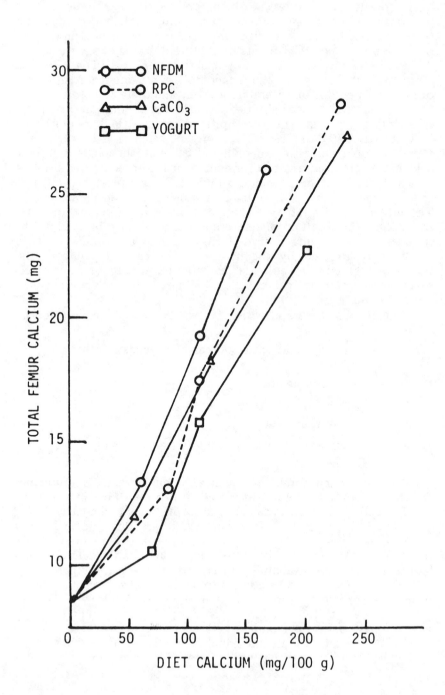

to decrease with increased age in both sexes, beginning at 45 years in women and 60 years in men. Emotional states of an individual such as stress, tension, anxiety, grief and boredom have been shown to decrease intestinal calcium absorption. Physical activity improves calcium absorption while immobilization which occurs during a prolonged illness requiring bed rest, decreases calcium absorption.

Medications such as antacids, tetracyclines, laxatives, diuretics and heparin have been demonstrated to impede calcium absorption. For example, antacids containing aluminum hydroxide, magnesium hydroxide, or a combination of the two, prescribed for patients with active ulcers, have been shown to result in calcium loss. Tetracyclines react with divalent or trivalent cations such as calcium, forming insoluble chelates which are not absorbed. There is some dispute regarding the impact of hormones, particularly the parathyroid hormone (PTH), the major endocrine control of blood calcium level, on calcium absorption. The effect, if any, appears to be minor.

INTERACTION OF DIETARY CONSTITUENTS

In addition to the aforementioned endogenous and exogenous factors efficiency of bone synthesis depends in a large measure on the maintenance of food quality and intake which affords optimal biochemical interrelationship of dietary components. These include: vitamin D, phosphorus, fat, protein, lactose, fiber, fluoride and magnesium.

Vitamin D: The observation that the efficiency of calcium absorption from bowel fluctuates inversely with the degree of skeletal mineralization led, some 20 years ago, to the suggestion that bone elaborated a hormone that informed the gut of skeletal needs.[10] Such a skeletal hormone has never been detected. It has, however, become apparent from many studies of its actions, that vitamin D is the single most significant influence on calcium absorption and proper bone mineralization. But unlike the postulated hormone, vitamin D is not synthesized by bone. Instead, the precursor is synthesized in the skin (cholecalciferol or D_3) or ingested in food (ergocalciferol or D_2) and then shuttled from organ to organ as it proceeds through an elaborate biochemical assembly line to its final, fully activated state (Figure 2). The odyssey begins in the liver where vitamin D is hydroxylated to 25-hydroxy D_3. It then is transported to the kidney, where a second hydroxyl group is attached to form 1,25-dihydroxy D_3. Clinical evidence suggests that it is this form of vitamin D that is most active and fulfills its role in the bloodstream as a skeletal guardian. It does this by working with PTH (Parathyroid hormone) in enhancing calcium and phosphate absorption from the bowel and reabsorption of these ions from bone. In the kidney, however, it opposes PTH by enhancing rather than diminishing, the tubular reabsorption of phosphate. Apparently this action occurs only in the presence of PTH which perhaps ensures that not too much phosphate is reabsorbed. It has been suggested, furthermore, that rising serum and cellular phosphate levels may suppress renal activation of vitamin D.

Figure 2

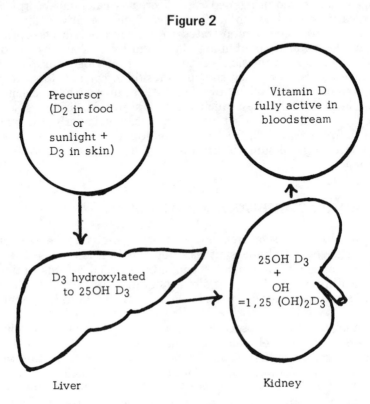

Liver   Kidney

By its actions, vitamin D maintains normal serum levels of both calcium and phosphate, thereby providing for bone adequate materials for construction of new osteons. In addition, vitamin D may stimulate a calcification front by directly enhancing hydroxyapatite formation. When vitamin D levels drop below normal—as in malabsorption states—bone becomes demineralized; this is seen clinically as osteomalacia in adults and rickets in children.

There is no information concerning the precise requirements of vitamin D in children and adults. Although the requirement in the normal adult can be met by exposure to sunlight, persons, especially the elderly, whose activities limit their exposure to sunlight should have an adequate source of vitamin D_2 (RDA 400 I.U. per day) either from the dietary and/or supplements.

Excessive levels of vitamin D—most likely to occur as a result of overdosage—may precipitate hypercalcemia with its attendant risks of metastatic calcification and renal lithiasis. The condition is difficult to treat since vitamin D has a long half-life of a month or more, and the hypercalcemia may therefore persist for weeks after cessation of vitamin D administration.

Thyrocalcitonin: This substance is phylogenetically probably the oldest known of the calcium-controlling hormones. The hormone's structure has been elucidated and its actions intensely explored. And it has also been put to some therapeutic use.[11] Produced in the parafollicular cells of the thyroid gland, thyrocalcitonin opposes the actions of parathyroid hormones. It rapidly lowers serum calcium and phosphate levels by inhibiting bone resorption. Thyrocalcitonin also reduces the number and activity of osteoclasts and inhibits osteocytic osteolysis. During childhood and adolescence the actions of thyrocalcitonin are profound, and it is tempting to draw the inference that the hormone is supporting the expansion of the skeletal mass. At maturity and beyond, however, thyrocalcitonin finds bone progressively unresponsive to its actions.

Whether in adult life it continues to play some role in the control of serum calcium levels is uncertain. What clinical evidence there is suggests an insignificant physiologic role of thyrocalcitonin in adults. For example, there may be no change in serum calcium levels following either excessive production of thyrocalcitonin by a carcinoma of the parafollicular cells (medullary carcinoma) of the thyroid, or total removal of thyrocalcitonin by thyroidectomy. The uncertainties surrounding the physiologic role of thyrocalcitonin in the adult have not precluded its therapeutic use. Thyrocalcitonin may be effective in treating conditions in which there is excessive breakdown of bone by osteoclasts. It has been used in lowering high serum calcium levels due to multiple myeloma, skeletal metastases and primary hyperparathyroidism. It has also proved useful in Paget's disease, in which the excessive remodeling of bone involves hyperactivity of bone osteoblasts and osteoclasts. Thyrocalcitonin has been of less value, however, in skeletal disorders unrelated to excessive activity, such as osteoporosis.

Calcium:Phosphorus Ratio: The amount of phosphorus in American diets is generally greater than that of calcium. If the diet contains adequate calcium, large variations in dietary phosphorus may not influence calcium absorption. Serum calcium concentration is influenced more by the dietary Ca:P ratio than by the absolute intake of dietary calcium. Skeletal changes in animals have been induced by excess dietary phosphorus or insufficient dietary calcium. It has not been firmly established whether the consumption of excess phosphorus induces mild chronic secondary hyperparathyroidism resulting in bone loss in man. The recommended Ca:P ratio is 1:1. However, the Ca:P ratio in the U.S. diets has been estimated to be 1:2.8 (Figures 3a and 3b). A recent dietary survey of 45 male and 45 female college students showed Ca:P ratios ranging

Figure 3

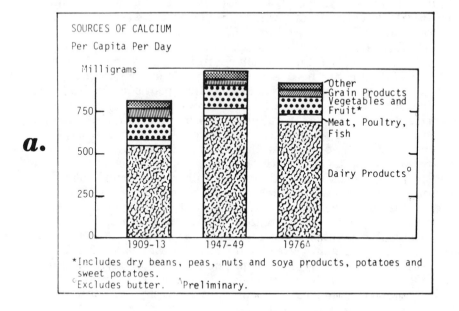

a.

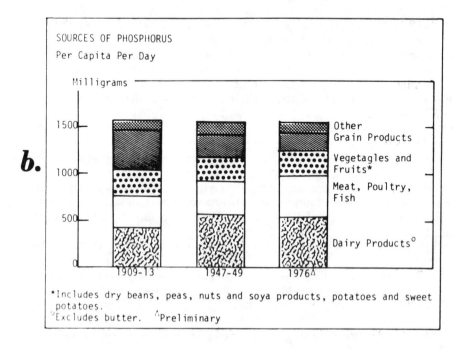

b.

from 1:1.2 to 1:3.6. Similar results were found in 105 normal healthy males and females 18 to 23 years of age.[5] These surveys disclose large intakes of phosphorus derived from processed foods, meat and other animal products and soft drinks frequently associated with a low consumption of milk and milk products of Ca:P ratio of 1:1.

Protein: Urinary calcium excretion is increased in subjects fed high protein diets. However, the effect of dietary protein on the apparent absorption and/or retention of calcium as a parameter of bond formation remains to be established. Metabolic balance studies with adult males have demonstrated that high protein intakes of 142 g and 800 mg of calcium per day resulted in a loss of substantial amounts of calcium; but calcium was retained when the subjects consumed 47 or 94 g of protein per day.[8] Although the possible osteoporotic effect of high protein intake has been considered, further investigations are indicated.

Fats: Calcium absorption may be influenced by dietary fats as follows:

(1) a low fat intake induces a decreased secretion of bile and bile salts with a consequent increase in intestinal fatty acid content which leads to the formation of insoluble calcium soaps, thus decreasing the availability of calcium for absorption;

(2) bile secretion, which is influenced by fat intake, is a major route for the secretion of calcium into the intestinal tract; and

(3) normal bile secretion is necessary for the optimal absorption of exogenous cholecalciferol.

A low fat intake, by decreasing the secretion of bile and bile salts, could retard exogenous cholecalciferol absorption which, in turn, could negatively impact calcium absorption. However, the relationship between the processes involved in fat digestion and absorption by bile and bile salts with the processes involved in calcium absorption is not clearly established. The type of dietary fat may also influence calcium metabolism. There is some evidence that calcium absorption is greater in the presence of milkfat than in the presence of hydrogenated vegetable fats. The latter have a greater tendency to form salts with calcium, resulting in a low calcium balance.

Lactose: The presence of lactose (a disaccharide composed of the monosaccharides glucose and galactose) found in the milk of most mammals has been shown by some investigators to enhance the utilization of proteins and minerals. Lactose ingestion improves calcium absorption in lactose-tolerant but not lactose-intolerant individuals. In lactose fed rats there was an increase in calcium retention as well as improved bone calcification; the beneficial effect of lactose was confined to the absorptive process of calcium and was more pronounced when dietary calcium was limited.[3] It has been suggested that the acidic environment resulting from the formation of lactic acid from lactose by intestinal microflora could possibly favor calcium absorption. Long-term studies in man are needed to determine (a) if the effect of lactose on calcium absorption is transitory or permanent and (b) specific biochemical mechanisms involved.

Lactase Deficiency: An intriguing variation is the prevalence of lactase deficiency has been noted in different parts of the world and among various ethnic groups. The prevalence ranges from 1-2% among Scandinavians to 99-100% among Orientals. With several exceptions, the only population that maintains normal lactase levels are Caucasians living in northern and western Europe and their overseas descendants. Almost all other ethnic groups lose lactase activity in childhood. This means that with a high prevalence of lactase deficiency among Blacks, Orientals, Jews, American Indians, and others, at least 70% of the world's population lose lactase activity in childhood.[5]

Among persons with lactase deficiency, the range of tolerance to lactose is extremely wide. Between 80-100% of persons with lactase deficiency are intolerant to an unphysiologic load of 50 g of lactose, which is equivalent to 1 quart of milk. At the other end of the spectrum, milligram amounts of lactose (as in tablets) occasionally have been reported to cause symptoms, but these anecdotal accounts are not substantiated by objective data. The critical question is: "What proportion of persons with lactase deficiency are symptomatic after a physiologic load of 12 g of lactose—an equivalent to one glass of milk?" Although the literature is controversial on this point, most studies have concluded that persons with lactase deficiency generally can drink 1 glass of milk without having significant symptoms, especially if the milk is taken with a meal, which would delay gastric emptying of lactose. For reasons that are unclear, children with lactase deficiency tend to be more tolerant of lactose than do adults with lactase deficiency. The tolerance to lactose may increase after ingestion of milk for several weeks. It is not uncommon for subjects with inherited delayed-onset lactase deficiency to first become symptomatic in adulthood, even though the enzyme loss occurs in early childhood. The reason for this delay in symptoms is not known. In some instances, a change in diet, which might include more milk as in pregnancy or in peptic ulcer treatment, brings out covert symptoms. Also, patients with lactase deficiency may become symptomatic after gastric surgery, which would result in accelerated emptying of the stomach.

Fiber: Dietary fiber has long been considered an inert and insignificant part of man's diet because it was believed to contribute little nutritionally. Recent epidemiological studies suggest that a habitual decline in dietary fiber intake may be associated with the etiology of gastrointestinal and cardiovascular diseases. Currently commercial exploitation of the limited data has resulted in supplementation of processed cereal foods with fiber.[4] Unfortunately, a high fiber diet may reduce calcium absorption by the formation of insoluble calcium salts within the intestinal lumen. Phytic acid found in the outer husks of cereal grains and whole meal flour and oxalic acid present in rhubarb, spinach, chard and beet greens can reduce calcium availability by formation of chelates. These effects are not of practical importance in the usual U.S. diet, but do pose a serious nutritional problem for the rapidly-growing number of vegetarians in our population.[13]

Other Nutrients: Reported effects of magnesium on calcium metabolism are controversial depending on the experimental conditions. Some investigators have shown that magnesium increases absorption, balance and retention of calcium, whereas others have found no effect of dietary magnesium on calcium utilization. There is evidence that magnesium is not incorporated into the apatite crystal lattice of bones, but is a surface limited ion. Fluoride, which constitutes 0.03% of bone mass, has been claimed by some investigators to promote calcium absorption and employed in the treatment of metabolic bone disease. However, kinetic studies on patients with osteoporosis showed no significant change in the size of calcium pool or accretion rate following fluoride administration. Contrary to subjective observations, radiodensitometric data obtained in studies with some 100 male and female adults (47-65 years) treated with fluoride and an equal number of controls showed no significant difference in range of bone density.

Blood Calcium: Factors involved in controlling blood calcium level are illustrated in Figure 4. Calcium in the blood is that which is actively transported through the walls of the intestine, as well as that mobilized from bone. The concentration of blood calcium is remarkably constant at about 10 milligrams (mg) calcium per 100 milliliters (ml) plasma with a diurnal fluctuation in man of 3%. Plasma calcium exists either as the free ion, or bound to proteins, or complexed with organic (for example, citrate) and inorganic (phosphate) acids.

The mechanisms responsible for maintaining the constancy of blood calcium at a normal level are not completely understood. The major endocrine control is PTH. The question of whether parathyroid regulation is due to a single hormone with hypercalcemic characteristics, or to two hormones, one hypocalcemic (calcitonin) and one hypercalcemic (PTH) is not resolved.

When the dietary intake of calcium is low, PTH stimulates the synthesis of the vitamin D metabolite, 1,25-dihydroxycholecalciferol. This vitamin D metabolite together with PTH functions to mobilize calcium from previously formed bone as well as to stimulate intestinal calcium absorption.

With the return of blood calcium to normal concentration,

PTH secretion and 1,25-dihydroxycholecalciferol synthesis are decreased. Thus, when calcium intake is insufficient, the body is capable of maintaining the blood calcium level, although at the expense of skeletal calcium. The above relationship underscores the importance of an adequate supply of dietary calcium.

The ratio of Ca:P in the diet is a prime determinant of blood calcium. An elevated intake of phosphorus relative to calcium can stimulate the parathyroid gland through a depression of serum calcium. The result is a decrease in calcium excretion and an increase in phosphate excretion, a state of secondary hyperparathyroidism.

Figure 4

Factors Involved in Control of the Plasma Calcium

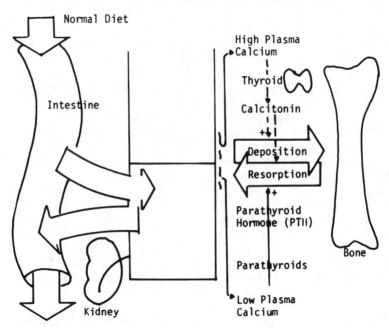

Calcium Excretion: Calcium in excess of body needs can be excreted via three routes:

(1) stool,

(2) urine, and

(3) sweat or perspiration.

Calcium in the stool is unabsorbed food calcium and nonresorbed digestive juice calcium (fecal endogenous calcium). Fecal calcium ranges from about 140 to 180 mg daily although it is highly correlated with calcium intake. In man, the proportion of fecal endogenous calcium to urinary calcium is about 1:1. Urinary calcium excretion can vary widely among individuals but under most conditions it appears to be relatively constant in amount for any given person. The normal loss of urinary calcium ranges from about 100 to 125 mg daily.

The incidence of kidney stone disease is greater than average in certain areas of the world. These include the southeastern and southwestern, and to a lesser extent the midwestern, parts of the U.S. In the southeastern U.S., as many as 15 of every 1,000 hospital admissions are due to calculous disease. Diet, water, climate and geologic formations have been mentioned as etiologic factors in stone disease, but for none of these has such an association been definitely established. As early as 1931, Jolly[7] found that stones were most common in areas where the inhabitants subsisted on diets composed mostly of cereal grains and were least common in areas where dairy farming was done and the inhabitants could be assumed to consume milk and milk products. The rising standard of living may be important in this regard, as an adequate and balanced diet may create a situation less favorable to stone disease. The incidence of stone disease among families of stone formers is higher by a factor of 10 than the incidence among the population in general.

Stone composition is extremely important in the elucidation of the problem. Tagasaki[14] analyzed 735 stones from 700 patients and found that 72.9% were composed of calcium oxalate or calcium phosphate, or both; 22.2% were composed of magnesium ammonium phosphate, mixed with calcium oxalate or calcium phosphate; 3.9% were composed of uric acid or urate; and 1.0% were composed of cystine. In the largest group, calcium oxalate and calcium phosphate stones, the cause is unknown in the majority of cases and the overall recurrence rate is 41.2%. Marshall and associates[8] found the recurrence rate for this type of stone to be 40.3% in men and 30.2% in women. Ten years after identification of the first stone, the chance of formation of a second stone was still 3% in men and as much as 14% in women.

Most of what is known clinically about urolithiasis is best explained in terms of supersaturation of urine. Uric acid stones, cystine stones and stones due to infection with urea-splitting organisms are considered nonidiopathic. In recent years, evidence has been mounting that even so-called idiopathic calcium oxalate stones are genetically associated with excessive urine supersaturation.

SKELETAL BONE LOSS

Numerous surveys taken in homes for the aged, and of ambulatory individuals aged 45-95 years requiring rehabilitation care, have disclosed an incidence of bone loss ranging from 15 to 50%. Other estimates indicate that at least 10% of the population over 50 years of age has osteoporosis severe enough to cause vertebral, hip or longbone fractures. Osteoporosis is a major orthopedic disorder in about 25% of postmenopausal women. Women of Anglo-Saxon origin are particularly prone to vertebral atrophy. Of the approximately 6,000,000 spontaneous fractures due to osteoporosis which occur annually in the U.S., about 5,000,000 are sustained by postmenopausal women. Projected increases in the population of 64 years of age and over can be expected to lead to a greater incidence of fractures. In short, bone health is fast becoming a *major health problem* that deserves the early establishment of preventive measures.[2]

It is apparent from the foregoing that a practical approach to the management of osteoporosis requires an understanding of bone physiology, biochemistry and nutrition. At the outset it is necessary to distinguish between osteomalacia and osteoporosis. Osteomalacia is characterized by decreased bone density due primarily to loss of calcium content in the protein matrix: abnormal calcium-protein ratios. This defect (adult rickets) results from a lack of vitamin D, which is required to utilize calcium in bone formation. Osteomalacia occurs primarily in geographic areas with limited sunshine and/or in populations with poor vitamin D intake. Osteoporosis, on the other hand, is decreased bone density of total substance without change in chemical composition; that is, with normal calcium-protein ratios. Severe skeletal bone loss in mature men and women may exist long before it is manifested by symptoms or outward physical changes. In many instances, advanced osteoporosis is first revealed by the occurrence of spontaneous fractures of the hip, spine or long bones. In postmenopausal women, the first symptom is progressive and persistent pain in the lumbar spine, which seldom radiates. Progressive decrease in vertebral bone mass results in a gradual loss of height and eventual kyphosis (dowager's hump) in the years following menopause. These changes constitute late overt physical evidence of advanced osteoporosis.

Detection: One of the persistent obstacles in probing this disease has been the lack of simple objective methods to detect bone loss in patients before it becomes a clinical *fait accompli*. Isotope tracer methods, measurements of serum and urine calcium levels, and calcium balance determinations have proved less than adequate for the early detection of bone loss. Consideration of the subjective shortcomings of conventional visual examination of x-ray films for the diagnosis of osteoporosis and other biomedical parameters prompted us to develop a practical and proved radiographic method for quantitative evaluation of bone loss.[1] Our 15 year radiogrammetric survey of a "normal healthy" median to upper economic population of some 4,000 females and 1,000 males (10-95 years old) has shown that (a) bone loss is a closely age-related phenomenon, (b) after 25 years of age bone density in males is approximately 25% greater than in females and (c) significant subnormal bone density prevails in 1-15% as early as age 25 in both sexes (Figure 5). In addition to these apparent basic sex differences, the observed incidence of excessive subnormal bone density in females may result because:

(1) Females indulge in reducing diets, even at an early age. When there is weight loss there is also bone loss.
(2) Childbearing takes a toll. The fetus requires 400 mg of calcium per day. Supplements recommended by obstetricians are generally not enough to cover the needs of the mother and fetus.
(3) During breast-feeding, the infant takes 300 mg of calcium per day from the mother.
(4) Changes in hormonal balance of menopause accelerate bone loss.
(5) More women live longer than men.

Figure 5

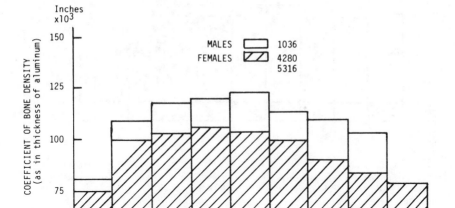

Among common foods, milk and cheese are the richest sources of calcium. Most other foods contribute much smaller amounts. The 1959 USDA survey of 5500 "normal" females showed that in the age group of 45+ years the estimated calcium consumption average approximated 450 mg per day—about 50% below the USRDA of 1000 mg per day.[15] In our studies we have found a high incidence of subnormal coefficients of bone density and spontaneous fractures for 313 females over 55 years of age. A significant relationship of inadequate calcium intake to the incidence of osteoporosis resulting in disabling fractures of long limbs and hips of postmenopausal women is indicated by the data shown in Figure 6.

Dietary Calcium: The data assembled in Figure 6 suggest that the occurrence of bone loss and fracture incidence in postmenopausal women could be related to an habitual consumption significantly below the USRDA of 1000 mg calcium per day.

Figure 6

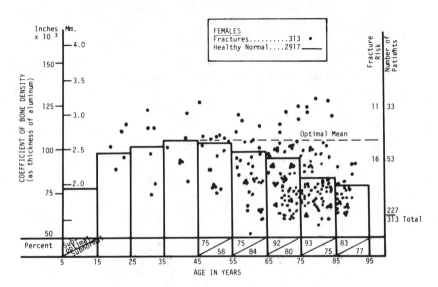

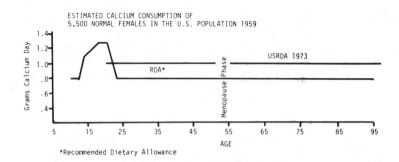

In order to validate this relationship additional tests were undertaken to probe the effects of dietary calcium intake and bone density of 52 "normal healthy" postmenopausal women (Figure 7). It is clear from the graphic representation of the results that bone density increases with intake of calcium and that dairy products constitute a major source of calcium. These findings indicate that a daily intake of 800-1000 mg of calcium is necessary to maintain normal or optimal bone health in postmenopausal women. To achieve this desirable goal the women would have to consume a quart of whole or skim milk (approximately 1000 mg of calcium) per day or an equivalent amount of cheese products. To avoid serious bone loss and its sequelae the alternative is a daily calcium supplement.

Figure 7

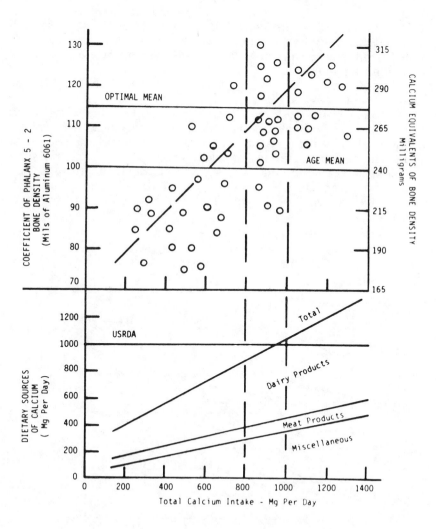

Calcium Supplements: To test the application of this therapeutic modality studies were undertaken to determine effects of daily supplements which provided 750 mg of calcium and 375 I.U. of vitamin D_2 on the bone density coefficients of 12 women residents of the Osborn Home, who ranged in age from 79-89. Their normal diet contained approximately 400 mg of calcium per day which with supplement supplied a total of some 1000 mg of calcium per day. At the same time, a control group which did not receive the supplement and whose average age was 82, continued receiving about 450 mg from their daily diets. Even though they were three years older than when they started taking the additional calcium, the bone density of the women taking the supplement was 12% higher than the control group. These results suggest that under conditions of low calcium intake, due to inadequate consumption of dairy products bone loss in elderly women may be decelerated or reversed by taking a calcium supplement.

Subsequently this evaluation was extended to include 23 female hospital employees (36-62 years) who volunteered for the study and were divided into a test and control group. All of these individuals lived at home, ate conventional meals and were engaged in normal activities of daily living. After a period of 36 months the test group of 14 who consumed 750 mg of calcium and 375 I.U. vitamin D_2 per day along with their regular diet showed an average increase of 11% in bone density. During the same period the placebo group of 9 showed no change in bone density. Typical results are shown in Figure 8.

Figure 8

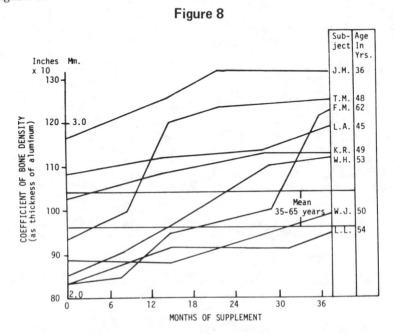

It will be noted that the pattern of change varies greatly and that generalizations as to individual mode of action could not be made in the absence of quantitative radiographic measurements. On the whole, little or no improvement was observed during the first six to nine months of supplementation. Attention is called to the fact that with 36 months of supplementation the bone density of six of the test subjects rose to the age level found in males of comparable age which is usually associated with a far lower fracture risk. Longitudinal measurements of eight test volunteers who stopped the supplements on their own initiative and later reentered the study showed significant increases in bone density during the initial supplementation period, deterioration after stopping the supplements and improvement in bone density after resuming the supplements (Figure 9).

Figure 9

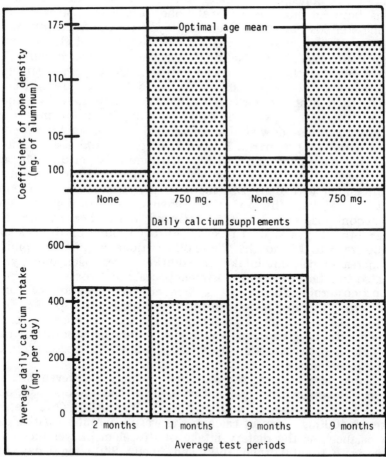

Average effects of calcium supplementation and no supplementation in eight "normal, healthy" women aged 46 to 59.

The sum-total of these and similar assays with 326 postmenopausal and elderly women who have participated in this program for periods of 2 to 8 years have amply confirmed the counter-osteoporotic efficacy of continuous calcium supplementation in conditions of habitual low calcium diets. Significantly these long term trials have revealed no adverse kidney effects.

To conclude, it has been reported that bone loss is an age related phenomenon which coincides with reduced physical activity of the later years of life in both sexes, vigorously aggravated by age associated hormonal changes in women and chronic inadequate intake of calcium. Optimum bone density in women is reached at about age 40. After that the wasting process begins. In the absence of preventative dietary measures osteoporosis may become a major health problem for America's fast growing senior female population.

Estrogen Therapy: Some 35 years ago, Albright and associates advanced the concept that bone formation is dependent on interactions between anabolic and catabolic hormones. Estrogens and anabolic steroids have long been employed for the management of bone loss in menopausal women. Since the results of clinical trials remain controversial, the National Academy of Science and FDA recently reached the opinion that estrogen therapy in postmenopausal osteoporosis is probably effective only when used in conjunction with other important therapeutic measures, such as diet, calcium, physiotherapy and other health-promoting measures. Our investigations showed that prescribed administration of conjugated estrogens (Premarin, 1.25 or 2.5 mg/day) and a supplement of 750 mg of calcium and 375 I.U. of vitamin D_2 per day (Os-Cal) or the calcium-vitamin D_2 supplement alone for 7 to 17 months was associated with 5 to 12% increases in the coefficient of bone density of 26 "healthy normal" women of 47 to 76 years. Age-related bone losses of 3 to 9% continued in 13 "healthy normal" women (45-72 years) who received estrogen alone or the calcium placebo for 6 to 15 months. These observations support the opinion that increased calcium intake, in addition to estrogen therapy, is indicated to retard or reverse postmenopausal bone loss.

Corticosteroid Therapy: A frequent complication of long-term steroid therapy is osteoporosis. Ordinarily, the spine is first involved, just as occurs in postmenopausal and senile osteoporosis. As the process progresses, the long bones become osteoporotic. However, not all patients under prolonged therapy with corticosteroids develop osteoporosis.

Calcium loss in corticosteroid therapy may be preventable. A 69 year old woman with rheumatoid arthritis and osteoporosis, had been on long-term therapy with prednisone (Figure 10). On a dietary intake of 400 mg a day of calcium she was in definite negative calcium balance. As the dietary intake of this element was increased progressively to 800, then 1600 and finally 2400 mg a day, her balance became more positive progressively with good storage of calcium.

Magnesium, Phosphorus and Calcium

Figure 10

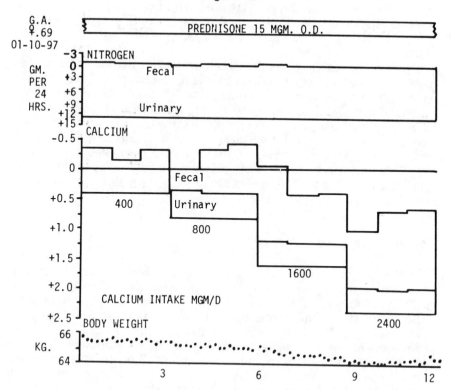

Figure 11

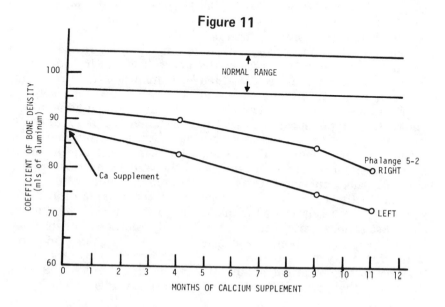

Studies on six patients over a period of 2-3 years of calcium supplementation of 750 mg per day concurrent with prescribed corticosteroid therapy showed progressive bone losses of about 5% per year beyond the initial subnormal level. A typical result is shown in Figure 11.

Stroke Patients: Several authors have documented the high frequency of osteoporosis and spontaneous fractures on the affected side in hemiplegic patients which can be attributed to metabolic effect of disuse phenomenon.[20,21] We found that a 10-25% bone loss is a frequent, early and readily detectable complication in hemiplegic patients. Quantitative measurements in 29 male and female stroke patients over periods of 3-20 months revealed that a 13% improvement in bone density was associated with calcium supplemented diets which provided these patients with about 1000 mg of calcium per day.

ALVEOLAR BONE

The physical form, histological structure, chemical composition, time of eruption and maturation process of teeth are determined by the combined influences of diet, nutritional state and heredity.[2] The integrity of the tooth depends on the pre- and post-eruptive development of a tooth with a stable apatite lattice. Certain nutrients are important during the period of tooth development. Namely, an adequate intake of calcium, phosphorus, protein, magnesium, fluoride and vitamins A, C, and D are needed during the formative period for the development of a tooth structure with increased resistance to oral disease. The calcification of teeth and bones is impaired in children whose diets are deficient in calcium, phosphorus and other essential nutrients.

The bulk of the tooth consists of dentin and the crown is covered by enamel, the hardest tissue in the body. The dentin and enamel of teeth do not readily release their calcium when body needs are not met by diet. Although the mature tooth is not metabolically inert, available evidence suggests that the fully erupted adult tooth, unlike the supporting structure, alveolar bone, is not significantly subject to modification in structure or calcification by changes in calcium or phosphorus metabolism. Alveolar bone, on the other hand, experiences active exchange with the vascular nutrient supply, as do all other bone and soft tissues elsewhere in the body.

The teeth bearing-alveolar processes of the maxilla (upper jaw) and the alveolar border of the mandible (lower jaw) are the thickest and most spongy part of these two skeletal structures. The insidious loss of periodontal bone frequently begins in the young and continues for many years before loss of teeth occurs. This process has been estimated to affect about 80% of the adults in the U.S. with the result that over 50% of the population is edentulous by 60

Figure 12

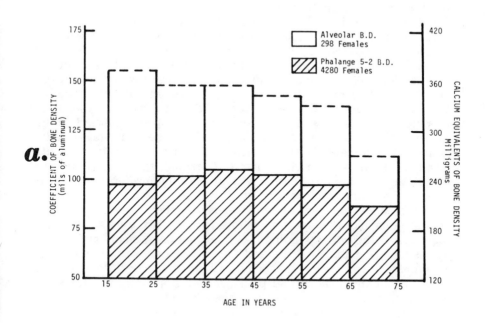

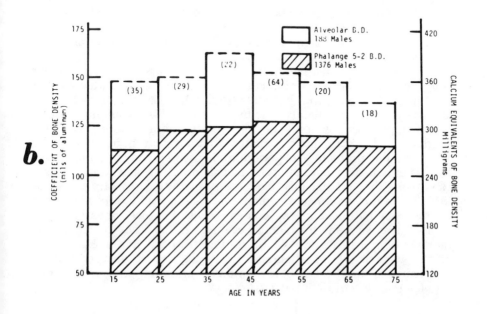

years of age. In other world areas the incidence is greater and loss of teeth occurs at much earlier ages. The fact that poor dentition is one of the principal causes of malnutrition of the elderly stresses the need to examine modalities which may minimize tooth loss. Difficulties in chewing with dentures often cause people to favor consumptions of less nutritious foods than those eaten by persons with healthy natural teeth.

A careful consideration of the foregoing clinical relationships emphasizes the need for a convenient procedure which would provide not only a means for detection of incipient alveolar bone loss but also a means of evaluating the efficacy of available dietary or therapeutic modalities. To this end, we developed a method similar in principle to that previously described for the quantitative radiographic determination of phalange 5-2 density. Briefly stated, dental x-rays of the left and right molar 1-2 area are taken at a fixed distance, angle and exposure time with an aluminum reference standard attached to the film. Density of the periodontal alveolar bone is then measured by means of a microdensitometer. The results showed that alveolar like skeletal bone values for both sexes are significantly higher than the corresponding 5-2 phalange coefficients and that an age associated pattern of rise and fall of bone density of both bone sites prevails for males and females (Figures 12a and 12b). The coefficients of the alveolar and phalange 5-2 bone densities for males and females at all ages reflect bone mass differences.

Dietary Calcium: Having satisfied ourselves regarding the reproducibility and accuracy of the alveolar bone measurements, we initiated an investigation on the relation of habitual daily intake of calcium and alveolar bone status. Although it has been inferred from available animal experiments that inadequate calcium intake may be a major factor in alveolar bone loss, direct objective evidence for man is lacking. In an attempt to probe this relationship calcium intake and source determined from tabular analysis of the dietaries of 22 "healthy normal" males and females (20-70 years) and their alveolar bone densities were determined as described above. The data collected in Figure 13 indicate that optimal alveolar, like skeletal, bone density is closely related to dietary calcium intake of 800-1000 mg per day which apparently can only be achieved by ample amounts of dairy products. Glickman has postulated that alveolar bone loss is the "crux of the problem" in periodontal disease. Though animal, and some human, data lend support to this point of view, dentists as a group feel that further investigations in man are necessary to establish this important concept. Quantitative radiography during periodic dental examinations appears to be a means of choice for early detection of alveolar bone loss and to evaluate the efficacy of dietary or therapeutic measures. It is interesting to note here, that many of the women who participated in our clinical studies reported a tightening of their teeth 3 to 4 months after initiation of calcium supplementation of their habitual diets.

Magnesium, Phosphorus and Calcium

Figure 13

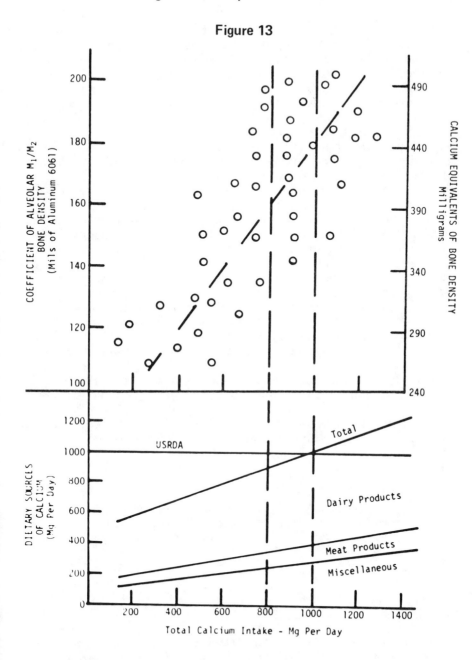

SUMMARY

The need for a consistent and reliable source of dietary calcium for skeletal growth and maintenance, as well as for several life essential processes has been documented. The present and other studies point to the likely adverse skeletal and alveolar bone health effects of inadequate calcium intake and the need for preventative and therapeutic measures either by dietaries or supplements which will provide optimal amounts of this mineral. Unfortunately, there remain many unknowns regarding the precise amounts of calcium necessary at various critical stages of the human life cycle. The development of simple quantitative radiographic methods for determination of skeletal and alveolar bone densities now permits early detection of bone loss and objective long-term evaluation of therapeutic measures to retard or reverse bone loss.

The results of dietary surveys disclose that calcium is one nutrient likely to be consumed in less than adequate amounts, particularly in the habitual diets of women aged 29 years and over. The inclusion of milk and other dairy products in the diet is important to supply a significant quantity of calcium in a readily available form. Clinical studies show that bone loss, characteristic of the aging process, can be retarded by increasing the calcium intake. There is growing support for a calcium intake of at least one gram daily throughout the life span to maintain normal, if not optimal bone density. This recommendation has particular bearing on nutritional efforts to minimize the incidence of bone loss and its consequences. Available knowledge of increased calcium needs for individuals predisposed to osteoporosis and those who cannot adapt to suboptimal calcium intake should and must not be overlooked. As of the present there remain considerable gaps in our understanding of the nutritional aspects of bone metabolism which can only be bridged by continued long-range research into the interrelationship of the structural components, calcium and protein, and other nutrients indispensable to maintenance of bone health.

> The author wishes to acknowledge the invaluable cooperation and assistance of Edward J. Lorenze, Jr., M.D., A. Herbert Edelson, D.D.S., Evelyn H. Wein and Lynne Carroll in the conduct of the investigations recorded herewith.

REFERENCES

(1) Albanese, A.A., Edelson, A.H., Lorenze, E.J., and Wein, E.H. 1969. Quantitative radiographic survey technique for detection of bone loss. *Am. Geriatr. Soc.* 17:142-154.
(2) Albanese, A.A. 1977. *Bone Loss: Causes, Detection and Therapy.* Vol. I. Current Topics in Nutrition and Diseases, A.A. Albanese and D. Kritchevsky (eds.). Alan R. Liss, Inc., New York.

(3) Ali, R.A.M., and Evans, J.L. 1971. Body composition in the growing rat as affected by dietary lactose, calcium, buffering capacity and EDTA. *J. Anim. Sci.* 33:765-770.
(4) Anonymous. 1975. The role of fiber in the diet. *Dairy Council Digest.* 46:1-4.
(5) Emiola, L., and O'Shea, J.P. 1978. Effects of physical activity and nutrition on bone density measured by radiography techniques. *Nutr. Rpt. Intl.* 17:669-681.
(6) Harrison, H.E 1976. Phosphorus. In 4th Ed. *Present Knowledge in Nutrition*, pp 241-246. Nutrition Foundation, Inc., New York.
(7) Jolly, J.S. 1931. *Stone and Calculus Disease in Urinary Organs.* C.V. Mosby Co., St. Louis.
(8) Marshall, V., White, R.H., Chaput de Saintonge, M., Tresidder, G.C., and Blandy, J.P. 1975. Natural history of renal and ureteric calcula. *Br. J. Urol.* 47:117-124.
(9) Newcomer, A.D 1979. Lactase deficiency. *Contemporary Nutrition* Vol. 4, No. 4.
(10) Omdahl, J.L., and DeLuca, H.F. 1973. Vitamin D. In R.S.. Goodhart and M.E. Shirls (eds.) *Modern Nutrition in Health and Disease*, 5th Ed., pp 158-165. Lea and Febiger, Philadelphia.
(11) Schrier, R.W. 1975. Derangements of calcium and phosphorus metabolism. In *The Sea Within Us*, pp 47-59. Science and Medicine Publ. Co., New York.
(12) Shils, M.E. 1976. Magnesium. In *Present Knowledge in Nutrition*, 4th Ed., pp 247-258. Nutrition Foundation Inc., New York.
(13) Southgate, D.A.T. 1973. Fibre and other unavailable carbohydrates and their effects on the energy value of the diet. *Proc. Nutr. Soc.* 32:131-136.
(14) Tagasaki, E. 1975. An observation on the composition and recurrence of urinary calcula. *Urol. Intl.* 30:228-236.
(15) USRDA-FDA Drug Bulletin, 1979. Washington, D.C.
(16) Walker, R.M., and Linkswiler, H.M. 1972. Calcium retention in the adult human male as affected by protein intake. *J. Nutr.* 102:1297-1302.
(17) Wong, N.P., and LaCroix, D.E. 1980. Biological availability of calcium in dairy products. *Nutr. Rept. Intl.* 21:673-680.

–13–
Fluoride Metabolism and Aging

Herta Spencer, Dace Osis and Menahem Lender

INTRODUCTION

Changes of fluoride metabolism as a function of age may have relevance to the important problem of bone loss with aging. Fluoride enters the human body with the diet and drinking water and is rapidly incorporated into bone.[6,31,32] A close relationship has been demonstrated between the fluoride content of the drinking water and the retention of fluoride in bones.[12,15,49,65] In view of the reported beneficial effects of fluoride on bone, retention of a certain amount of fluoride may be necessary to maintain the normal bone structure with aging. Fluoride has been shown to improve the crystallinity of the hydroxyapatite crystal,[66] to decrease the solubility of bone,[3] to decrease bone resorption,[18] to increase bone formation during fluoride therapy for osteoporosis,[7,11,20,21,29,39,43] and to decrease the fracture incidence in osteoporosis.[40,60] Also, a survey reported a lower incidence of osteoporosis in high fluoride areas than in areas where the fluoride content of the water was low.[4]

There is ample evidence[1,2,9,17,30,34,36,45,46] for bone loss with aging and for the high incidence of asymptomatic osteoporosis.[50] As fluoride is beneficial in osteoporosis treatment,[7,11,18,20,21,29,39,43] the use of sodium fluoride may therefore be considered a rational type of therapy for persons of the older age group afflicted with this condition.

Information is available on fluoride levels in bone with aging in man,[8,12,13,15,16,47,49] and some data have been reported on the changes of the plasma fluoride level with age.[14,38] The fluoride content in bone has been reported to increase with age[12,15,24,64] and to reach a plateau at about the age of 55 years,[12,15] suggesting an equilibrium between bone fluoride and tissue fluoride at that time. However, with advancing age the storage of fluoride in bone may decrease both in animals[12,15,24,35,61,62,64] and in man.[8,13,16] the total amount of fluoride in the skeleton of persons in the older age group may be low because of mobilization of fluoride from the skeleton due to increased bone resorption with aging.

METHODOLOGY

The excretion and retention of fluoride was determined by fluoride balances which were carried out under strictly controlled conditions. The subjects were fully ambulatory adult males who were observed in the Metabolic Research Unit. The diet was constant and was analyzed for fluoride and calcium serially throughout all studies. The average dietary fluoride intake ranged from 1.2 to 1.5 mg/day; however, the total daily fluoride intake was greater and depended on the amount of fluoridated drinking water consumed which contained about 1 ppm fluoride. The fluid intake was kept constant throughout the studies. The total fluoride intake ranged from 3.8 to 5.2 mg/day and averaged 4.3 mg/day. Complete collections of urine and stool were obtained throughout all studies. Fluoride and calcium balances were determined for several weeks by analyzing aliquots of 6-day pools of urine and of stool and by analyzing aliquots of the diet in each 6-day metabolic study period. In studying the effect of added fluoride these supplements were given as sodium fluoride at three dose levels of 10, 20 and 45 mg fluoride per day. The average duration of the fluoride balance studies in the control phase was 28 days. The duration of the high fluoride studies, using the 10 mg/day dose was 30 days, that of the 20 mg dose was 90 days, and that of the 45 mg dose was 84 days. The 45 mg dose of fluoride was given to patients as a form of treatment for osteoporosis. The net absorption of fluoride was determined by the following formula:

$$\text{Net Absorption, \%} = \frac{\text{Fluoride Intake} - \text{Fecal Fluoride}}{\text{Fluoride Intake}} \times 100$$

In order to determine whether excess fluoride is excreted after the discontinuation of the fluoride supplements, urinary and fecal fluoride excretions were determined for several weeks after the administration of fluoride. The difference between these excretions and the urinary fluoride excretions prior to the intake of the fluoride supplements permitted the calculation of the release of the previously retained fluoride.

The effect of several inorganic elements such as calcium, phosphorus, and magnesium on the fluoride balance was investigated. In another study the effect of aluminum on fluoride metabolism was investigated by giving the antacid aluminum hydroxide in a dose of 30 ml three times per day, containing a total of 1.8 g aluminum.

Fluoride in the diet, urine, stool, plasma, in the sodium fluoride tablets, and in the drinking water was analyzed as total fluoride by a modification[37] of the method of Singer and Armstrong.[44] Calcium in the diet, urine, and stool was analyzed by atomic absorption spectroscopy[63] and phosphorus by a modification of the method of Fiske and SubbaRow.[10]

The data were analyzed statistically using the paired Student's t-test.[51]

RESULTS

Table 1 shows fluoride balance data of six patients studied under controlled dietary conditions during a similar fluoride intake which was due to the fluoride content of the diet and drinking water. The data show that the main pathway of fluoride excretion is via the kidney and that the excretion of fluoride via the intestine is very low. Variability of the urinary fluoride excretion was noted which ranged from 56.8 to 86.8% of the fluoride intake and averaged 73.1%. The fecal fluoride varied by a factor of 3, however, these excretions are very low, ranging from 0.1 to 0.3 mg/day, corresponding to a range of 2.6 to 7.5% of the fluoride intake and averaging 4.2%. The fluoride balances of all patients were positive, the average balance being 1.0±0.23 mg/day, corresponding to an overall retention of 23.3% of the fluoride intake. However, this fluoride balance is a maximal balance as the loss of fluoride in perspiration has not been considered in this determination. Due to the very low fecal fluoride excretion, the net absorption of fluoride was very high, averaging 95.8% of the fluoride intake.

Table 2 shows data of fluoride balances during different fluoride intakes. The baseline fluoride balance studies show again that on a similar fluoride intake of approximately 4 mg/day, urinary fluoride ranged from 2.5 to 2.9 mg/day, corresponding to an excretion ranging from 57 to 73% of the fluoride intake. The stool fluoride in the three studies was again very low, ranging from 0.1 to 0.3 mg/day and corresponding to a range of 2.3 to 6.8% of the fluoride intake. The fluoride balance was similar in the three studies, ranging from 0.9 to 1.6 mg/day, corresponding to a fluoride retention ranging from 22.5 to 36.4% of the fluoride intake. Increasing the fluoride intake by a factor of 2.5, 5 or 10 by adding 10, 20 or 45 mg of fluoride as sodium fluoride per day was reflected by a marked increase in urinary fluoride excretion. This excretion increased threefold on addition of the 10-mg dose of fluoride; it increased more than four-fold on adding the 20-mg dose, and increased ten-fold on adding the 45-mg dose of fluoride per day. The increase in stool fluoride varied during the higher fluoride intake; it increased threefold during the intake of the 10- and 20-mg intake of fluoride but increased eleven-fold on increasing the intake by 45 mg/day. There

Table 1: Fluoride Balances and Absorption During a Normal Fluoride Intake

Case	Age	Days	Fluoride (mg/day)				Fluoride Excretion (% of intake)		Net Absorption of Fluoride* (%)
			Intake	Urine	Stool	Balance	Urine	Stool	
1	63	36	4.5	2.7	0.2	+1.6	60.0	4.4	95.6
2	48	18	5.2	4.3	0.2	+0.7	82.7	3.8	96.2
3	51	36	3.8	3.3	0.1	+0.4	86.8	2.6	97.3
4	50	54	4.0	3.2	0.3	+0.5	80.0	7.5	92.5
5	64	30	4.4	2.5	0.2	+1.7	56.8	4.5	95.4
6	58	30	4.0	2.9	0.1	+0.9	72.5	2.5	97.5
Average	56	34	4.3	3.2	0.2	+1.0	73.1	4.2	95.8
SEM	±2.83	±4.81	±0.21	±0.26	±0.03	±0.23	±5.05	±0.74	±0.74

*Net absorption determined from fluoride intake and fecal fluoride excretions, see text.

Table 2: Effect of Higher Intakes of Fluoride on Fluoride Excretions and Retention in Man

Type of Study*	Added Fluoride (mg/day)	Study Days	Fluoride (mg/day**)				Fluoride (% of intake)		
			Intake	Urine	Stool	Balance	Urine	Stool	Balance
Control	—	30	4.4	2.5	0.3	+1.6	56.8	6.8	36.4
Fluoride	10	30	13.7	7.5	0.9	+5.3	54.7	6.6	38.7
Control	—	30	4.4	2.9	0.1	+1.4	65.9	2.3	31.8
Fluoride	20	90	23.7	14.2	0.3	+9.2	59.9	1.3	38.8
Control	—	30	4.0	2.9	0.2	+0.9	72.5	5.0	22.5
Fluoride	45	90	49.7	34.1	2.2	+13.4	68.6	4.4	27.0

*The fluoride intake in the control study was due to the fluoride content of the diet and water. The higher fluoride intakes were due to the addition of 10, 20 and 45 mg fluoride as sodium fluoride, respectively.

**Values are averages for the number of study days.

was a progressive increase in fluoride retention on increasing the fluoride supplements from 10 to 45 mg/day and the fluoride balance changed from a baseline value of about 1 mg/day to 5, 9 and 13 mg/day as the fluoride intake was increased from 4.4 mg/day in the control study to 13.7, 23.7 and 49.7 mg/day, respectively. The values for the urinary and fecal fluoride excretion, expressed as percent of the fluoride intake, as well as the fluoride balances, remained in a similar range during the low and higher fluoride intake reflecting the increasing retention of fluoride with increasing fluoride intake.

Figure 1 shows graphically the positive correlation between the fluoride intake and the fluoride balance in man.

Figure 1

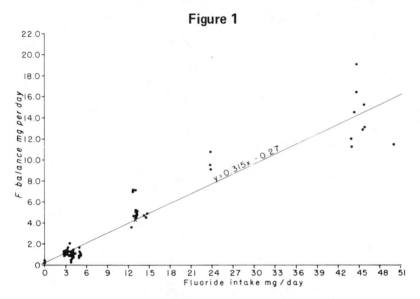

Fluoride intake in control study (4 mg/day) was due to the fluoride content of the diet and water. During the high fluoride intake, fluoride, as sodium fluoride, was added. Values are averages for each 6-day period.

Table 3 shows data of the excretion and retention of fluoride during a prolonged high fluoride intake given for the treatment of osteoporosis and after the discontinuation of these large amounts of fluoride. In the 18-day control study, the daily fluoride intake averaged 3.7 mg/day and 70% of the intake was excreted in urine and a very small amount, 5.4%, was excreted in stool. The retention of fluoride was 0.9 mg/day or a total of 16.2 mg/day during the 18 days of this "native" fluoride intake. When the fluoride intake was increased by adding sodium fluoride to the constant dietary fluoride intake, the pattern of excretion was the same, i.e., the major

Table 3: Fluoride Balances, Retention and Release After Discontinuation of Fluoride Supplements

Study	Days of Fluoride Intake	Fluoride (mg/day) Intake	Fluoride (mg/day) Urine	Fluoride (mg/day) Stool	Fluoride (mg/day) Balance	Total Fluoride Retained (mg/day)	Excess Excretion of Fluoride (mg)	Percent of Retained Fluoride
Control*	18	3.7	2.6	0.2	+0.9	16.2	—	—
High F**	76	48.5	32.7	4.6	+11.2	851.2	16.7	2***

*Fluoride intake due to fluoride content of diet and water.
**High fluoride intake due to the addition of 45 mg fluoride as sodium fluoride.
***Excess fluoride excreted in first 8 days after the discontinuation of the fluoride supplements.

Table 4: Effect of Fluoride Supplementation on Calcium Excretion and Retention in Osteoporosis and in Paget's Disease of Bone

Patient	Diagnosis	Study Condition	Study Days	Calcium (mg/day) Intake	Calcium (mg/day) Urine	Calcium (mg/day) Stool	Calcium (mg/day) Balance
1	osteoporosis	Control	60	198 ± 15	101 ± 23	220 ± 27	−124 ± 35
		F (45 mg/day)	54	175 ± 6	76 ± 6*	202 ± 15	−104 ± 19
2	Paget's disease of bone	Control	30	403 ± 9	395 ± 47	203 ± 31	−195 ± 75
		F (45 mg/day)	12	397 ± 0	333 ± 6**	220 ± 15	−157 ± 8

Note: Values are average ± SD.
*$p < 0.01$.   **$p < 0.05$.

pathway of excretion was via the kidney; 67.5% of the intake was excreted in urine and the fecal excretion, although relatively low in relation to the intake, corresponded to 9.6% of the high fluoride intake. The fluoride retention was high, 11.2 mg/day, or a total of 851.2 mg during the 76 days of the high fluoride intake. Following discontinuation of the high fluoride intake, only a total of 16.7 mg fluoride was excreted corresponding to 2% of the retained fluoride. This amount of excess fluoride was excreted in the first 8 days after the high fluoride intake was discontinued and, thereafter, there was no excess fluoride excretion, neither via the kidney nor via the intestine.

Table 4 shows data of the effect of a high fluoride intake on the calcium balance in a patient with osteoporosis and in a patient with Paget's disease of the bone. In Patient 1, the patient with osteoporosis, the calcium balance was slightly negative in the control study during the low fluoride intake of about 4 mg/day. This is expected because of the low calcium intake. During fluoride supplementation of 45 mg fluoride per day, given as sodium fluoride, for 54 days, the urinary calcium decreased significantly, $P < 0.01$, the stool calcium remained in the same range, and the calcium balance was only slightly less negative, the change not being significant. Similar observations were made in Patient 2, the patient with Paget's disease of bone, who received a somewhat higher calcium intake than Patient 1. The urinary calcium of the patient with Paget's disease of bone was very high in relation to the intake in the control study and the calcium balance was distinctly negative. During the addition of 45 mg fluoride per day, given as sodium fluoride for only 12 days, the urinary calcium decreased distinctly and significantly, $P < 0.05$. There was no change of the stool calcium during the high fluoride intake. Due to the decrease of urinary calcium excretion the calcium balance became less negative but the difference was not significant.

Figure 2 shows data of the retention of fluoride during 54 days of fluoride supplementation. The concomitant changes in urinary calcium are also shown. In the four 6-day study periods in the control study the fluoride retention, determined by fluoride balances, ranged from 0.6 to 1.3 mg/day. In the nine 6-day periods, or 54 days of fluoride supplementation of 45 mg fluoride per day, given as sodium fluoride, the retention of fluoride was high and ranged from 12 to 16 mg/day. Although fluctuations of the fluoride balances were noted, the fluoride retention was quite uniform during this relatively prolonged intake of added fluoride, except for the slightly higher retention in the first 6-day study period and the higher fluoride retention in the fourth period of the high fluoride intake. There was a significant decrease of the urinary calcium excretion during the high fluoride intake with time, $P < 0.01$.

Table 5 shows data of the effect of relatively small doses of the antacid aluminum hydroxide on the fecal fluoride excretion. In each of the four patients studied there was a significant increase of the fecal fluoride excretion, the factor of increase ranging from 9 to 27. This increase reflects a significant decrease of the intestinal absorption of fluoride induced by aluminum hydroxide.

Figure 2

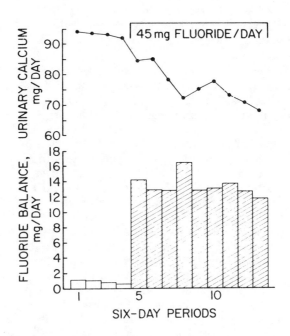

Each point depicting the urinary calcium excretion and each bar showing the fluoride balance represent averages for a 6-day study period.

Table 5: Effect of Aluminum Hydroxide on Fecal Fluoride Excretion in Man

Patient	Fecal Fluoride, mg/day	
	Control	Aluminum Hydroxide*
1	0.1	2.7
2	0.3	2.6
3	0.3	2.7
4	0.2	3.2

*Dose of aluminum hydroxide = 30 ml/day, containing 1.8 g aluminum.

DISCUSSION

The observation that the main pathway of fluoride excretion is via the kidney has been reported many years ago.[27,28] These findings were confirmed in controlled studies carried out in this Research Unit during both a low fluoride intake and during a high fluoride intake.[56] It was found that 50 to 60% of the fluoride ingested with food and with fluoridated water was excreted in urine during a low fluoride intake averaging 4.2 mg/day. This range of fluoride excretion also applied when the intake of fluoride was increased to 14 mg fluoride per day.[56] When the intake of fluoride was even greater due to the administration of fluoride supplements which were used as a form of treatment for patients with osteoporosis[55] and the total fluoride intake was increased to 50 mg/day, the urinary fluoride excretion also corresponded to about 60% of the fluoride intake.[56] The fecal fluoride excretion was low during all fluoride intakes, particularly during the low fluoride intake. This low excretion of fluoride via the intestine indicates the efficient absorption of fluoride[56] and therefore the fluoride retention or fluoride balance primarily depends on the urinary fluoride excretion. These balances are usually positive, even during low fluoride intake. The positive balances have to be considered as maximal balances as the loss of fluoride through the skin, i.e., the secretion of fluoride in sweat, has not been measured in these studies. It is possible that the ingested fluoride is almost completely absorbed from the intestine and that the small amounts excreted in stool have been reexcreted into the intestine following the absorption of ingested fluoride, that is, that these excretions are of endogenous origin. To our knowledge, no studies have been reported on endogenous fecal fluoride excretion in animals or in man. This high net absorption of fluoride is in agreement with observations of others.[28]

The problem whether fluoride is well retained during the intake of fluoride and following the discontinuation of fluoride supplements is relevant when using fluoride therapeutically for osteoporosis. Studies carried out in this Research Unit have shown that once fluoride is absorbed it is well retained. This applies to the period of fluoride administration, even if it extends for a relatively long period of intake and relatively large amounts of fluoride are used, primarily for the treatment of osteoporosis.[55,59] Other studies in this Research Unit have also demonstrated that the retention of fluoride continues after the discontinuation of large doses of fluoride which were given for several weeks or months[53,55,58] and that only very small amounts of the retained fluoride are excreted for a very short period of time following its use. The observation of the effective retention of fluoride and the low excess excretion of fluoride after the discontinuation of fluoride supplements is in contrast to reports of others stating that the excretions of fluoride remain high for a prolonged period of time following the discontinuation of a high fluoride intake.[22,25] The difference of these observations may be due to the fact that the subjects in the latter reports had resided in high fluoride areas for many years and the skeleton of these persons may have been saturated with fluoride.

The intestinal absorption of fluoride may depend on the form in which fluoride is given[23] as the absorption of fluoride from calcium fluoride, cryolite, and from bone meal was lower than from sodium fluoride.[33] However, in a study carried out in this Research Unit the absorption of fluoride was in the same range whether it was given during a low calcium intake or during a high calcium intake for several weeks.[57] In the latter study the high calcium intake was given in three or four divided doses during the day. In a study of others in which fluoride and calcium, given as calcium carbonate, simultaneously[19] the plasma fluoride levels were lower than during the intake of fluoride without calcium. These authors concluded that calcium decreases the intestinal absorption of fluoride.

Fluoride that is retained in the body is primarily and rapidly taken up by bone and this incorporation of fluoride into bone has been demonstrated in man by the use of radioactive fluoride.[6] Due to the rapid uptake of fluoride in the skeleton[64] the use of the fluoride isotope, ^{18}F, has been introduced for bone scanning for diagnostic purposes.[5] A considerable percentage of large amounts of ingested fluoride has also been reported to be deposited in bone in a short period of time.[31,32,47,48]

With regard to the therapeutic use of fluoride for osteoporosis, this type of treatment is indicated because of the pathophysiologic change in osteoporosis and the effects of fluoride on bone. Osteoporosis occurs with greatest frequency in persons of the older age group, particularly in older females and the major alteration in bone physiology in this condition is increased bone resorption in relation to the rate of bone formation. The incorporation of fluoride in bone leads to improvement of the crystal structure of bone,[66] and to the decreased bone resorption.[18] In view of these effects, the use of fluoride appears to be a reasonable therapeutic measure for osteoporosis.[20,41,42,55] It also appears that the life-long intake of certain amounts of fluoride may protect against the development of osteoporosis as a survey reported the incidence of osteoporosis to be lower in naturally high fluoride areas than in areas where the fluoride content of the water is low.[4] The decrease of the urinary calcium excretion, induced by fluoride therapy in patients with osteoporosis, is most likely a result of decreased solubility of the bone crystal and the decreased bone resorption.[66] This decrease is the single most important change in bone metabolism induced by fluoride in patients with osteoporosis treated with sodium fluoride. It is our experience that the best results with fluoride can be achieved in patients with osteoporosis when fluoride is given during a high calcium intake. Other investigators suggested for the treatment of patients with osteoporosis the prolonged use of fluoride in combination with the use of a high calcium intake and large doses of vitamin D.[20] These investigators suggest that fluoride leads to increased bone formation and that the added calcium would promote calcification of the newly formed bone matrix. However, it is not clear whether the addition of large doses of vitamin D is beneficial as the effect of vitamin D, given in conjunction with sodium fluoride, has not been investigated with regard to the intestinal absorption of calcium during fluoride therapy. The clarification of this aspect is of importance as large doses of vitamin D may increase the urinary calcium in the absence of increased intestinal absorption of calcium.[52] It would, therefore, be safer to recommend a small dose of vitamin D, 400 IU, the recommended daily allowance, during fluoride therapy.

The effect of aluminum-containing antacids on mineral and fluoride metabolism are usually not considered when these antacids are used in clinical medicine. The studies carried out in this Research Unit have shown that aluminum hydroxide increases significantly the fecal fluoride excretion and thereby markedly inhibits the intestinal absorption of fluoride.[54] This effect of aluminum hydroxide together with the adverse effects on phosphorus and calcium metabolism[26,54] may contribute to the development of skeletal demineralization. As studies in this Research Unit have also shown that aluminum hydroxide has a similar inhibitory effect on fluoride metabolism during a high fluoride intake, the use of this antacid during fluoride therapy for osteoporosis may interfere with the beneficial therapeutic effect of this type of treatment.[18,20,41,42,55]

(This work was supported by U.S. Public Health Services Grant DE-02486.)

REFERENCES

(1) Arnold, J.S., Bartley, M.H., Tont, S.A. and Jenkins, O.P. 1966. Skeletal changes in aging and disease. *Clin. Orthop.* 49:17-38.
(2) Atkinson, P.J., Weatherell, J.A. and Weidmann, S.M. 1963. Changes in density of the human femoral cortex with age. *J. Bone Joint Surg.* 448:496-502.
(3) Baylink, D., Wergedal, J., Stauffer, M. and Rich, C. 1970. Effects of fluoride on bone formation, mineralization, and resorption in the rat. In T.L. Vischer (ed.) *Fluoride in Medicine*, p. 37-69. Hans Huber, Bern Stuttgart, Vienna.
(4) Bernstein, D., Sadowsky, N., Hegsted, D.M., Guri, C. and Stare, F. 1966. Prevalence of osteoporosis in high- and low-fluoride areas in North Dakota. *JAMA* 198:499-504.
(5) Blau, M., Nagler, W., and Bender, M.A. 1962. Fluorine-18: A new isotope for bone scanning. *J. Nucl. Med.* 3:332-334.
(6) Carlson, C.H., Armstrong, W.D. and Singer, L. 1960. Distribution and excretion of radiofluoride in the human. *Proc. Soc. Exp. Biol. Med.* 104:235 239.

(7) Cass, R.M., Croft, J.D., Jr., Perkins, P., Nye, W., Waterhouse, C. and Terry, R. 1966. New bone formation in osteoporosis following treatment with sodium fluoride. *Arch. Intern. Med.* 118:111-116.
(8) Ellis, G. and Maynard, L.A. 1936. Effect of low levels of fluoride intake on bones and teeth. *Proc. Soc. Exp. Biol. Med.* 35:12-16.
(9) Epker, B.N. 1967. A quantitative microscopic study of bone-remodeling and balance in a human with skeletal fluorosis. *Clin. Orthopaed. Rel. Res.* 55:87-93.
(10) Fiske, C.H. and SubbaRow, Y.T. 1925. The colorimetric determination of phosphorus. *J. Biol. Chem.* 66:375-400.
(11) Franke, J., Rempel, H. and Franke, M. 1974. Three years' experience with sodium fluoride therapy of osteoporosis. *Acta Orthop. Scand.* 45:1-20.
(12) Geever, E.F., McCann, H.G., McClure, F.J., Lee, W.A. and Schiffmann, E. 1971. Fluoridated water, skeletal structure, and chemistry. *Health Services and Mental Health Admin. Reports* 86:820-828.
(13) Glock, G.E., Lowater, F. and Murray, M.M. 1941. The retention and elimination of fluorine in bones. *Biochem. J.* 35:1235-1239.
(14) Husdan, H., Vogl, R., Oreopoulos, D., Gryfe, C. and Rapoport, A. 1976. Serum ionic fluoride: normal range and relationship to age and sex. *Clin. Chem.* 22:1884-1887.
(15) Jackson, D. and Weidmann, S.M. 1958. Fluorine in human bone related to age and the water supply of different regions. *J. Path. Bact.* 76: 451-459.
(16) Jenkins, G.N. 1967. Fluoride. *World Review of Nutr. and Dietetics* 7:138-203.
(17) Johnston, C.C., Jr., Norton, J.A., Jr., Khairi, R.A. and Longcope, C. 1979. Age-related bone loss. In U.S. Barzel (ed.) *Osteoporosis-II*, p. 91-100. Grune and Stratton, New York.
(18) Jowsey, J. and Kelly, P.J. 1968. Effect of fluoride treatment in a patient with osteoporosis. *Mayo Clin. Proc.* 43:435-443.
(19) Jowsey, J. and Riggs, B.L. 1978. Effect of concurrent calcium ingestion on intestinal absorption of fluoride. *Metabolism* 27:971-974.
(20) Jowsey, J., Riggs, B.L., Kelly, P.J. and Hoffman, D.L. 1972. Effect of combined therapy with sodium fluoride, vitamin D and calcium in osteoporosis. *Am. J. Med.* 53:43-49.
(21) Kuhlencordt, F., Kruse, H.P., Eckermeier, L. and Lozano-Tonkin, C. 1970. The histological evaluation of bone in fluoride treated osteoporosis. In T.L. Vischer (ed.) *Fluoride in Medicine*, p. 169-174. Hans Huber, Bern Stuttgart, Vienna.
(22) Largent, E.J. 1952. Rates of elimination of fluoride stored in the tissues of man. *Am. Med. Assoc. Arch. Indust. Hygiene* 6:37-42.
(23) Largent, E.J. 1961. Ingestion, inhalation, excretion and storage of fluoride by man. In E.J. Largent (ed.) *Fluorosis*, p. 22. Ohio State University Press, Columbus, Ohio.
(24) Lawrenz, M., Mitchell, H.H. and Ruth, W.A. 1940. Adaptation of growing rat to ingestion of constant concentration of fluorine in diet. *J. Nutr.* 19:531-546.
(25) Likins, R.C., McClure, F.J. and Steere, A.C. 1962. Urinary excretion of fluoride following defluoridation of a water supply. In F.J. McClure (ed.) *Fluoride Drinking Water.* U.S. Government Printing Office, Washington, DC, Public Health Services Publication No. 825.

(26) Lotz, M., Zisman, E. and Bartter, F.C. 1968. Evidence for a phosphorus-depletion syndrome in man. *New Engl. J. Med.* 278:409-415.
(27) Machle, W. and Largent, E.J. 1943. The absorption and excretion of fluoride. II. The metabolism at high levels of intake. *J. Indust. Hyg. Toxicol.* 25:112-123.
(28) Machle, W., Scott, E.W. and Largent, E.J. 1942. The absorption and excretion of fluorides. I. The normal fluoride balance. *J. Indust. Hyg. Toxicol.* 24:199-204.
(29) Manzke, E., Rawley, R., Vose, G., Roginsky, M., Rader, J.I. and Baylink, D.J. 1977. Effect of fluoride therapy on nondialyzable urinary hydroxyproline, serum alkaline phosphatase, parathyroid hormone, and 25-hydroxyvitamin D. *Metabolism* 26:1005-1010.
(30) Maziere, B., Kuntz, D., Comar, D. and Ryckewaert, A. 1979. In vivo analysis of bone calcium by local neutron activation of the hand: results in normal and osteoporotic subjects. *J. Nucl. Med.* 20:85-91.
(31) McCann, H.G. and Bullock, F.A. 1957. The effect of fluoride ingestion on the composition and solubility of mineralized tissues of the rat. *J. Dent. Res.* 36:391-393.
(32) McClure, F.J. 1939. Fluoride in food and drinking water. *Natl. Inst. Health Bull.* 172:1-53.
(33) McClure, F.J., Mitchell, H.H., Hamilton, T.S. and Kinser, C.A. 1945. Balances of fluorine ingested from various sources in food and water by five young men. *J. Indust. Hyg. Toxicol.* 27:159-170.
(34) Meema, H.E. and Meema, S. 1974. Involutional (physiologic) bone loss in women and the feasibility of preventing structural failure. *J. Am. Geriat. Soc.* 22:443-452.
(35) Miller, R.F. and Phillips, P.H. 1956. The effect of age on the level and metabolism of fluorine in the bones of the fluoridated rat. *J. Nutr.* 59:425-433.
(36) Newton-John, H.F. and Morgan, D.B. 1970. The loss of bone with age, osteoporosis, and fractures. *Clin. Orthopaed. Rel. Res.* 71:229-252.
(37) Osis, D., Waitrowski, E., Samachson, J. and Spencer, H. 1974. Fluoride analysis of the human diet and of biological samples. *Clin. Chim. Acta* 51:211-216.
(38) Parkins, F.M., Tinanoff, N., Moutinho, M., Anstey, M.B. and Waziri, M.H. 1974. Relationships of human plasma fluoride and bone fluoride to age. *Calc. Tis. Res.* 16:335-338.
(39) Parsons, V., Mitchell, C.J., Reeve, J. and Hesp, R. 1977. The use of sodium fluoride, vitamin D and calcium supplements in the treatment of patients with axial osteoporosis. *Calc. Tis. Res.* 22:236-240.
(40) Reutter, F.W. and Olah, A.J. 1970. Bone biopsy findings and clinical observations in longterm treatment of osteoporosis with sodium fluoride and vitamin D_3. In T.L. Vischer (ed.) *Fluoride in Medicine,* p. 249-255. Hans Huber, Bern Stuttgart, Vienna.
(41) Rich, C., Ensinck, J. and Ivanovich, P. 1964. The effects of sodium fluoride on calcium metabolism of subjects with metabolic bone disease. *J. Clin. Invest.* 43:545-556.
(42) Rose, G.A. 1965. A study of the treatment of osteoporosis with fluoride therapy. *Proc. Roy. Soc. Med.* 58:436-440.
(43) Schenk, R.K., Merz, W.A. and Reutter, F.W. 1970. Fluoride in osteoporosis. In T.L. Vischer (ed.) *Fluoride in Medicine,* p. 153-168. Hans Huber, Bern, Stuttgart, Vienna.

(44) Singer, L. and Armstrong, W.D. 1965. Determination of fluoride. *Anal. Biochem.* 10:495-500.
(45) Smith, D.M., Khairi, M.R.A., Norton, J. and Johnston, C.C., Jr. 1976a. Age and activity effects on rate of bone mineral loss. *J. Clin. Invest.* 58:716-721.
(46) Smith, D.M., Norton, J.A., Jr., Khairi, R. and Johnston, C.C. 1976b. The measurement of rates of mineral loss with aging. *J. Lab. Clin. Med.* 87:882-892.
(47) Smith, F.A. (ed.) 1966. Metabolism of inorganic fluoride. In *Pharmacology of Fluoride*. Springer-Verlag, New York.
(48) Smith, F.A., Gardner, D.E. and Hodge, H.C. 1952. Skeletal storage of fluorine in the human. Atomic Energy Comm. Quarterly Tech. Report (University of Rochester), 200:12-14.
(49) Smith, F.A., Gardner, D.E. and Hodge, H.C. 1953. Age increase in fluoride content in human bone. *Fed. Proc.* 12:368.
(50) Smith, R.W. and Frame, B. 1965. Concurrent axial and appendicular osteoporosis. Its relation to calcium consumption. *New Eng. J. Med.* 273:73-78.
(51) Snedecor, G.W. 1959. *Statistical Methods*, 5th edition. Iowa State College Press, Ames, Iowa.
(52) Spencer, H., Kramer, L., Gatza, C. and Osis, D. 1977. Calcium absorption and calcium balances in man during vitamin D intake. In A.W. Norman, K. Schaefer, J.W. Coburn, H.F. DeLuca, D. Fraser, H.G. Grigoleit and D. v. Herrath (eds.) *Vitamin D. Biochemical, Chemical and Clinical Aspects Related to Calcium Metabolism*, p. 611-613. Walter deGruyter, Berlin.
(53) Spencer, H., Kramer, L., Osis, D. and Waitrowski, E. 1975a. Excretion of retained fluoride in man. *J. Applied Physiol.* 38:282-287.
(54) Spencer, H. and Lender, M. 1979. Adverse effects of aluminum-containing antacids on mineral metabolism. *Gastroenterology* 76:603-606.
(55) Spencer, H., Lewin, I., Osis D. and Samachson, J. 1970a. Studies of fluoride and calcium metabolism in patients with osteoporosis. *Am. J. Med.* 49:814-822.
(56) Spencer, H., Lewin, I., Waitrowski, E. and Samachson, J. 1970b. Fluoride metabolism in man. *Am. J. Med.* 49:807-813.
(57) Spencer, H., Osis, D., Kramer, L., Waitrowski, E. and Norris, C. 1975b. Effect of calcium and phosphorus on fluoride metabolism in man. *J. Nutr.* 105:733-740.
(58) Spencer, H., Osis, D. and Waitrowski, E. 1974. Studies of fluoride metabolism in man. In D.D. Hemphill (ed.) *Trace Substances in Environmental Health–VII*, p. 289-294. University of Missouri, Columbia, Missouri.
(59) Spencer, H., Osis, D. and Waitrowski, E. 1975c. Retention of fluoride with time in man. *Clin. Chem.* 21:613-618.
(60) Vose, G.P., Keele, D.K., Milner, A.M., Rawley, R., Roach, T.L. and Sprinkle, E.E., III. 1978. Effect of sodium fluoride, inorganic phosphate, and oxymetholone therapies in osteoporosis: a six-year progress report. *J. Gerontol.* 33:204-212.
(61) Wallace-Durbin, P. 1954. The metabolism of fluorine in the rat using F-18 as a tracer. *J. Dent. Res.* 33:789-800.
(62) Weddle, D.A. and Muhler, J.C. 1956. Frequency of fluoride administration in relation to skeletal storage. *J. Dent. Res.* 35:65-68.

(63) Willis, J.B. 1961. Determination of calcium and magnesium in urine by atomic absorption spectroscopy. *Analyt. Chem.* 33:556-559.
(64) Zipkin, I. and McClure, F.J. 1952. Deposition of fluorine in the bones and teeth of the growing rat. *J. Nutr.* 47:611-620.
(65) Zipkin, I., McClure, J.F., Leone, N.C. and Lee, W.H. 1958. Fluoride deposition in human bones after prolonged ingestion of fluoride in drinking water. *Public Health Report* 73:732-740.
(66) Zipkin, I., Posner, A.S. and Eanes, E.D. 1962. The effect of fluoride on the x-ray diffraction pattern of the apatite of human bone. *Biochim. Biophys. Acta* 59:255-258.

–14–
Nutrition Related Diseases of the Aged

Alfred H. Lawton

INTRODUCTION

Nutrition related diseases usually have not been considered as forming a unique combination for medical specialization. Neither are these diseases grouped in textbooks of medicine or of geriatrics. On the other hand, diseases that have nutritional components in their etiologies, that have nutritional complications in their courses, or that have nutritional measures useful in their therapies occur in all specialties of medicine.

There are no nutrition related diseases that are limited to those humans who are considered to be aged. No nutrition related disease is fully characteristic of the aging process. No nutrition related diseases nor any collection of nutrition factors have been identified as causative of nor preventive of the human aging process.

Many diseases known to have nutrition components manifest more different characteristics when the diseases involve aged subjects than do the same diseases when they affect more youthful hosts. It is the body of knowledge about nutrition related diseases of the aged that will be considered in this chapter.

DIABETES MELLITUS AS A REPRESENTATIVE DISEASE

Diabetes mellitus frequently appears to cause premature aging in its victims.[20] For this reason, the disease has often been studied by those seeking to find the cause of the aging process. This aspect is not to be considered, rather diabetes mellitus is being reviewed here as a disease that has well-known nutritional factors and manifests very different characteristics and follows a very different course when it occurs in an elderly patient than it does when it involves a younger person.

Diabetes mellitus is a pleomorphic disease and it can involve any or all of the organ systems and tissues. The pathologic complications of diabetes are many, varied, and are often more serious than diabetes itself. As a consequence, definitions of the disease become descriptive statements either covering the more common signs and symptoms of diabetes or describing the various facets of the physiopathology of the condition.[46]

Diabetes mellitus occurring in youths is usually sudden in onset, brittle in its course, and often progresses rapidly to complications such as acidosis, coma, and other serious metabolic derangements. The disease tends to have a more insidious onset and course in older persons. As a result, the diagnosis is often delayed or missed and, if the wrong diagnostic criteria are used, diabetes may be diagnosed too frequently among aging subjects.

It is generally accepted that glucose tolerance decreases[19] and fasting and postprandial blood glucose levels rise progressively with age although a few investigators disagree.[48] Many elderly patients have hyperglycemia without glycosuria and these individuals are frequently not suffering from textbook diabetes. The standard diagnostic criteria for the diagnosis of diabetes mellitus cannot always be applied rigidly in medical studies of aged patients.[36]

The diet of aging persons changes for many reasons. These include subtle alterations in the senses of taste and of smell which ultimately leave the taste sensation of sweetness one of the few remaining positive pleasure factors accompanying eating so that the diet becomes filled with sugars. Other age related dietary changes may be brought about because of the loss of teeth, by bad habits of eating, by economic pressures limiting food choices, by loneliness and other stresses of living, and by complicating medical conditions. Unfortunately, these changes all seem to bring about the intake of foods high in carbohydrates at the expense of other nutriments in the diets of the elderly.

In persons of all ages, the more common presenting signs and symptoms of diabetes mellitus are generalized weakness, loss of weight in spite of satisfactory food intake, increased thirst, increased urinary output and hyperglycemia.[8] On the other hand, the onset of diabetes in the elderly may be heralded by blurred vision, pruritus vulvae, or some other symptom complex. Medical and laboratory studies of elderly patients presenting themselves with various neurologic syndromes including peripheral neuritis, with various acute abdominal symptoms, with infectious disease processes, with congestive heart failure, with pulmonary disease, with renal complications, or in coma may reveal that the basic disease process is diabetes mellitus.

The best known nutritional disturbance of diabetes is that of an abnormal carbohydrate metabolism. Not only is the carbohydrate metabolism deranged but it is accompanied by an altered fat metabolism. This latter derangement is thought to be one of the causative factors for the enhanced development of atherosclerosis which accompanies diabetes.[14,3,24] Abnormal fat metabolism also alters hepatic lipid production, bile acid synthesis, and may be involved in the formation of pigmented gallstones. Derangement of the metabolic pathways for the utilization of glucose, potassium, and insulin by diabetes in turn produces myocardial metabolic changes manifested by electrocardiographic abnormalities.[52]

Mineral metabolism is disturbed in diabetics. As plasma glucose and cholesterol levels rise magnesium and sodium levels fall, the magnesium being lost in the urine.[26,51] Chromium has been found to be integral in the glucose tolerance factor[45] so this metal must be involved in carbohydrate metabolism. Since both healthy elderly subjects and aged diabetics have decreased tissue chromium concentrations as reflected by their urinary excretion patterns[9] these investigators suggest that the diabetic patients have lost their ability to convert chromium into glucose tolerance factor.

CHRONIC DISEASE, AGE, AND NUTRITIONAL FACTORS

The above glimpse at a few of the nutritional elements involved in diabetes mellitus serves to introduce the fact that many chronic diseases have nutritional components. Simply because they have lived longer lives, the aged have had the opportunity to become the victims of one or more chronic diseases. So, a consideration of chronic diseases is usually also a consideration of the diseases of the elderly, i.e., geriatrics. Diabetes mellitus, atherosclerosis, obesity, biliary tract disease, cirrhosis, iron deficiency anemia, pernicious anemia, osteoporosis, and pressure and stasis ulcers are among the nutritionally related diseases of the aging. See Table 1.

Table 1: Chronic Diseases with a Dietary Component

Disease	Nutritional Factor
Anemia	Minerals, folate, vitamin B_{12}
Atherosclerosis	Carbohydrate, fat, vitamins, minerals, contaminants
Bone disorders	Protein, minerals, vitamin D
Cirrhosis	Ethanol, carbohydrate, minerals
Diabetes mellitus	Carbohydrate, minerals
Obesity	Calories
Renal disorders	Protein, fluids, electrolytes, urates
Ulcers of skin	Zinc
Vascular disorders	Minerals

Source: Modified from Jernigan.[27]

Nutrition Related Diseases of the Aged

Most gastrointestinal diseases have nutritional components either in their etiology, their complications, or both. Most gastrointestinal diseases occur at higher rates and produce more symptoms in older subjects than in patients of any other age group. These symptoms and diseases, in turn, are among the factors that often lead the elderly to further modify their dietary patterns.[56]

Malnutrition, by itself, is infrequently a problem among the elderly. When it is present in highly industrialized societies, the malnourishment is much more apt to be obesity than it is starvation. Undernourishment, as opposed to malnutrition, is fairly common among those who are extremely old,[58] particularly those suffering from dementia or advanced senility. Frank failure to take adequate carbohydrate, fat, or protein in the diet or actual starvation is a rare cause of nutritionally related disease of the aging when other intercurrent factors are not present. Most of the malnutrition of the elderly is failure to maintain adequate intake or failure of metabolic utilization of vitamins, minerals, amino acids and other specific food factors required for a complete and nutritional diet.

The factors which may lead to inadequate nutrition in the aging are many and varied. These include apathy, loneliness, physical disability, poverty, unsuitable living conditions, drugs, alcohol, inadequate dietary knowledge, mental deterioration, malabsorption, and disease.[57] Malnutrition and undernourishment need to be carefully differentiated, for an increasing number of diseases are being found to relate to dietary variables resulting from abnormal diets.[54]

Malnutrition presents an added risk when surgical procedures are required. Undernourishment with its advanced stage of cachexia will often result in rapid postoperative deterioration with death in a few days even though the patient at first appeared to withstand the surgical procedure well.[21] Protein deficiency, caloric lack, or a combination of these two factors always increase surgical risk but these states of malnourishment are sufficiently responsive to dietary supplementation with carbohydrate, protein, vitamins, minerals and fluids so that a patient can be adequately prepared to withstand the needed operation.

Obesity is definitely a nutritionally related disease. Most adults tend to overeat and the excesses are stored as fat. Even those who are aged and state that they have only maintained their body weight constant since young adulthood are storing fat for it is known that with age muscle and tissue mass shrinks and is replaced by fat if food intake remains constant. The overabundance of stored fat predisposes to a variety of illnesses including diabetes and arteriosclerotic related diseases such as coronary occlusion with myocardial infarction and cerebrovascular accidents.

It has been demonstrated that there is a significant decrease in average weight of successive five year age groups between the ages of 65 and 94 years.[35] At least a part of this decrease in average weights is caused by a progressive reduction in the numbers of overweight persons because of the relatively high mortality rate for the obese when compared with lean individuals. This differential in life expectancy is not surprising for excess fatty tissue produces increased bodily stresses such as the increased workload placed upon the heart by the network of extra capillaries needed to supply blood to redundant adipose tissue.

In general, healthy elderly individuals show no loss of efficiency in the processes of digestion and absorption.[49] Yet, metabolic processes may go awry. For example, the osteoporosis which accompanies aging may lead to hypercalcemia and this becomes a factor in the formation of calcium gallstones[5] and the accompanying hypercalciuria may be a factor in the production of urinary tract stones.[44]

Diarrhea and constipation are symptoms, not diseases, but both should be considered because both have nutritionally related factors. These may be etiologic or may be complications following the diarrhea or constipation. The causes of diarrhea differ little between various age groups. Moderately prolonged diarrheas may become serious because of the resulting complications, chiefly deficiencies in one or more of the essential food factors, including the vitamins and minerals. Among the elderly, the sudden onset of a prolonged diarrhea should immediately lead to detailed medical, x-ray, and laboratory studies to ascertain the cause so that specific treatment can be instituted. Too often, some forms of cancer will be found to be causing the diarrhea and frequently these have already advanced so far that only paliative therapeutic measures can be utilized.

Constipation sometimes seems to be ubiquitous among the elderly for the present group of aged persons grew up in an era of poor bowel habits, highly purified diets, inadequate fluid intake, and habituation to laxatives. The laxative habits resulted from an accumulation of misinformation including stress on the idea that a day without a bowel movement could lead to all manner of dire results and that the only way to avoid these disastrous events was to take a certain laxative.

Prolonged constipation is difficult to treat and is itself of little danger. The addition of roughage and liquids to the diet, the establishment of good bowel habits, and the weaning of the patient from laxatives is usually sufficient to slowly produce improvement. The sudden onset of constipation is a more serious matter for its abrupt appearance can herald the onset of underlying serious disease. Like diarrhea, this new symptom, or an alternation of diarrhea and constipation should lead to a detailed examination to elucidate the cause and institute appropriate treatment.

Aging combined with prolonged or heavy alcoholic ingestion can produce hepatic disease[30] and cirrhosis. Patients with alcoholic related cirrhosis have been found to have low serum and liver zinc and excess secretion of zinc in the urine.[53] It is not known whether this abnormal zinc metabolism is related to the etiology of the disease or that it is a zinc deficiency resulting from the alcohol ingestion or the cirrhosis. It is more commonly known that high serum copper, high ceruloplasmin copper concentrations, high liver and brain copper and high urinary copper excretion are found in patients afflicted with hepatolenticular degeneration or Wilson's disease.[2] The copper accumulation is somehow a factor in the development of the cirrhosis and the neurologic alterations that characterize this disease.

It has frequently been suggested that nutritional factors might be related to the central nervous system changes and mental illnesses that seem to be specific for aged people.[33,31,38] Recent observations[34] substantiate that there is relationship between nutritional factors and degenerative changes in the aging human brain. Minor changes in nutrient quality and quantity were observed to produce changes in certain brain neurotransmitters. These alterations, in turn, might change a variety of physiological responses mediated by the release of neurotransmitters. One particular chemical contained in the brain which increases after the ingestion of appropriate foods containing tryptophan is the neurotransmitter seratonin.[57] The investigators that established this information point out that the prolonged consumption of a diet deficient in protein could lead to a modification of the chemical composition of the brain and that this chemical change in turn could be responsible for behavioral and learning deficits.

There is suggestive evidence that tardive dyskenesia can be successfully treated by feeding diets rich in choline.[42] Tardive dyskenesia is an abnormal condition in which the patient manifests a variety of hyperkinetic movements. This abnormality frequently follows treatment of individuals with phenothiazine, butyrophenone, and possibly other classes of drugs. Low levels of the neurotransmitter, acetylcholine, are a finding in patients with tardive dyskenesia. The levels of acetylcholine can be raised by feeding the patients choline per se or by feeding them a diet high in lecithin which is a rich source of choline.

Other neurotransmitters are being studied. Many of these may be dependent on dietary constituents so deficits or excesses of certain nutritional substances may prove to be factors in diverse diseases. Depression and mania are two mental states under intensive study for neurohormone factors. This new area of research raises exciting possibilities that unexpected relationships will be discovered between nutrition, disease, and aging.

Thus far, specific diseases including diabetes mellitus, hypertension, atherosclerosis, iron deficiency anemia, and obesity have more responsibility for producing nervous system disease in the aged than any known chemicophysiologic alteration.[38] Stroke or cerebrovascular accident is the most common nervous system disease of the elderly and its origins arise in the above named disease states. All of these disease states in turn have their nutritionally related origins.

There has been prolonged controversy about the role of excessive calories, excess saturated fat, refined sugar, cholesterol, triglycerides, table salt, alcohol, caffein-containing beverages, or tobacco in the causation of athero- and arteriosclerosis and its special concomitants, stroke and myocardial infarction. The human aorta shows an increase in lipids and esterified cholesterol[47] and other human tissues show elevated levels of cholesterol and its esters with advancing age.[11,40] Unfortunately, increased cholesterol levels are not only related to advancing age but also appear to be related to sex, height, obesity, smoking, elevated blood pressure, exercise, stress, season of the year, and uric acid levels. Much information is still required before specific relationships between nutrition, age, and sclerotic vascular disease is understood. Probably several of these factors cooperate to produce the pathology of sclerotic blood vessels, stroke, and myocardial occlusion and infarction.

An epidemiologic study[16] suggested that cardiac risks from smoking are only manifest when associated with other risk factors. A study of autopsied males[50] indicated that atherosclerotic involvement of the aorta and the coronary arteries was greatest in heavy smokers and least in nonsmokers. Another epidemiologic study[17] concluded that cardiovascular mortality, including sudden death, was one result of the high consumption of alcohol. A fourth report[43] concluded that alcohol, cigarettes, and coffee used separately or used together in moderation did not disturb the heart in otherwise normal subjects but the author did suggest that in patients with an unexplained cardiac irritability or an arrhythmia associated with mitral valve prolapse or hypertrophic subaortic stenosis the casual relationship to alcohol and tobacco should be sought and, if present, the use of these agents should be moderated. Many, many other studies of the possible role of alcohol, tobacco, caffein beverages in cardiovascular disease have been done with equally tangential relationships and resultant confusion about these pseudodietary factors in causing sclerotic diseases for the aging. At this point, it can only be assumed that moderation in most nutritional, dietary, and pseudodietary factors is helpful for those who aspire to live long relatively free of disease.

Even water is possibly a dietary factor in cardiovascular disease. A study of water supplies and coronary heart disease[41] indicated that deaths from coronary heart disease were associated with low chromium and copper levels in the water supplies. Calcium and magnesium values appeared to have no relationship to the coronary heart disease. This report seems to continue an old argument about the relative values of hard versus soft water in the etiology of vascular scleroses.

It was mentioned earlier that increased uric acid levels bore a relationship to the increase of cholesterol levels with age. However, there is little to differentiate gout and purine malmetabolism occurring in an elderly patient from that which occurs in a youth. Gout is a frequent associate with other chronic diseases such as obesity, atherosclerosis, and diabetes mellitus. The elderly can develop hyperuricemia following courses of diuretics for either cardiac or renal disease.[4] Gout and the hyperuricemic states which accompany polycythemia and the myeloproliferative disorders may lead to the formation of uric acid renal stones.[44]

The development of calcium renal stones in the elderly may be related to the increased calcium absorption accompanying vitamin D intoxication or appear in milk-drinkers syndrome. The appearance of hypercalciuria and/or of renal stones should raise a suspicion of the presence of some condition causing dissolution of bones. These conditions include hyperparathyroidism, multiple myeloma, metastatic cancer, Paget's disease, chronic pulmonary disease, renal tubular acidosis, osteoporosis, long time consumption of corticosteroid drugs, or immobilization.[44] Only vitamin D intoxication, milk-drinkers syndrome, osteoporosis, and renal tubular acidosis have well-known nutritional etiologic relationships.

Advancing age involves a shift in the distribution of mineral salts in the body.[39] Calcium is a mineral in which this age variation is prominent. This shift of calcium may be manifest as osteoporosis. This is especially apparent in otherwise normal elderly women in whom the bone loss begins with the menopause. In the extreme elderly of both sexes, osteoporosis of some degree is an almost constant finding. Senile osteoporosis may be directly nutritionally related, for a diet deficient in calcium which has extended over many years can result in osteoporosis. In addition, the consumption of highly acidic diets can increase calcium and hydroxyproline in the urine. This loss is at the expense of the skeleton and enhances osteoporosis.[55,12] but the mechanisms responsible for this interrelationship have not been elucidated.

Vitamin D is an important factor in bone physiology. It is essential in the metabolic pathways of parathormone and of calcium. A long-standing dietary deficiency of vitamin D, especially when the patient is simultaneously exposed to sunlight,[1] can cause bone demineralization. This loss of mineral salts leaves a demineralized matrix and results in the disease, osteomalacia, which is a very different condition from osteoporosis.[55]

Osteoporosis is a major factor making the elderly susceptible to compression fractures of the vertebrae, femoral fractures, fractures of the long bones, and fractures of the flat bones of the pelvis.[39] In a sense, these fractures can be classified among the nutritional related diseases of the aged.

It has been stated that the best predictors of mortality in the aged of both sexes are higher age and proteinuria. Among males, low vitamin C intake was a significant predictor of mortality.[23] Scurvy has become a rare disease among young adults but it is increasingly present, in a subclinical or prescorbutic form, among the aged and especially in aged widowers.

Vitamin A levels tend to fall with advancing age largely as the result of limited dietary intake. As a result, nyctalopia or night blindness, is more frequently discovered in aged patients. This condition plus the development of cataracts and other chronic eye diseases causes the loss of visual acuity and explains the inability of many older persons to walk or to drive a car in the dark.

In addition to age related loss of vitamins A, C, and D, vitamin K deficiency among the elderly has been reported.[22] Absorption of the fat soluble vitamin requires the presence of bile so certain liver diseases or obstructive jaundice can result in a serious deficiency of this vitamin. Vitamin K levels may be low in patients manifesting malabsorption syndrome, being maintained on long-term antibiotics therapy, or receiving long-term anticoagulant treatment.[25]

There is also a slow decline in vitamin B_{12} levels with advancing age.[18,37] In spite of this, few new cases of pernicious anemia are diagnosed among the elderly. It is possible that these lowered vitamin B_{12} levels may bear some causal relationship to the diminishing of proprioceptive senses that results in gait abnormalities and difficulty in maintaining balance observed in the elderly. The oral administration of vitamin B_{12} will raise serum vitamin B_{12} temporarily but does not relieve any of the symptoms and the vitamin B_{12} values slowly return to pretreatment levels when the supplementation is stopped.[13] Peripheral vascular thrombosis may be a complicating factor in aged patients being treated with vitamin B_{12} for pernicious anemia or any other condition. The thromboses usually follow the reticulocyte peak that the treatment brings about.[27]

There are conflicting reports about the possible roles of vitamins either causing or being therapeutically beneficial for neoplastic diseases. A recent review of the relationships between vitamins and cancer[15] indicates that no final recommendations can be made relative to vitamins, antivitamins, and cancer but raises hope that dietary manipulation will ultimately minimize the incidence of metastatic disease.

Growing old is accompanied by an increasing incidence of neoplastic disease. Presumably everyone could ultimately have cancer if they lived long enough. Certainly cancer increases morbidity and mortality for the aged. The complications produced by the appearance and spread of cancer are many and varied. Unchecked cancer ultimately will produce cachexia and will terminate in death. The cachexia of advanced carcinoma results in anorexia, weakness, weight loss, and alterations in the body's immune system similar to the symptoms manifest in persons suffering from advanced malnutrition or from starvation.[10] In part, the cachexia of malignancy is because of poor food intake, involvement of some part of the gastrointestinal tract, or other interruption of normal metabolic pathways. However, there are additional nutritional abnormalities complicating advanced metastatic disease not evident in simple malnutrition or starvation. Many of the therapeutics measures utilized in cancer treatment by themselves can cause or intensify the cachexia of malignancy.

Numerous dietary and pseudodietary factors have been accused of being cancer-causing. The intake of coffee, tea, alcohol, tobacco, diets high in either animal fat, protein, red meat, or carbohydrates, and diets either high or deficient in various vitamins or minerals have all received attention as possible causes of cancer. Presently, the contamination of foodstuffs by chemical additives or by environmental pollutants is the popular nutritionally-related area being searched for carcinogens. A single example will suffice to indicate the confusion abounding in relating nutrition and cancer. Selenium has been blamed for the rise in the numbers of human cancer[32] and has been found to be present in high levels in the soil of those areas of the United States having very low human cancer mortality.[6]

Many metals, present in the body in trace amounts, are suspected in their deficiencies or by their excesses to have a role in causing one or another disease. It is accepted that cobalt is a factor in vitamin B_{12} metabolism and that copper, along with iron, is a factor in hemoglobin synthesis. These metals in deficient amounts or misutilized by the body can cause specific anemias.

Deficiency of zinc is suspected as being a factor in the development of pressure and stasis ulcers which complicate many chronic diseases. The metal seems to have been efficacious in treating such ulcers among aged patients. Zinc and chromium both have been suggested as etiologic agents in vascular disorders and chromium is mentioned as a suspect in the etiology of diabetes mellitus. Silica is being studied for its role in bone metabolism. Much is yet to be learned about the interrelationship of metals, nutrition, metabolism, chronic diseases, and aging.

Low nutrient intake results in energy lack and in turn these have been associated with provoking mental illness in women and causing dyspnea and deafness in men.[33] Low potassium in the diets of older persons can leave them vulnerable to frank potassium deficiency if they become afflicted with an acute disease like influenza or with chronic diseases like cardiac failure or if they have to receive diuretics as a therapeutic measure. Potassium deficiency in the elderly, alone, may result in muscular weakness, paralysis, hypotension, ileus, and various cardiac and pulmonary abnormalities so that potassium deficiency can be mistaken for any one of a number of chronic diseases.

CONCLUDING REMARKS

It is evident that there are many diseases that occur among the elderly that have nutritionally related components. It is even more apparent that there is confusion, misinformation, and lack of knowledge about nutrition related diseases of the aged. It is, therefore, a challenging area for careful observation and for laboratory study.

The current status is indicated in the results of a comparison of senators elected between 1789 and 1860 with those elected between

1931 and 1966 which revealed that modern nutritional advances, scientific progress, and therapeutic advances had added only one-half year to their life expectancy.[7]

If a one sentence summary was offered it would be a paraphrase of Johnson in *Human Paleopathology*.[28] This statement would be that there is no precise one-to-one relationship between any etiology and any disease process, but disease is a geometric function of the patient's combined physical, pathological, chemical, and nutritional state.

In the aged, diseases are seldom single but are usually multiple. Their diseases tend to be chronic, persistent, and progressive. Their diseases arise from multiple etiologic factors, which factors are not limited to physical pathology, anatomic alterations, chemical malfunctions, and metabolic processes but also include social pressures, psychological problems, economic deficiency, and religious confusion.

REFERENCES

(1) Anderson, W.F. 1974. Preventative aspects of geriatric medicine. *J. Amer. Geriat. Soc.* 22:385-392.
(2) Bearn, A.G. 1953. Genetic and biomedical aspects of Wilson's disease. *Amer. J. Med.* 15:442-449.
(3) Bennion, L.J. and Grundy, S.M. 1977. Effects of diabetes mellitus on cholesterol metabolism in man. *New Eng. J. Med.* 296:1365-1371.
(4) Bienenstock, H. and Fernando, K.R. 1976. Arthritis in the elderly: an overview. *Med. Clin. No. Amer.* 60:1178-1189.
(5) Boudher, I.A.D. 1976. Gallstones. *Br. Med. J.* 2:870-872.
(6) Broghamer, W.L., et al. 1976. Relationship between serum selenium levels and patients with carcinoma. *Cancer* 37:1384-1388.
(7) Calloway, N.O. 1974. Modern nutrition adds only six months to life. *Med. Trib.* p. 2, Apr. 3, 1974 citing from *Amer. Fam. Physician* 10:146.
(8) Campbell, I.W. Non-fatal keto-acidosis in a 94 year old diabetic patient. *J. Amer. Geriat. Soc.* 22:473-474.
(9) Canfield, W.K. and Doisy, R.J. 1976. Chromium and diabetes in the aged. In J.M. Hsu, R.L. Davis and R.W. Neithamer (eds.), *The Biomedical Role of Trace Elements in Aging*. Eckerd College, St. Petersburg, FL.
(10) Costa, G. and Donaldson, S.S. 1979. Current concepts in cancer: Effects of cancer and cancer treatment on the nutrition of the host. *New Eng. J. Med.* 300:1471-1474.
(11) Crouse, R.J., Grundy, S.M. and Ahrens, E.H. Jr. 1976. Cholesterol distribution in the bulk tissues of man: Variation with age. *J. Clin. Invest.* 51:1292-1296.
(12) Daniel, H.W. 1976. Osteoporosis in the slender smoker. *Arch. Int. Med.* 136:298-304.
(13) Davis, R.L., Lawton, A.H., Prouty, R., et al. 1965. The absorption of oral vitamin B_{12} by an aged population. *J. Geront.* 20:169-172.

(14) DeLang, D.J., Vivier, F.S. and Davis, W.H. 1965. Observations on carbohydrate and fat metabolism in different age and racial groups. *S. Afr. Med. J.* 39:119-122.
(15) DiPalma, J.R. and McMichael, B.S. 1979. The interaction of vitamins with cancer chemotherapy. *Cancer* 29:280-286.
(16) Dick, T.B.S. and Stone, M.C. 1973. Prevalence of three cardinal risk factors in a random sample of men and in patients with ischemic heart disease. *Brit. Heart J.* 35:381-385.
(17) Dyer, A.R., Stamler, J., Paul, O., et al. 1977. Alcohol consumption, cardiovascular risk factors and mortality in two Chicago epidemiologic studies. *Circulation* 56:1067-1074.
(18) Gaffney, G.W., Horonick, A., Okuda, K., et al. 1957. Vitamin B_{12} serum concentrations in 528 "healthy" human subjects of ages 12-94. *J. Gerontol.* 12:32-38.
(19) Hadnagy, C., Horvath, E., Elekes, I., et al. 1964. Effect of vitamin B_{12} and cortisone on the glucose tolerance of aged persons. *Gerontologia* 9:71-77.
(20) Hamlin, C.R., Kohn, P.R. and Luschin, J.H. 1975. Apparent accelerated aging of human collagen in diabetes mellitus. *Diabetes* 24:902-904.
(21) Harken, D.E. 1977. Malnutrition: A poorly understood surgical risk factor in aged cardiac patients. *Geriatrics* 32:83-85.
(22) Hazell, K. and Baloch, K.H. 1970. Vitamin K deficiency in the elderly. *Geront. Clin.* 12:10.
(23) Hodkinson, H.M. and Exton-Smith, A.N. 1976. Factors predicting mortality in the elderly in the community. *Age & Aging* 5:110-115.
(24) Houcke, E., Houcke, M. and Leblois, J. 1964. Histologic study of the senile pancreas. *Presse Med.* 72:1887-1892.
(25) Hussain, S. 1976. Disorder of hemostasis and thrombosis in the aged. *Med. Clin. N. Amer.* 60:1273-1287.
(26) Jackson, E.C.H. and Meier, D.W. 1968. Routine serum magnesium analysis. *Ann. Int. Med.* 69:743-748.
(27) Jernigan, J.A. 1979 (December). Personal communication.
(28) Johnson, L.C. 1976. The principle of structural analysis. In S. Jarcho (ed.), *Human Paleopathology*. Yale Univ. Press, New Haven, CT.
(29) Klemetti, L. 1964. Is the vitamin B_{12} treatment of pernicious anemia a predisposing factor for thromboses in aged persons? *Acta. Med. Scand.* 176:121-122.
(30) Kramer, K., Kuller, L. and Fisher, R. 1968. The increasing mortality attributed to cirrhosis and fatty liver in Baltimore (1957-1966). *Ann. Intern. Med.* 69:273-282.
(31) Krehl, W.A. 1974. The influence of nutritional environment on aging. *Geriatrics* 29:65-76.
(32) Lipkin, M. 1979. Dietary, environmental, and hereditary factors in the development of colorectal cancer. *Cancer* 29:291-299.
(33) Lonergan, M.E., Milne, J.S., Manle, M.M., et al. 1975. A dietary survey of old people in Edinburg. *Brit. J. Nutr.* 34:517-527.
(34) Lytle, L.D. and Altar, A. 1979. Diet, central nervous system, and aging. *Fed. Proc.* 38:2017-2022.
(35) Master, A.M., Lasser, R.B. and Beckman, G. 1959. Analysis of weight and height of apparently healthy populations, ages 65 to 94 years. *Proc. Soc. Exp. Biol. Med.* 102:367-370.
(36) Moss, J.M. 1976. Pitfalls to avoid in diagnosing diabetes in elderly patients. *Geriatrics* 31:52-55.

(37) Nyberg, W., Eriksson, A., Forsuis, H., et al. 1964. Serum vitamin B_{12} levels in an isolated island population. *Acta. Med. Scand.* 127:79-82.
(38) Ostfeld, A.M. 1976. Nutritional aspects of stroke, particularly in the elderly. In M. Rockstein and M.L. Sussman (eds.), *Nutrition, Longevity, and Aging*. Academic Press, New York.
(39) Piersol, G.M. 1943. Medical considerations of some geriatric problems. *Arch. Ophth.* 29:26-35.
(40) Pincheile, G. 1971. Factors affecting the mean serum cholesterol. *J. Chron. Dis.* 24:289-297.
(41) Punsar, S. 1976. Research on water quality and CHD in two Finnish male cohorts. Hardness, Drinking Water. *Public Health. Proc. Eur. Sci. Colloq.* 359-368, 1975. Comm. Eur. Communities, EUR.
(42) Rawls, R.L. 1978. Diet can influence the functioning of the brain. *Chem. Engr. News* 56:27-30.
(43) Regan, T.J. 1979. Of beverages, cigarettes, and cardiac arrhythmias. *New Eng. J. Med.* 301:1060-1061.
(44) Rosen, H. 1976. Renal disease in the elderly. *Med. Clin. No. Amer.* 60:1105-1119.
(45) Schwartz, K. and Mertz, W. 1959. Chromium III and the glucose tolerance factor. *Arch. Biochem. Biophys.* 85:292-295.
(46) Shagan, B.P. 1976. Diabetes in the elderly patient. *Med. Clin. No. Amer.* 60:1191-1208.
(47) Smith, E.B. 1965. The influence of age and atherosclerosis on the chemistry of aortic intima. *J. Atherosclerosis Res.* 5:224-240.
(48) Smith, L.E. and Shock, N.W. 1949. Intravenous glucose tolerance tests in aged males. *J. Gerontol.* 4:27-33.
(49) Southgate, D.A.T. and Durnin, J.V.G.A. 1970. Caloric conversion factors: An experimental reassessment of the factors used in the calculation of the energy value of human diets. *Brit. J. Nutr.* 24:517-535.
(50) Strong, J.P. 1976. Cigarette smoking and atherosclerosis in autopsied men. *Atherosclerosis* 23:451-476.
(51) Stutzman, F.L. and Amatuzio, D.S. 1953. Blood serum magnesium in portal cirrhosis and diabetes mellitus. *J. Lab. Clin. Med.* 41:215-219.
(52) Timaffy, M. 1974. Electrocardiographic changes during intravenous loading in the aged. *J. Amer. Geriat. Soc.* 22:429-431.
(53) Vallee, B.L., Wacker, W.E.C., Bartholomay, A.F., et al. 1957. Zinc metabolism in hepatic dysfunction. *New Eng. J. Med.* 257:1055-1065.
(54) Weindruch, R.H., Kristie, J.A., Cheney, K.E., et al. 1979. Influence of controlled dietary restriction on immunologic function and aging. *Fed. Proc.* 38:2007-2016.
(55) Wheeler, M. 1976. Osteoporosis. *Med. Clin. No. Amer.* 60:1213-1224.
(56) Wilson, R.W. 1974. Prevalence of selected chronic digestive conditions— United States, July-December 1968. *Vital Health Stat.*, Series 10, No. 83.
(57) Wurtman, R.J. and Fernstrom, J.D. 1974. Effect of diet on brain neurotransmitters. *Nutr. Rev.* 32:193-200.
(58) Young, V.R. 1979. Diet as a modulator of aging and longevity. *Fed. Proc.* 38:1994-2000.

–15–
Dietary Fiber and the Aging Processes

Robert F. Borgman

INTRODUCTION

Often we consider many of the degenerative diseases in our industrialized society, i.e., cardiovascular disease, stroke, neoplasms and diabetes, as part of the aging process. Since these diseases are less common in comparable age groups in the less developed societies, there is some question of whether they should be considered in the true aging process.

Certainly, the aging process and development of degenerative diseases can be hastened by environmental factors, life-style and diet. The excessive use of alcohol and tobacco, stresses of urban life, environmental pollutants, gluttony and dietary imbalances have known life-shortening influences.

The diet of man has changed markedly the few short millennia that man has dominated this planet. Until a few centuries ago, man was probably primarily an ovo-lacto-vegetarian. More recently in the industrialized societies, he has changed to a more carnivorous type of diet which is relatively purified, nonfibrous, and high in caloric density. In this relatively short period, the questions must be asked, "Has man had the time to adjust for this radical dietary change? Is this lack of adjustment responsible for the high incidence of degenerative diseases in our industrialized societies?"

Since the incidence of degenerative diseases in the industrialized societies has increased with time, and since these diseases are less common in the undeveloped societies, dietary changes and environmental differences should contain some clues as to the etiologies of these diseases. The amount and type of dietary fiber in various societies is one of several logical topics of investigation.

In general, the amount of dietary fiber in the past was much greater than now in the industrialized societies. Dietary fiber is presently much greater in the rural societies in the undeveloped world than in the industrialized societies. The daily intake of fiber approximates 5 grams in the United States and most other industrialized countries, whereas in rural Africa the level is about 25 grams per day. Dr. D.P. Burkitt, an English physician, has pioneered investigations on the levels of dietary fiber in rural Africa vs industrialized societies and has compared incidences of certain degenerative diseases with most interesting results.

The etiologies of degenerative diseases are generally recognized to be multicasual situations, and the diet is only part of the picture. Furthermore, the dietary constituents all have some role in the etiology of degenerative diseases. A diet containing appreciable amounts of dietary fiber must also have less purified carbohydrate, protein and fat than a diet with a small amount of fiber. This raises a logical question, "Is a low incidence of a particular degenerative disease in a society the result of sufficient amounts of dietary fiber or the result of less of the other dietary constituents mentioned above?"

Considering the physiological and biochemical effects of dietary fiber, there appears to be logical reasons why it can be effective in the prevention of certain degenerative diseases. However, the role of dietary fiber must be considered in the matrix of all of the causative factors. Increasing the level of dietary fiber must not be considered a panacea for the ills of mankind.

CHEMICAL AND PHYSIOLOGICAL PROPERTIES OF DIETARY FIBER

The term "dietary fiber" is actually a catchall for a number of materials which are not absorbed from the digestive tract and therefore have no nutritive value for man. These materials may be altered in the tract by the action of the intestinal microflora. Older terms for dietary fiber included nonnutritive bulk, roughage, undigestible dietary materials, and undigestible carbohydrates and lignin.

The sources of dietary fiber are almost exclusively plant materials which form the cell walls and supportive structure in the plants, except for the plant gums. The milling of grains removes much of the fiber. Animal food products contain very little dietary fiber. Nutrient materials such as sucrose, fats and alcoholic beverages contain no fiber.

Chemical determinations of foods can reveal the contents of the various types of fiber, and these analyses can predict the actions within the digestive tract.[31] The chemical nature of fibers and their ability to bind water and other materials have been reviewed by Anderson and Chen.[2] Much of this information is outlined in Table 1 below.

Table 1: Chemistry and Properties of Dietary Fiber

Fiber	Chemical Nature	Action	Ability to Produce Bulk	Complexing Ability for Bile Acids, etc.
Cellulose	Polymer of glucose about 3,000 units	Binds about 0.4 g water/g	Good	Insignificant
Methylcellulose	Similar to cellulose	Similar to cellulose	Good	Insignificant
Lignin	Phenylpropane polymers	Binds bile acids and other organic materials	Poor*	Good
Hemicelluloses	About 250 different types, polymers of pentose and hexose sugars	Binds water and cations	Good	Fair to Good
Pectins	Mainly polymers of galacturonic acids	Binds water, bile acids, and cations	Good	Good
Gums and similar plant materials	Branched polysaccharide polymers of a number of hexoses and pentoses	Binds water, bile acids and organic materials	Poor to Good*	Good
Silicates	Oxides of silicon	Binds cations and bile acids	None*	Good

*Usually only small amounts are present in foods, and therefore they produce little bulk.

It may be noted from Table 1 that fibers have the ability to produce bulk by binding water. Some fibers are also able to bind bile acids and other materials. Not every type of fiber has both actions. The water binding capacity retains water within the ingesta aiding in peristalsis. In the colon where considerable water absorption occurs, dietary fiber retains water in the rejecta and the effort needed for voiding the soft fecal mass is less than for voiding a hard fecal mass found when the dietary fiber level is low. This bulking action stimulates peristalsis making bowel transit time less and also dilutes any toxins present.

As noted in Table 1, certain of the dietary fibers have the ability to complex bile acids and other potentially toxic organic materials. This binding action tends to keep these materials away from the intestinal walls lessening the change of chemical insult and of reabsorption.

The binding of bile acids prevents reabsorption into the enterohepatic pathways. Since the bile acids are formed from cholesterol, the net effect is a reduction of the serum cholesterol level.

Considering these chemical and physiological effects of dietary fiber, a logical hypothesis can be made for the role of fiber in the prevention of many of the degenerative diseases generally associated with the aging process.

ROLE OF DIETARY FIBER IN PREVENTION OF DISEASES AFFECTING THE INTESTINAL TRACT

Constipation

The intestinal transit time is reduced by increasing the amount of dietary fiber. In Africa, Ugandan villages consuming about 25 grams of dietary fiber daily had transit times averaging 35 hours and stool weights averaging 470 grams per day. English students consuming approximately 5 grams of fiber daily had transit times averaging 69 hours and daily stool weights averaging 107 grams.[10] The decreased transit times and increased stool weights are attributed to the water binding capacity of the fiber which in turn increased bulk.[33]

In patients where constipation was found to be a problem, the addition of citrus pectin and cellulose to bread was helpful. These patients were confined to a care facility, and some were confined to bed or a wheelchair.[34] Citrus pectin can be added to moist baked goods without loss of flavor or texture.[3]

Diverticulitis

Diverticulitis, sometimes called "left-side appendicitis" is a common malady affecting the colon of the elderly. The pathogenesis of the disease is thought to result from a ballooning of the weak portions of the colon to form blind pockets which impede the flow of ingesta.

This disease has increased in incidence to become a major problem in the Western countries during the last 50 years as the level of dietary fiber has decreased.[10,17] This disease is almost unknown in rural Africa.

Surgery for diverticulitis is only partially effective, and there are approximately half a million operations per year for this malady in the United States. Many more cases are treated without surgery.[32]

The development of pouches in the colon is thought to result from increased intraluminal pressures which occur when hard fecal masses are propelled by strong peristaltic contractions. Increased dietary fiber increases bulk making the fecal material more fluid and more easily propelled without as great intraluminal pressures as with the hard fecal masses. This is illustrated in Figure 1.

Studies with increased levels of dietary fiber in patients with this disease yielded beneficial results. The addition of 20 grams of fiber alleviated the condition while reducing intraluminal pressures in the distal colon, shortening transit times, resulting in a better mixing of fecal solid and liquid phases and increasing stool weights.[15,19]

Figure 1

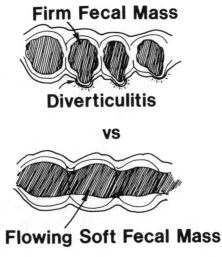

Firm Fecal Mass

Diverticulitis

vs

Flowing Soft Fecal Mass

When there is sufficient dietary fiber (bottom), water is bound resulting in a fecal mass which is soft in consistency. Relatively mild peristaltic contractions with modest increase in intraluminal pressure are sufficient to move the fecal material. With insufficient fiber (top), fecal masses are difficult to propel resulting in excessive intraluminal pressure. This pressure causes ballooning of weak portions resulting in blind pouches.

Appendicitis

Appendicitis is one of the most frequent causes of abdominal surgery in the United States, and yet this disease is virtually unknown in rural Africa. This disease became common in England after 1880.[10]

Insufficient dietary fiber levels are thought to have a role in the pathogenesis of appendicitis since many cases of this disease are the result of an impacted hard fecal mass in the appendix. This hard mass causes chronic irritation, and is the result of insufficient fiber in the diet, as illustrated in Figure 2.

Figure 2

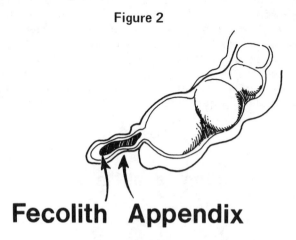

Fecolith   Appendix

A hard fecal mass is lodged in the appendix resulting in appendicitis.

Ulcerative Colitis

Ulcerative colitis is common in the Western countries, yet is uncommon in most rural undeveloped countries of the world. A probable cause of this disease would be the prolonged contact of irritating materials with the intestinal mucosa. Sufficient dietary fiber speeds transit of fecal material, dilutes toxins by bulk production, and binds toxins preventing prolonged contact with the mucosa.

Hemorrhoids

Hemorrhoids are not respectors of persons in the United States, affecting both presidents and laborers. This malady is present in half the persons over 50 in the United States but is uncommon in rural Africa.[10]

Hemorrhoids are varicose veins in the anal region and the cause appears to be a combination of straining during defecation to evacuate a hard fecal mass, and the mechanical irritation from this mass.

Dietary Fiber and the Aging Processes

These effects are illustrated in Figures 3 and 6. Sufficient fiber in the diet retains water in the fecal mass rendering the mass somewhat soft in consistency.

Figure 3

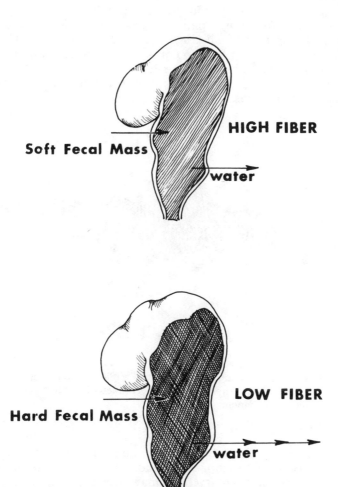

The soft fecal mass (top) resulted from sufficient dietary fiber preventing excessive water absorption in the colon. The hard fecal mass (bottom) resulted from insufficient water retention, and this mass would be more difficult to evacuate.

Colon Cancer

In the United States, colon cancer is a major problem, and follows lung cancer as the second most fatal form of cancer. It caused 47,000 deaths in the U.S. in 1972.[32] In general, colon cancer is common only in societies where the mean daily fecal mass is lower than 150 grams.[7]

The role of dietary fiber in preventing colon cancer may be one or more of the following: binding carcinogens thereby preventing contact with the colon wall (Figure 4); decreasing transit time thus lessening the time of carcinogen contact with the colon wall; increasing the stool volume subsequently reducing carcinogen concentration (Figure 5); or altering the microflora in a manner which lessens the synthesis of carcinogens.

Figure 4

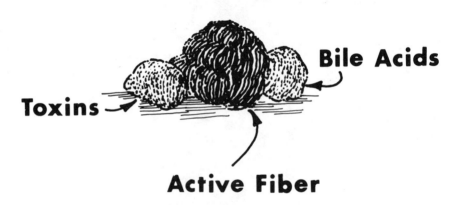

Certain dietary fibers have the ability to bind bile acids and toxins thereby lessening contact with intestinal walls and enhancing excretion.

Some workers[13,14] are of the opinion that dietary protein in excess may result in carcinogen production, i.e., nitrogenous wastes, in the colon and that dietary fiber may alter the process. However, when they added 31 grams of wheat bran to human subjects consuming high or low protein diets, they did not find fecal ammonia concentration was influenced by the dietary protein level, but increasing dietary fiber increased fecal weight and shortened transit time.

Some of the bile acids, particularly deoxycholic acid, have been suspected to be weak carcinogens. The amount of fecal deoxycholic acid excretion in English adults consuming low fiber diets was greater than that of Nigerian adults consuming a high fiber diet.[22]

Figure 5

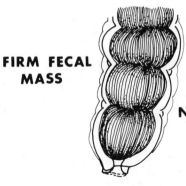

FIRM FECAL MASS — Slow movement, long contact of concentrated toxins with tissue = Neoplasms & Ulcerative Colitis

vs

SOFT FECAL MASS — Rapid movement and Short contact of diluted toxins with tissue = Less Disease

Dietary fiber increases bulk and binds water. When dietary fiber is insufficient, excessive amount of water are removed resulting in a hard fecal mass (top), while sufficient fiber holds water and results in a softer fecal mass with a lower concentration of toxic materials (bottom).

DISEASES ASSOCIATED WITH STRAINING DURING DEFECATION (HERNIAS AND VARICOSE VEINS)

Hard stools require more straining during evacuation, and thus the intra-abdominal pressure is excessive at that time resulting in hernias and varicose veins. Softer stools resulting when dietary fiber is sufficient do not require this amount of exertion during evacuation.

The incidence of varicose veins and hiatal hernias has increased in the Western world during the past century. These maladies are relatively rare in the rural areas of the undeveloped countries where the intake of dietary fiber is high.

The relationship of varicose veins to a low fiber intake was first reported by Cleave.[11] The rationale that dietary fiber has a role in the etiology of varicose veins is that straining excessively during defecation to evacuate hard stools results in excessive intra-abdominal pressure. This pressure prevents venous return and the pooling of blood eventually disrupts the valves which prevent back flow. Repeated insults to the veins in this matter results in permanent dilation, venous stasis and varicosity of the superficial and deep veins. Venous thrombosis may follow.[8] See Figure 6.

Figure 6

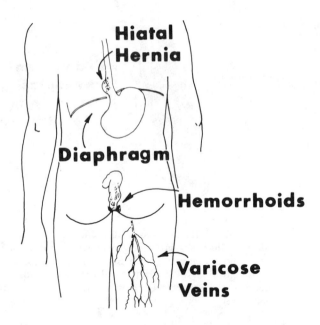

Straining During Defecation

Hemorrhoids (a type of varicose vein) and varicose veins in the legs result when the flow back to the heart is prevented for long periods of excessive intra-abdominal pressure. Increased pressure also forced a portion of the stomach through a weak portion of the diaphram resulting in a hiatal hernia.

Hiatal hernias are more common in societies where dietary fiber is insufficient, and the excessive intra-abdominal pressure during evacuation of hard stools is thought to contribute.[9] See Figure 6.

DISEASES ASSOCIATED WITH ATHEROSCLEROSIS (ISCHEMIC HEART DISEASE, STROKE, ETC.)

Over half of the deaths over age 45 in the Western world result from diseases associated with atherosclerosis.[33] This group includes diseases of the heart (ischemic heart disease, coronary artery disease, myocardial infarction and angina pectoris), stroke, aortic aneurism, and intermittent claudation.

While the exact role of cholesterol in the pathogenesis of atherosclerotic lesions is a matter of debate, it has long been observed that societies with low serum cholesterol levels and consuming high levels of dietary fiber have less incidence of diseases associated with atherosclerosis than found in the Western world.

Comparisons of serum cholesterol levels and dietary fiber intake in several societies have been made, and invariably those societies with high intakes have low levels.[20,21,27]

The hypocholesterolemic effects of plant fibers has been reviewed by Anderson and Chen.[2] In most human studies, all of the plant fibers had some effect, and the soluble fibers such as pectin, guar gum and bengal gum had the greatest hypocholesterolemic effect. Effects were noted from both purified fibers and the plant materials containing these fibers. In young men, Mathur et al.[28] found bengal gum to be particularly efficient in lower serum cholesterol levels. In healthy adults, Palmer and Dixon[29] found modest hypocholesterolemic effects from feeding 15 grams daily. Guar gum was given to patients with type II hyperlipidemia by Jenkins et al.,[23] and the serum triglyceride levels were unaffected but there was a modest reduction in the serum cholesterol levels.

Dietary fibers are effective in their native form within the plant foods. Carrots, containing a mixture of fibers, were fed at the rate of 200 grams daily for 3 weeks to normal adults. The serum cholesterol levels were modestly reduced concurrent with increased fecal weights, and increased excretion of bile acids and fats.[30] In another study, feeding a diet containing considerable amounts of various vegetable fibers in the native state to seven young American adults resulted in lower serum concentrations of cholesterol and triglyceride.[1]

The method whereby dietary fiber increases cholesterol removal and lowers serum cholesterol levels and ultimately reduces those diseases associated with atherosclerosis is thought to be by enhancement of the fecal excretion of bile acids. Since bile acids are synthesized from cholesterol, the net effect of increased excretion of bile acids would be the removal of cholesterol from the body. See Figure 7.

Figure 7

Entero Hepatic Pathway

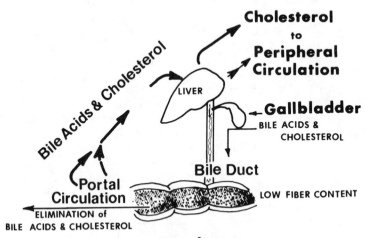

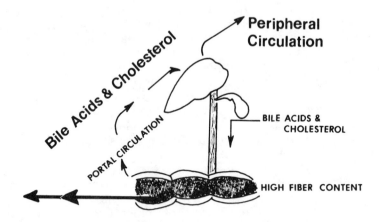

When dietary fiber is insufficient (top), less bile acids and cholesterol are excreted and more return to the liver via the enterohepatic pathway, and then to the peripheral circulation. When dietary fiber is sufficient (bottom), much of the bile acids are bound to the dietary fiber and fecal excretion of these acids and cholesterol is enhanced. This leaves less bile acids and cholesterol to be returned to the liver, and therefore less to enter the peripheral circulation.

Evidence of increased bile acid excretion and reduction of serum lipids by increasing dietary fiber has been found in humans. When six American males were changed from a Western type diet to a high fiber diet common in rural Guatemala or when oat bran was added to the Western type diet, increased excretion of bile acids resulted.[25]

There appears to be some influence of dietary fiber on the type of bile acids secreted, apparently by influence upon the intestinal microflora. Increasing the fiber content of experimental animal rations increased the ratio of the primary bile acids (cholic and chenodeoxycholic acids) to the secondary bile acids (deoxycholic and lithocholic acids). Since the secondary bile acids are formed by the intestinal microflora and enter the enterohepatic pathway, it may be assumed that the microflora were influenced.[26]

CHOLELITHIASIS

Gallstones are relatively common in the developed countries, and are a major cause of abdominal surgery in the United States. Since approximately half the people in their middle seventies have gallstones in the United States, a time when surgery is difficult and sometimes fatal, methods of prevention would seem most desirable. Gallstones are relatively rare in rural Africa where the content of dietary fiber is much higher than in the United States.

Eastwood[16] has reviewed the influence of dietary fiber upon gallstone formation both in man and in experimental animal models and he is of the opinion that fiber prevents stone formation by increasing fecal elimination of bile acids and by increasing the ratio of cholic and chenodeoxycholic acids to deoxycholic acid. Chenodeoxycholic acid is thought to keep cholesterol suspended in the bile more than deoxycholic acid thereby preventing cholesterol stone formation.

Rabbits formed gallstones when fed a purified ration containing considerable fat within 4 to 12 weeks.[6] When rabbits with formed stones were returned from the purified ration to a commercial rabbit ration, the stones disappeared.[4] The addition of pectin to a purified ration will cause formed stones to disappear.[5] Also, stone formation was prevented when rabbits were fed the purified ration with added pectin, and the serum cholesterol concentrations were lower in the pectin-fed group.[6]

Longitudinal studies of cholelithiasis in humans would be difficult, but if the information gathered in the animal models can be applied, it appears logical that increasing dietary fiber would lower the incidence of cholelithiasis in the aged.

OBESITY

Obesity is the scourge of the sedentary gluttonous Western

society. In the United States over 25% of the population is overweight or obese. In Africa, obesity is relatively rare in the rural areas but becomes more common as the population eats more Western type foods.[10] The role of dietary fiber in preventing obesity may be in reduction of the caloric density of foodstuffs. Foodstuffs containing sufficient fiber will produce a volume of food resulting in satiety without adding excess calories.

DIABETES MELLITUS

The role of dietary fiber in the prevention of diabetes mellitus is not clear, although the disease is more common in the Western countries than in the underdeveloped countries. Dietary fiber reduces the caloric density of foods, and Cleave et al.[12] have noted that the removal of fiber from foodstuffs may lead to overconsumption of food. Also, the refined foods are more rapidly absorbed than high fiber nutrients. Excess of refined cereals and sucrose may play a role in the etiology of diabetes mellitus.

Anderson and Chen[2] have reviewed the relationship of dietary fiber to diabetes. Some studies indicated that high fiber diets improved glucose tolerance in normal persons, while other studies did not. In diabetic patients, increasing the fiber content of the diet reduced postprandial glycemia, and in some patients less insulin dosage was required.

POSSIBLE PROBLEMS WITH INCREASING DIETARY FIBER

The average daily consumption of dietary fiber in the United States is about 5 grams, while in rural Africa it is 20 to 25 grams. Probably within this range there should be little detrimental effect upon any age group. However, it must be remembered that dietary fiber may lessen nutrient absorption by diluting the ingesta and speeding transit. An elderly person consuming a nutritionally marginal diet may develop a deficiency if fiber is consumed to excess, although the more logical step would be to improve the diet.

Humans have been fed diets high in plant fibers and compared to diets where less fiber was fed.[24] Daily balances of calcium, magnesium and iron were positive on the low fiber diet but negative on the high fiber diet.

Another complication which might be expected to result from a high fiber diet is flatulence. A direct relationship has been found between the level of dietary fiber and the amount of flatus in humans.[18]

SOURCES OF DIETARY FIBER

The common trend in modern nutrition is to seek a special

source of a specific nutrient, fortify a common food material, and generally utilize the specialty shops dealing with "food fad items." In the case of dietary fiber, this approach is completely unnecessary. Dietary fiber can be obtained economically and easily at the ordinary market place.

In seeking values for the dietary fiber content of various plant foods, one finds crude fiber and dietary fiber values. Usually, the dietary fiber values for a foodstuff are higher than are the crude fiber values because crude fiber is only the lignin and cellulose portion, whereas the dietary fiber includes pectins and hemicelluloses.

Dietary fiber values given in the literature are given both on a wet weight basis and a dry weight basis. The fiber content of lettuce may be listed as 1.5%, but on a dry weight basis this figure would be about 10 times this amount. Also, the consumption of lettuce would be considerable in a salad, and thus would make a valuable contribution to the daily fiber intake.

Most all plant foods make a contribution to the daily intake of fiber, except for purified materials such as white flour and other purified grains. Bran is particularly high in fiber content. Citrus materials, apples, peaches, bananas, pears and tomatoes are particularly high in pectin.

Perhaps, in suggesting ways to increase the dietary fiber levels, suggestions should be made to increase unrefined plant foods at the expense of refined plant foods and animal foods. In this manner, a variety of functional fibers would be ingested.

PROSPECTUS ON THE USE OF DIETARY FIBER TO PREVENT PREMATURE AGING

Increasing the level of dietary fiber in the Western world to prevent those degenerative diseases generally observed with the aging process in the developed countries should be viewed with other aspects of diet and life style. Many preventative measures are known, and many known measures have application in the early stages of degenerative diseases. Changes, such as increasing the level of dietary fiber, must be accomplished many years before the clinical phases of degenerative disease in order to have the optimal effect.

Some caution should be observed in a radical change in the dietary fiber level in the aged with a degenerative disease. A sudden increase in dietary fiber might be unwise in a patient with diverticulitis, and each case must be individually assessed.

REFERENCES

(1) Albrink, M.J., Newman, M.D., and Davidson, P.C. 1979. Effect on high and low-fiber diets on plasma lipids and insulin. *Am. J. Clin. Nutr.* 32:1486-1491.

(2) Anderson, J.W., and Chen, W.L. 1979. Plant fiber. Carbohydrate and lipid metabolism. *Am. J. Clin. Nutr.* 32:346-363.
(3) Borgman, R.F., Acton, J.C., Nesbitt, J.C. Jr., Patterson, M.Y., and Toler, J.E. 1977. The influence of citrus pectin upon appetite and hunger in humans. S.C. Agric. Exp. Sta. Bull. No. 599.
(4) Borgman, R.F., and Haselden, F.H. 1968. Cholelithiasis in rabbits: effects of diet upon formation and dissolution of gallstones. *Am. J. Vet. Res.* 29:1287-1292.
(5) Borgman, R.F., and Haselden, F.H. 1969. Cholelithiasis in rabbits: effects of several treatments on formation and dissolution of gallstones. *Am. J. Vet. Res.* 30:1979-1984.
(6) Borgman, R.F., and Wardlaw, F.B. 1974. Serum cholesterol and cholelithiasis in rabbits treated with pectin and cholestyramine. *Am. J. Vet. Res.* 35:1445-1447.
(7) Burkitt, D.P. 1971. Epidemiology of cancer in the colon and rectum. *Cancer* 28:3-13.
(8) Burkitt, D.P. 1972. Varicose veins, deep vein thrombosis, and haemorrhoids: epidemiology and suggested aetiology. *Br. Med. J.* 2:556-561.
(9) Burkitt, D.P., and James, P.A. 1973. Low-residue diets and hiatus hernia. *Lancet* 2:128-130.
(10) Burkitt, D.P., Walker, A.R.P., and Painter, N.S. 1974. Dietary fiber and disease. *JAMA* 229:1068-1074.
(11) Cleave, T.L. 1960. *On the Causation of Varicose Veins.* John Wright and Sons, Ltd., Bristol, England.
(12) Cleave, T.L., Campbell, G.C., and Painter, N.S. 1969. *Diabetes, Coronary Thrombosis and the Saccharine Disease,* 2nd ed. John Wright and Sons, Ltd., Bristol, England.
(13) Cummings, J.H., Hill, M.J., Jivraj, T., Houston, H., Branch, W.J., and Jenkins, D.J.A. 1979. The effect of meat protein and dietary fiber on colonic function and metabolism. I. Changes in bowel habits, bile acid excretion, and calcium absorption. *Am. J. Clin. Nutr.* 32:2086-2093.
(14) Cummings, J.H., Hill, M.J., Bone, E.S., Branch, W.J., and Jenkins, D.J.A. 1979. The effect of meat protein and dietary fiber on caloric function and metabolism. II. Bacterial metabolites in feces and urine. *Am. J. Clin. Nutr.* 32:2094-2101.
(15) Eastwood, M.A. 1974. Dietary fibre in human nutrition. *J. Sci. Fd. Agric.* 25:1-5.
(16) Eastwood, M.A. 1975. The role of vegetable dietary fibre in human nutrition. *Medical Hypotheses* 1:46-53.
(17) Eastwood, M.A., Fisher, N., Greenwood, C.T., and Hutchinson, J.B. 1974. Perspectives on the bran hypothesis. *Lancet* 1:1029-1033.
(18) Eastwood, M.A., and Terry, S.I. 1974. Diet and gastroenterology. *Br. J. Hosp. Med.* 12:713-721.
(19) Findlay, J.M., Smith, A.N., Mitchell, W.D., Anderson, A.J.B., and Eastwood, M.A. 1974. Effects of unprocessed bran on colon function in normal subjects and in diverticular disease. *Lancet* 1:146-149.
(20) Groen, J.J., Tijong, K.B., Koster, M., Willebrands, A.F., Verdonck, G., and Pierloot, M. 1962. The influence of nutrition and ways of life on blood cholesterol and the prevalence of hypertension and coronary heart disease among Trappist and Benedictine monks. *Am. J. Clin. Nutr.*, 10:456-470.

(21) Hardinge, M.G., Chambers, A.C., Crooks, H., and Stare, F.J. 1958. Nutritional studies of vegetarians. III. Dietary levels of fiber. *Am. J. Clin. Nutr.* 6:523-525.
(22) Hill, M.J., Crowther, J.S., Drasar, B.S., Hawkesworth, G., Aries, V., and Williams, R.E.O. 1971. Bacteria and etiology of cancer of large bowel. *Lancet* 1:95-100.
(23) Jenkins, D.J.A., Leeds, A.R., Slavin, B., Mann, J., and Jepson, E.M. 1979. Dietary fiber and blood lipids: reduction of serum cholesterol in Type II hyperlipidemia by guar gum. *Am. J. Clin. Nutr.* 32:16-18.
(24) Kelsay, J.L., Behall, K.M., and Prather, E.S. 1979. Effect of fiber from fruits and vegetables on metabolic responses of human subjects. II. Calcium, magnesium, iron, and silicon balances. *Am. J. Clin. Nutr.* 32:1876-1880.
(25) Kretsch, M.J., Crawford, L., and Calloway, D.H. 1979. Some aspects of bile acid and urobilinogen excretion and fecal elimination in men given a rural Guatemalan diet and egg formulas with and without added oat bran. *Am. J. Clin. Nutr.* 32:1492-1496.
(26) Kritchevsky, D., Tepper, S.A., and Story, J.A. 1975. Symposium: Nutritional Perspectives and Atherosclerosis. Nonnutritive Fiber and Lipid Metabolism. *J. Food Sci.* 40:8-11.
(27) Lyken, R., and Janse, A.A.J. 1960. The cholesterol level in blood serum of some population groups in New Guinea. *Trop. Geogr. Med.* 12:145-148.
(28) Mathur, K.S., Khan, M.A., and Sharma, R.D. 1968. Hypocholesterolemic effect on bengal gum: a long term study in man. *Brit. Med. J.* 1:30-31.
(29) Palmer, G.H., and Dixon, D.G. 1966. Effect of pectin dose on serum cholesterol levels. *Am. J. Clin. Nutr.* 18:437-442.
(30) Robertson, J., Brydon, W.G., Tadesse, K., Wenham, P., Walls, A., and Eastwood, M.A. 1979. The effect of raw carrot on serum lipids and colon function. *Am. J. Clin. Nutr.* 32:1889-1892.
(31) Saunders, R.M., and Hautala, E. 1979. Relationships among crude fiber, neutral detergent fiber, in vitro dietary fiber, and in vivo (rats) dietary fiber in wheat foods. *Am. J. Clin. Nutr.* 32:1188-1191.
(32) Scala, J. 1974. Fiber. The forgotten nutrient. *Food Tech.* 28:34-36.
(33) Scala, J. 1975. The physiological effects of dietary fiber, in A. Jeanes and J. Hodge (eds.), *Physiological Effects of Food Carbohydrates*, pp. 325-335. Am. Chem. Soc. Symp. Ser. No. 15.
(34) Weingarten, B., Simon, M., Zimetbaum, M., Weiss, J., and Smith, L.W. 1962. The use of pectin-cellulose bread as an adjuvant in dietary regulation and treatment of functional constipation. *Am. J. Gastroent.* 37:301-310.

–16–
Nutritional Hazards of Retirement

Audrey K. Davis

INTRODUCTION

Retirement with its full-time leisure life-style appears to carry with it some hazards which can affect nutritional status. Leisure itself may contribute to or create some nutritional problems.

Retirement is a comparatively recent phenomenon. During the 17th century, less than 20% of Americans survived to the age of 70, and retirement as we know it today was extremely rare. People did not plan for retirement because there were no retirement programs. Men remained active and worked as long as they were able, or until overcome by disease or accident. Women also worked and continued their family duties indefinitely throughout life.[11]

Today there are retirement programs of some sort available to almost everyone, and the majority of men and women expect to live long enough to retire at about 65 years of age. The Bureau of Census 1974-1975 report indicated that one-tenth of the United States population was of retirement age and the retired age group is now a large and growing segment of our population.[37]

The work ethic has been and will continue to be of paramount importance. Society has, of necessity, been dominated by the need to engage in productive work, but work may not necessarily be the only important life function. Ideally, retirement and the leisure phase of the life cycle should also be respectable and satisfying.[3] Since this period of life often covers a span of years longer than childhood and adolescence combined, recognition of the need for good health, or at least good nutritional management of disease, is essential.

Health is a retirement resource. Adequate finances, friends, varied interests and other types of support all help to make the retirement years satisfactory, but health status overrides all other factors in dictating the limitations imposed upon the individual and the extent to which the expectations of retirement will be realized.[33] It is reality that among those 60 years of age almost one-half have been forced into retirement because of deteriorating health and a majority of these have one or more conditions that requires some type of nutritional therapy treatment.

The most prevalent medical problems reported among those of retirement age include obesity, heart disease, hypertension, gastrointestinal problems, fractures, osteoporosis, kidney problems, gallbladder disease, cancer and diabetes. These are the common diseases associated with aging and all have nutritional aspects which, if not properly managed, can affect the course of the disease and the quality of life of the individual during retirement. The fact that the conditions are usually chronic means that long-term dietary management is indicated and must be continued indefinitely as part of the treatment program.

Retirement is perceived as free—free from work, free from responsibility, and free to do as one desires. Possibly involvement in work suppresses acknowledgement of the physiological changes inevitably associated with growing older, but all too often the realities of aging, particularly where food is concerned, seem to come as a surprise just prior to or soon after retirement. For many, the anticipated freedom of retirement does not include freedom to eat and drink as they please. It is not only adequacy of nutritional intake that is important to the health of the retired population. Nutritional management of one or more disease conditions is also of concern to a majority of retired persons and the economic and social conditions of retirement influence greatly how successfully nutritional problems will be handled and how effective disease management will be.

LOW INCOME

Retirement income level has a direct effect on the nutritional adequacy of diet and nutritional status.[4] Income is usually reduced as a result of retirement, but inflationary trends and spiraling medical costs erode spendable income even further. The United States Bureau of Census in 1974-1975 found approximately one-third of the population 60 years of age and older were below the poverty level and the number has probably increased since that time.

When income is low, essential expenses such as housing, utilities, and medicine cannot be easily adjusted. Food is the one large expense item that can be reduced when the cost of living increases beyond expectations. The lower the retirement income, the less money will be spent for food, the greater the nutritional hazard, and the more likelihood that food intake will be inadequate to maintain optimum health. When food budgets are low, the first items to be reduced or omitted are milk and meats.[14] These are the high-cost foods, but they are the important sources of protein, calcium, iron and riboflavin. Milk and milk products are the only really dependable food sources of calcium and the need for this mineral does not decrease with age.[26]

Studies have shown that the lower the income level, the greater the risk that nutritional intake will fall below the recommended dietary allowances.[25] The Ten-State Nutrition Survey which studied the needs of older people in lower economic groups showed that these people 60 years of age and older did not consume a quantity of food sufficient to provide the recommended quantities of essential nutrients.[35]

POOR DENTITION

Poor dentition is a nutritional hazard that may account for more nutritional problems among retirees and older people than is generally admitted. The Ten-State Survey reported 90% of those 67 to 74 years of age suffered from gum disease with subsequent loss of teeth, and that dentition problems increase with age.[35] It would seem practical to prepare for retirement and be sure that teeth or dentures are in the best possible condition. The cost of dental work often seems exorbitant and if the problems do not show there is a tendency to procrastinate and "take care of it later." However, the cost of dental care will certainly be no less as time goes on and, if anything, can be expected to increase. Disturbing as the expense may seem before retirement, the impact will probably be worse later when income is reduced.

Good dental care is important to prevent or correct mouth conditions that limit the kind or consistency of food that can be consumed.[12] When chewing is difficult or painful, there is a tendency to omit meat fresh fruits and vegetables from the diet, and to eat only soft, easily chewed foods.[34] Such foods are usually high in carbohydrate (starches and sweets) and low in fiber. This type of diet contributes to constipation and other gastrointestinal problems associated with a low fiber intake.[6] Long-term restriction of meat, which is a major source of dietary iron, could result in iron deficiency anemia as well as a lack of riboflavin.[28] Anemia has been identified as a common problem among older adults. While it is possible to omit meats and have a nutritionally adequate intake, such diets require careful planning and are even more difficult when there are chewing problems. Restricted use of fruits and vegetables reduces mineral and vitamin intake.

Poor dentition leads to consumption of a diet that is limited to foods of soft consistency. Some other foods may also be omitted because reduced sense of taste and smell cause them to seem unpalatable.[31] A diet of limited variety and poor quality over an extended period of time increases susceptibility to infection, slows healing after surgery or bone fractures, and can result in lack of control, progression, or premature disability from disease conditions in which diet is an important part of management such as diabetes, hypertension, and heart disease.

The use of soft food, or food mills or blenders to produce food of a soft consistency is a poor way to compensate for dental problems that could be corrected, yet many people resort to such strategies when there are chewing problems.

Well cared for teeth or properly fitting dentures are desirable for proper nutrition and also for cosmetic reasons. It is very true that a good personal appearance promotes self esteem as well as social acceptability. People who are edentulous look old and may be treated accordingly. It is a reality that younger people are inclined to have negative attitudes toward those who exhibit characteristics associated with advancing age.[26]

The retiree needs to be aware of the nutritional and social hazards of dentition problems. Any physical limitation, including dentition which results in loss of function, contributes to feelings of inadequacy, invalidism and social withdrawal and should be avoided if at all possible. Dental work is something people usually must do for themselves; it is a very personal problem and no one else can fully appreciate the extent to which poor mouth conditions limit food intake or socialization opportunities.

BOREDOM

Retirement that is devoid of meaningful responsibilities and creativity is not comfortable for many people. Boredom is a hazard to nutritional health because food is used in many ways to meet emotional needs, and it is often used to compensate for frustrating situations. When there seems to be nothing to do, boredom sets in and many people eat. Behavior modification techniques for control of weight identify "cues" that are perceived as objects, events, or associations that trigger eating.[16,29] One of the powerful cues to eating is boredom itself. Other cues are the television and the refrigerator. Television and eating seem to be synonymous for some people and many who are trying to conquer a weight problem report that they find themselves opening the refrigerator simply because it is there. After retirement, the refrigerator is available as it never was while working. It is available at all times of the day and the night and it contains food. It may also contain beer or other alcoholic beverages. The consumption of food or drink can become the chief activity of the day although hunger is not involved in this type of compulsive eating behavior. Rather, eating becomes an escape from an otherwise dull, uninteresting existence.[24]

Food is also used as a means to manipulate others and to express disapproval, to threaten, to gain attention, or to set one apart by creating an identity that has special needs and desires. For the disabled, eating may be important because it may be one of the last pleasures and it is sometimes one of few remaining functions that can be managed independently. The disabled retiree who visits the corner bar every afternoon may not really have a desire for alcohol; what he may be looking for is companionship and relief from boredom.

Food that is used as an escape all too often results in obesity which increases the difficulty of managing other medical problems. Boredom promotes feelings of futility concerning dietary controls of any kind. Consciously or unconsciously, food is sometimes a vehicle for self-destruction as evidenced by the grossly obese cardiac patient or the diabetic who refuses to make any effort toward dietary control.

ERRATIC FOOD CONSUMPTION PATTERNS

Work provides personal discipline and necessitates organization of individual and family life. Employment schedules usually dictate the framework within which meal times, eating habits and food intake become stabilized. During the working years, meals ordinarily take place at fairly regular and predictable times.

Leisure may have little impact on the nutritional status of those who make few changes in their way of living after retirement and continue long-established adequate food consumption patterns.[14] However, if food habits and nutritional intake were poor in the past and are perpetuated into and beyond retirement, there is increased likelihood that nutritional risks will be reflected in the development or progression of health problems. Retirement is often a time of many changes and adjustments that may include new meal schedules, reduced frequency of meals, and eating behavior that can be detrimental to health and effective management of disease conditions.

Absence of the work-structured day may result in a reduction in the number of meals consumed each day. A change in meal frequency may involve a shift from three meals a day to a routine of having breakfast late in the morning and a substantial meal sometime in the afternoon, but earlier than the traditional evening meal. This arrangement of only two recognized meals is invariably supplemented with snacks later in the evening which usually consist of large quantities of foods such as ice cream, cookies, desserts, crackers, soft drinks or alcoholic beverages. In effect, one main meal of the day has been replaced with a meal of high caloric snack foods of little nutritional value,[30] and usually excess calories.

Another version of changed meal schedules reduces the number of meals even further to only one a day. These people may have coffee for breakfast and one large meal at noon, which is often eaten in restaurants. This schedule also includes consumption of excessive quantities of snack foods throughout the afternoon and evening, resulting in a diet that is conducive to obesity and usually deficient in several essential nutrients. The practice of skipping meals can be expected to result in overeating during the meals that are consumed.

Retirees who revel in the absence of the alarm clock and establish a schedule of watching television late in the evening and sleeping late in the morning may face other problems related to meals. The spouse who does not change schedules with the retiree and continues to rise early and eat breakfast alone is often unreceptive to the idea of preparing a second breakfast when the late riser appears. The late sleeping retiree has a choice of either omitting breakfast altogether and waiting until lunch to eat with the spouse, or of having breakfast late in the morning and remaining on a different meal schedule from the spouse for the whole day. At best, they may join each other for the evening meal.

These simple sleeping-getting-up changes in routine which result from the spouse continuing to follow a preretirement meal schedule while the retiree adopts new routines can cause considerable conflict. Unless the problem is recognized, the retiree may find the number of meals reduced to two or even one each day, and the couple each preparing their own meals and eating alone on separate schedules most of the time.[39] Such meal schedules are conducive to reduced communication and loneliness as well as problems of weight control and nutritional intake.[19,40] The management of any disease condition under these circumstances would be highly unlikely.

The omission of meals promotes overconsumption of snack foods of higher calories, fat, cholesterol, and sugar than would probably have been contained in the meal omitted. For comparison, the caloric content of a breakfast of juice, egg and toast with margarine would provide approximately 235 calories. An afterdinner snack of ice cream, cookies, and soft drink or alcohol might easily provide 500-600 calories.[1]

RETIREE-SPOUSE RELATIONSHIPS

The relationship that exists between the retiree and spouse has a direct bearing upon the food and eating habits and nutritional status of the retiree. The spouse is usually the significant other person in the retired age group.

Although it is generally assumed that couples who have lived together many years and are together at the time of retirement enjoy a comfortable relationship, this is not always the case. A counselor to older people reports that "The divorce rate among retirees exceeds that of other age groups and that many other retired couples stay together in a stressful situation of emotional dissatisfaction and are unhappy with the relationship."[10]

There are several reasons that relationships may deteriorate at this stage of life. Retirement follows a long period of shared interests and responsibilities for home, children, careers, cars, recreation, community affairs, schools and involvement with others having similar interests. Eventually there comes a time when these activities are no longer of daily concern and interests of the couple may separate. While the man may be totally involved in work, the woman may develop her own activities and social support groups and there may be few mutual friends or concerns.

When work ceases to be of vital interest to either party, the couple may discover that they have very little in common or even topics to talk about. Communication may be difficult or break down to little exchange. Sometimes emotional separation occurs and although living together, each party seems to avoid the other and to live in separate worlds, even to the extent of preparing food and eating by themselves. Under such circumstances, meal time is far from a pleasurable social experience. When the atmosphere is strained or unpleasant and companionship is lacking and emotional needs are not met, appetite and interest in food declines, eating becomes mechanical, and it may seem better to avoid meals or to eat elsewhere than to eat at home. Emotional stress surrounding meals has been shown to be associated with poor food intake and nutritional deficiencies.[13]

Besides the possible role that stress plays in the development and progression of health problems, stressful relationships can have a catastrophic impact upon nutritional management of disease. Many persons who need nutritional therapy treatment also need assistance in the planning of meals, shopping and preparation of food. When there is little support or assistance on the part of the spouse, treatment programs that require special dietary considerations, such as sodium or fat restriction, weight control or diabetes management may be abandoned altogether. The person who shops for groceries and prepares the food controls to a great extent the kind of food that will be available. For example, the person who has diabetes and needs to restrict energy intake and sweet foods will have a difficult time following a diet if pastries are served with many meals and no alternative foods are available. In such cases, the diabetes may not be controlled. The attitude and support of the spouse in the management of special dietary needs can be an important factor in dietary compliance and effectiveness of treatment programs.

INFLUENCE OF ADVERTISING

The influence of persuasive advertising can be a hazard to the nutritional health as well as the pocketbook of retirees. The absence of work and its social and business contacts often reduces outside sources of available information. Advertising that is brought into the home by way of television, newspapers and magazine subscriptions takes on added importance and these sources of information may be influential in providing health information and determining products that the retiree will purchase and use. Retired people are apt to spend many more hours a day watching television than they did while working, and much advertising is aimed directly at this group of people. The media is well aware that persons 60 years of age and older are very interested in topics concerning health, food, nutrition and products that might affect the aging process. Some popular television programs specialize in nostalgic music, and types of entertainment known to be preferred by the retired generation. It is understandable that these programs are sponsored by advertisements for food supplements, laxatives, antacids and retirement homes. Talk shows also feature well-known personalities to promote exotic diets, "health" books and products which wrongly claim to be necessary or capable of erasing the physical evidence of aging.

"Nutrition" and "health" publications that are written to advance the sale of products advertised can be expected to present biased information. Slanted information caused people to buy products which they do not need and may interfere with good nutrition.[2]

Advertising has been influential in promoting the notion that medical problems of aging can be prevented or solved by taking pills. As a result, many people take self-prescribed substances that are ineffective and in some cases may actually be harmful.[8] Although the nutritional needs of most people can be met at lowest cost by using the regular foods from the supermarket, some people will always feel that expensive or special "health" foods provide nutrition of superior quality.

Antacids and laxatives are two widely advertised products that are used and abused by older people. Antacids interfere with the digestion and absorption of iron by changing the acidity of the stomach to alkaline. Also, some older people have reduced stomach acidity which results in decreased absorption and for this reason, ascorbic acid (vitamin C) is often given with iron supplements to insure the acid medium needed to promote absorption. Antacid abuse, low stomach acidity, or a diet low in iron could cause anemia, symptoms of which include apathy, listlessness and fatigue. The indiscriminate use of antacids could be a contributing factor in the occurrence of anemia, which is a major nutritional deficiency problem among older people.[35] Antacids also interfere with calcium absorption and could be associated with fractures and other bone problems.

Advertising has been largely responsible for publicity that has created concern that irregularity of bowel function is unhealthy and that use of laxatives and related products are necessary for health. Laxative misuse is common among older adults. In addition, people frequently omit milk from the diet in the belief that it causes constipation. The combination of low calcium intake resulting from omission of milk and the use of laxatives may account for some orthopedic problems among older people, since laxatives decrease transit time of food through the intestinal tract resulting in decreased absorption of nutrients.

The gastrointestinal tract of older adults ordinarily functions adequately in the absence of disease. Problems of constipation and related conditions are frequently the result of a diet that includes mostly soft, high carbohydrate foods that have little fiber. Increasing the dietary fiber intake can be remarkably effective in controlling chronic constipation without the need to resort to medications. Advertising that results in improper use of drugs can lead to serious nutritional problems.

OBESITY

Obesity has been identified as a nutritional hazard and a major health problem among 88% of those of retirement age.[35] Almost all chronic diseases associated with aging are more difficult to control and may progress faster when obesity is a contributing factor.[33] A person is considered to be overweight when 10% or more above desirable body weight; obesity is said to exist when weight exceeds desirable level by 20% or more. It is recognized that desirable weight determinations for any particular person are not exact, but such guidelines are helpful even though there may be justifiable exceptions. The control of weight is accepted as an important aspect of health management, but adjustments following retirement that include reduced income and social contacts, changes in daily activities and meal schedules, and sometimes even the place of living, may be reflected in eating behavior that makes weight control more difficult than at any previous time.

The problem of excess weight might be avoided to some extent if people were aware that as age advances a normal reduction in cell metabolism and metabolic needs along with decreased activity usually necessitates reduction of energy intake to maintain normal weight.[32] It has been estimated that after maturity age alone dictates a need to reduce caloric intake about 2% for each decade. A gradual loss of body weight can be expected after 70-75 years of age.[7,21] One of the problems of weight control is that very few middle-aged and older people know their desirable weight[25] and many who are frankly obese realize that they have cause for concern only after a medical crisis occurs and excess weight is identified as a contributing factor.[21]

Unexpected weight gain may also occur when smoking is stopped. Persons who have pulmonary disease, hypertension, diabetes, and cardiac conditions are usually advised to quit smoking. When such advice is followed, it is not uncommon for weight to increase 20 or 30 pounds in a surprisingly short period of time. Without the cover of smoke, food has more flavor and larger quantities may be consumed than previously. In addition, body circulation is improved and nutrients are utilized to a greater degree. Excess food may also serve as a form of oral gratification that was formerly provided by the cigarettes. Appropriate nutritional counselling directed at promoting an awareness of the potential for weight gain could prevent some people from experiencing the discouraging situation of overcoming the smoking problem only to be faced with a weight problem. Unfortunately, this exchange of one hazard for another frequently occurs among people who have serious medical conditions and for these individuals excess weight may present as great a risk as smoking.

The disabled retiree who must use a wheelchair is particulary apt to see food as a source of security and to consume excess quantities to compensate for limitations. Many elderly wheelchair persons surround themselves with food and keep supplies of cookies, candy and other snacks available in the pockets of the wheelchair and other places such as bedside tables, desks or tables where time is spent. These people have a difficult time at best controlling weight because of their low activity level; and the loss of a limb may further limit the need for energy intake due to reduced body mass. Obesity contributes to disability for these people by increasing mobility problems.

Simply stated, the person who is overweight has consumed more calories than the body needs and the excess has been converted to body fat. In order to lose weight, energy intake must be reduced below energy expenditure. When energy intake is balanced with energy expenditure, weight will be maintained at a constant level in the absence of abnormalities.

LACK OF EXERCISE

The value of exercise is well established and lack of activity has been shown to have some distinctly detrimental effects on nutritional status. Exercise is an important aspect of nutrition because it prevents muscle atrophy, promotes functioning of the cardiovascular system and increases circulation of nutrients to all tissues and organs of the body including the heart and brain. Exercise may also play a part in preventing the development of osteoporosis by promoting calcium deposits in the bones, thereby maintaining bone density.[9]

Retirement usually results in a decrease in energy expenditure due to absence of activities related to work. Any type of work, even the activities needed to get to and from work, requires a certain amount of energy expenditure. The more strenuous the work and the longer it is performed, the greater the energy expenditure and the greater the energy intake required.[22] People whose work involves much physical activity, such as carpenters, have a greater need to adjust energy intake after retirement than office workers. This reduction in activity level resulting from retirement is often the cause of a rapid unexpected weight gain that is not usually understood. For example, even if work activity requires only 300 calories per day, after retirement there could be an increase in weight of about 20 pounds in one year provided energy intake stays the same and new retirement activities do not replace former work-related activity. If such a trend in weight gain is not reversed, within a period of five years weight could increase by as much as 100 pounds.

While there are exceptions, of course, a low activity level is characteristic of most people at the time of retirement. Although studies have demonstrated that many older adults can safely exercise more than they think they can or should, exercise on a regular basis is seldom reported to be more strenuous than talking or moving the arms.[32] Some activity restrictions are self-imposed and others are the result of disability; but, on the whole, the activity level of older adults is usually described as sedentary and the individual whose weight is at desirable level seldom justifies energy intake of more than 25 calories per kilogram of body weight. Those who need to lose weight would need to reduce intake further.

NEED FOR SOCIALIZATION

Food represents friendship and plays an important part in practically all group social functions that are popular with retired people. Events that bring people together such as church functions, card playing, dancing, playing golf, organizational meetings and almost every other recreational activity, include "refreshments" of some sort. The need for social contacts and participation in available activities can present difficult decisions for those who are trying to control weight or follow some nutritional therapy program.

Dining out is one of the important and popular social activities among retired people. This form of socialization requires no particular exertion, fills a need to be with people, and also offers a variety of food that might not be prepared at home. Eating in restaurants is an activity that single older women can do without feeling conspicuous, providing the setting is carefully chosen and late evening meal hours are avoided. For some retirees, one restaurant meal each day may provide a major portion of the total food intake. Meals away from home are often understandably encouraged by women who are happy to retire, at least in part, from some of the tasks related to food preparation.

While restaurant meals promote socialization and serve to get people out of the house and relieve monotony, this popular activity may be disastrous as far as weight control or special dietary needs are concerned. Portions of food served in restaurants are invariably several times larger than most older people need, and many eating places rely heavily on the use of breaded and deep-fried foods, gravies, sauces and dressings which the public seems to want and which promotes the profit margin. This type of food also promotes obesity. Of course, people do not have to eat everything served to them, but human nature being what it is, restraint is difficult. Some older people solve the large portion size by taking part of the food home for a lunch the next day or dividing portions with someone else.

Salad bars are particularly deceptive to the person concerned about calories. These attractive displays of foods start with lettuce and other low-calorie vegetables, but end with an assortment of calorie-loaded items such as bacon chips, chopped olives and salad dressings. A salad of 1500-3000 calories can be assembled without much effort and this is usually in addition to the main course. It is also very difficult to resist the buffet type of service that offers an unlimited quantity of food for a set price. This form of food service is very popular with retired people and it is natural to justify overeating with the rationalization that "I've paid for it, I might as well get my money's worth."

Retirees who find themselves on the covered-dish or pot-luck meal circuit also face usually insurmountable odds on holding the line on calorie intake or other dietary restrictions. These social events are popular in retirement villages and mobile home parks and are an established tradition with many organizations. Such occasions can be counted upon to bring out the competitive instincts and gourmet talents of participants and overeating is the rule rather than the exception at these events. It is unfortunate that the common perception of "good food" is that which is covered with frosting, gravy, fried, breaded or otherwise laced with some form of fat or mixture that is high in energy content and well salted.

Another favorite after-retirement activity is travel and it, too, makes dietary management difficult. Travel by car usually means eating many meals in restaurants and in a vacation-type atmosphere food restrictions or dietary management programs are usually ignored. Cruises and organized travel tours are popular with the retired group and those in which meals are included as part of the package fare will almost surely result in disregard for dietary needs and overeating. Retirees who use their freedom to take extended trips to relatives and friends claim that since they are visiting they have no control over the food and drink consumed. Of course, the reverse is no better because when friends or relatives come to visit them they feel that they are expected to abandon their own dietary plans and return the entertainment; then both parties lament that the situation is beyond control.

The human need for socialization apparently takes precedence over medical needs.[18] Just as studies have shown that companionship at mealtime is more important than the food itself to older people, the need for a sense of belonging, to be accepted as part of a social group, can usually be expected to override any need for dietary control or adherance to any nutritional therapy plan.

NUTRITIONAL ASSESSMENT AND COUNSELING

Nutrition can be a vital force in health care when it becomes an integral part of medical evaluation and clinical practice. A routine part of health care of older patients in all medical settings, private or public, should include careful assessment of dietary intake as well as other factors that influence nutritional status. Assessment of nutritional needs should be followed by counseling programs to provide health and nutrition education directed not only toward treatment of existing problems, but also designed to promote the goals of health maintenance, prevention of disease and disability, and an acceptable quality of life and health during the retirement years.

Effective nutritional counseling for the older adult must recognize the influence of psycho-social needs as well as the obvious medical needs of the individual.[15,22,42] Due to the chronicity of health problems associated with age, most dietary management programs must, of necessity, be long term—often for the remainder of life. There will be greater likelihood of compliance with treatment goals if nutritional plans are realistic and recommended day-to-day routines are practical.[5] Older people are interested in nutrition and professionals can provide reliable information provided it is presented in a non-technical manner that is understandable and has some meaning to the individual.[3] Nutritional therapy plans that consider income level, cultural and religious traditions, availability of acceptable food supply and food preparation equipment, as well as the educational, physical and emotional capacity of the individual will be more acceptable than plans that ignore these factors.[23,38] Nutritional therapy prescribed for older people should restrict or modify long-established food consumption habits only to the extent believed necessary to result in improvement or control of threatening conditions. People make their own decisions about food and lack of dietary compliance should be appropriately met with a non-judgemental attitude on the part of the counselor to avoid adding feelings of guilt to existing problems of the older person.[20]

Understanding of the interrelationships of nutrition, medicine and geriatrics can result in a higher level of care for older people and should prove to be cost-effective by reducing disability and institutionalization as solutions for social problems of the elderly.[17]

SUMMARY

Retirement age is often accompanied by chronic disease conditions that require long-term nutritional management as part of treatment programs. Poor nutritional management may be the cause of some cases of involuntary retirement. Adjustments in income level, living habits, and eating behavior that frequently follow retirement can have a detrimental influence on food consumption, nutritional status, and disease control. Nutritional hazards associated with retirement include erratic food consumption patterns, boredom, poor retiree-spouse relationships, and the influence of advertising and lack of exercise. In addition, poor dentition may limit food intake and affect social relationships. Obesity is a nutritional hazard for many older adults and is associated with inadequate control of many disease conditions that can affect the quality of life of the retired person. Socialization needs can be expected to take precedence over needs for dietary restrictions prescribed for many conditions. Nutritional assessments and evaluation of psychosocial factors as well as dietary needs can provide valuable information that contributes to effective medical care of older people.

The views expressed in this paper are not necessarily those of the Veterans Administration.

REFERENCES

(1) Adams, K. 1975. Nutritive value of American foods. *Agriculture Handbook No. 456*, Washington, D.C.
(2) Barrett, S., and Knight, G. 1976. *The Health Robbers—How to Protect Your Money and Your Life.* George F. Stickney Co., Philadelphia.
(3) Boss, D. 1977. Reaching out: Diet and senior Americans. *Food Mgmt.* 12:70-73, 91-93.
(4) Brien, M., and Bauernfiend, J.C. 1978. Vitamin needs of the elderly. *Post Grad. Med.* 63:155-163.
(5) Brun, J.K, and Clancy, K.L. 1980. Low income and the elderly population. *J. Nutr. Educ.* 12:128-130.
(6) Burkitt, D.P, and Painter, N.S. 1974. Dietary fiber and disease. *JAMA* 229:1068-1074.
(7) Busse, E.W. 1980. Eating in late life: Physiologic and psychologic factors. *NY State J. Med.* 80:1496-1497.
(8) Comfort, A. 1976. *A Good Age*, pp 179-191. Simon and Schuster, New York.
(9) DeVries, H.A. 1970. Physiological effects of an exercise training regimen upon men aged 52-88. *J. Geront.* 25:325-326.
(10) Finkelhor, D.C. April-May 1980. Help for retirement marriages. *Mod. Maturity*, pp 79-82.
(11) Fischer, D.H. 1977. *Growing Old in America*, pp 3-25, 113-156. Oxford Univ. Press, New York.
(12) Gambert, S.R., and Guansing, A.R. 1980. Protein-calorie malnutrition in the elderly. *J. Amer. Geriat. Soc.* 28:272-275.
(13) Gifft, H.H., Washbon, M.B., and Harrison, G.G. 1972. *Nutrition, Behavior and Change*, pp 187-211. Prentice-Hall, Englewood Cliffs, NJ.
(14) Hamilton, E.M., and Whitney, E. 1979. *Nutrition Concepts and Controversies*, pp 453-479. West Publishing Co., St. Paul, MN.
(15) Hatten, A.M. 1976. The nutrition consultant in home care services. *J. Amer. Diet. Assoc.* 68:250-252.

(16) Jordan, H.A., and Levitz, L.S. 1975. A behavioral approach to the problem of obesity. *Obesity and Bariatric Med.* 4:58.

(17) Kane, R.L., and Kane, R.A. 1978. The care of the aged: Old problems in need of new solutions. *Science* 200:913-919.

(18) Lutwak, L. 1969. Symposium on osteoporosis: Nutritional aspects of osteoporosis. *J. Amer. Geriat. Soc.* 17:115-119.

(19) Lyons J.S., and Tralson, M.F. 1956. Food practices of older people living at home. *J. Geront.* 11:66-72.

(20) Maiman, L.A., Wang, V., Becker, M.H., Finlay, J., and Simon, M. 1979. Professionals' attitude toward obesity and the obese. *J. Amer. Diet. Assoc.* 72:331-336.

(21) Marx, J.L. 1974. Aging research. 1. Cellular theories of senescence. *Science* 186:1105-1107.

(22) Mason, M., Wenberg, B.G., and Welsch, P.K. 1979. *The Dynamics of Clinical Dietetics.* John Wiley and Sons, New York.

(23) McNutt, K.W., and Steinberg, L.H. 1980. Persons with diet related diseases. *J. Nutr. Educ.* 12:131-137.

(24) Miller, I., and Solomon, R. 1980. The development of group services for the elderly. *J. Geront. Soc. Work* 2:241-257.

(25) National Research Council 1980. *Recommended Dietary Allowances*, 9th ed. National Academy of Sciences, Washington, D.C.

(26) Newgarten, B.L. 1973. Patterns of aging: Past, present, and future. *Soc. Svc. Rev.* 47:571-580.

(27) Olson, R E. and Doisy, A.A. 1978. Clinical nutrition, and interface between human ecology and nutritional medicine. W.D. Atwater Memorial Lecture. *Nut. Rev.* 36:6.

(28) Rao, D.B. 1973. Problems of nutrition in the aged. *J. Amer. Geriat. Soc.* 21:362-367.

(29) Ritt, R.S., Jordan, H.A., and Levitz, L.S. 1979. Changes in nutrient intake with behavioral weight control. *J. Amer. Diet Assoc.* 72:325-330.

(30) Rockstein, M., and Sussman, M.L. 1976. *Food, Nutrition, Longevity, and Aging,* pp 1-7. Academic Press, New York.

(31) Scheffman, S.S. 1978. Changes in taste and smell in old persons. *Adv. Res.* 2:1-6.

(32) Sidney, K.H, Shepherd, R.G., and Harrison, J.E. 1977. Endurance training and body composition of the elderly. *Amer. J. Clin. Nutr.* 30:326-333.

(33) Stunkard, A.J. 1976. Nutrition, aging and obesity. In *Nutrition, Longevity and Aging,* pp 253-284. Academic Press, New York.

(34) Swanson, P. 1964. Adequacy in old age, Part II. Nutrition programs for the aging. *J. Home Econ.* 56:728-734.

(35) Ten-State Nutrition Survey. 1968-1970. DHEW Pub. No. (HSM) 72-8130, 8133. Center for Disease Control, Atlanta.

(36) U.S. Department of Agriculture. 1979. *Food,* U.S. Government Printing Office, Washington, D.C.

(37) U.S. Department of Commerce, Bureau of Census, 1975. Money, Income and Poverty Status of Families and Persons in the United States, 1974. U.S. Government Printing Office, Washington, D.C.

(38) Watkin, D.M. 1976. Biochemical impact of nutrition on the aging process. In M. Rockstein and M.L. Sussman (eds.), *Nutrition, Longevity and Aging,* pp 47-66. Academic Press, New York.

(39) Weinberg, J. 1972. Psychologic implications of the nutritional needs of the elderly. *J. Amer. Diet. Assoc.* 60:293-296.
(40) Williams, L.M. 1978. A concept of loneliness in the elderly. *J. Amer. Geriat. Soc.* 26:183-187.
(41) Young, C M. 1974. Nutritional counseling for better health. *Geriatrics* 29:83-91.

–17–
Modern Nutrition for Those Who Are Already Old

Donald M. Watkin

INTRODUCTION

The connotations on the words "modern nutrition" and "those who are already old" are sufficiently numerous to justify a few paragraphs devoted to definitions.

Modern nutrition is not some striking innovation. It is not a new fad, a new miracle diet. It is not a means of rolling back the clock, of eliminating pathology already *in situ*.

Modern nutrition is the application of known principles. It is the union into a single regimen for each person of the appropriate elements of the triad of nutrition, health and aging.[52] It is the process of optimizing the physical and mental health of each person through a combination of nutrition and appropriate health and social services. Finally, and very importantly, it is the collection, analysis and interpretation of data collected during the application of scientific principles in the optimization process and the transformation of that research into new applications beneficial to those who are already old.

In defining those who are already old, it is appropriate to start with the technical definition of the life span as it applies to human beings. Technically speaking, the life span is the oldest documented age to which any member of a given species has survived. According to the 1980 edition of the *Guinness Book of World Records*,[36] a Japanese gentleman died in 1979 exactly 114 years and zero days after his birth in 1865. The issue then becomes that of selecting a proportion of the life span of human beings (114 years) after which the term "already old" applies. Those 86 years of age and older have survived at least three-quarters of the human life span. Those 76 and older, about two-thirds. Those 66 and older, about 57 percent. Those 61 and over, about 53 percent.

As noted in the introductory chapter, only those in the last quarter of the technical life span for human beings are truly old in terms of chronological aging *per se*. However, diseases and disabilities impose the equivalence of old age on many whose chronological ages are appreciably younger. The instant aging associated with injuries to the cervical portion of the spinal cord (resulting in quadriplegia) are well documented.[49,46] The relative disabilities occasioned by myocardial infarcts and myocardiopathies among young middle-aged persons yield the health status erroneously associated only with persons much older. The same may be said for those disabilities associated with the residues of strokes, of neoplasms and of metabolic bone disease, to mention only a few. Heredity may shorten survival but also, as has been shown in families favored by high serum concentrations of high density lipoproteins,[23] may be characterized by great longevity.

With the possible exception of those persons living in the last quarter of the technical life span of human beings, the most sensible definition of old age is a functional, not a chronological one. The nutrition of a specific person must be tailored to meet the needs of the functional definition of old.

FUTILITY OF CONVENTIONAL PREVENTION BY NUTRITIONAL MEANS AMONG THE ELDERLY

Conventional prevention refers to the modification of environmental factors including nutrition to thwart the development of a disease or to defer its manifestations until the very end of the technical life span of man. The success of such conventional prevention is not at issue. The data of Belloc and Breslow[4] have shown that the application of seven life styles, at least four of which are nutrition-related, throughout life will lead to health status measured objectively among persons entering the last quarter of the technical life span of man which is equivalent to that of those persons at the beginning of the second quarter who have not practiced the same life styles up to that age (v.s., Introduction). What is at issue is that conventional prevention by nutritional means must be practiced throughout life and throughout life must be integrated with other preventive measures of which abstinence from tobacco and avoidance of alcohol abuse are classic examples. Pathology already *in situ* obviously can not be prevented.

In spite of this *res ipsa loquitur*, older persons are subjected daily to media meassages suggesting that they can prevent an array of diseases ranging from myocardial infarction to neoplasms by adopting some new dietary regimen. In spite of solid evidence[26] that dietary manipulations will be futile in those over 55 years of age, older persons continue to be advised by physicians, nonmedical nutritionists and others of less sophisticated backgrounds to undertake dietary regimens which are at variance with the older persons' dietary habits or are impractical from the viewpoints of availability, price or facilities required for preparation. Compliance is therefore low, but the level of concern manifest by feelings of anxiety and guilt remains high. Much evidence has been accumulated suggesting that those surviving beyond the first half of the technical life span of human beings have a genetic endowment which protects them from many hazards to which the nonelite succumb.[30-33]

Primary prevention among those already old does serve a useful purpose in certain specific foci. For example, prevention of accidents in the kitchen and elsewhere in the home is possible and is nutrition-related in that it may avert hospitalization and the deleterious and sometimes fatal effects of nosocomial infections, immobilization and therapy involving a variety of anesthetics and drugs. Primary prevention of food-borne infectious diseases is nutrition-related and quite possible. Prevention of drug and alcohol abuse among older persons who have not indulged in such abuse earlier in life is possible but requires the coordinated efforts of the full spectrum of professionals, subprofessionals and laymen seriously concerned in applying concepts inherent in the nutrition-health-aging triad.[52]

Although certain exceptions like those mentioned above do exist, conventional prevention of disease by nutritional means is not operational among the elderly. What is pertinent to those who are already old is the application of modern nutrition principles to those who are genetically nonelite and who have acquired specific disease entities present in old age, those regardless of genetic background who are the victims of acute illnesses and trauma and those not afflicted by acute or chronic diseases or trauma whose health status may be optimized via modern nutrition.

NUTRITIONAL MANAGEMENT OF THE GENETICALLY NON-ELITE

Improving the outlook for older persons who have survived in spite of genetic handicaps is a challenge not only in a humanitarian sense but also from an economic point of view. In the U.S. Veterans Administration Department of Medicine and Surgery nationwide a minority of potential beneficiaries is responsible for the vast majority of hospital admissions.

The key to the successful management of the older persons in this category is education.[47,50,51,53] Proper education requires personnel but also sufficient time for reinforcement through repetition. Participation in the education process by the older trainees or students is an effective approach. Once properly educated, the elderly have the potential of becoming multipliers of the information among their peers.

The classic example of inherited disease is diabetes mellitus. Whether acquired during childhood, middle age or old age, diabetes mellitus is manageable by control of the dietary intake whether the individual is insulin-dependent or noninsulin-dependent. In the case of the former, particular care must be taken to adjust food intake to correspond to peak insulin activity in order to avoid the life-threatening complication of hypoglycemic shock. In noninsulin-dependent diabetics, success lies in reducing food intake and weight and when possible in increasing the level of physical activity. Attaining compliance with such regimens is difficult but can be achieved if time and effort expended are equivalent to the task. The very diagnosis of diabetes among older persons is difficult, since carbohydrate intolerance as measured by conventional glucose tolerance tests is present in many who do not have genetically transmitted disease.[2]

Although not all instances of overweight are harmful (v.s., Introduction), there is no disagreement regarding the harm imposed by morbid obesity, i.e., a body weight equal to or greater than 130% of desirable weight by the Metropolitan Life Insurance Company's 1959 tables.[37] Such obesity may be difficult to treat by conventional calorie restriction not only because of lack of compliance with the dietary regimen but also because some persons are apparently able to maintain a weight even when rigidly adhering to a low energy diet. Until recently, such persons were complete conundrums. Now the possibility of inherited or acquired (at younger ages) differences in the amount of brown fat in human bodies[16,45,44] may offer some rational explanation. Weight loss by the integration of nutritional, exercise and medically supervised regimens is a necessary goal in the morbidly obese. The extreme difficulty of managing such persons when they become ill with or without the requirement of surgery is in itself justification for the effort.

Familial hyperlipoproteinemia is estimated to be present in less than 1% of the North American population.[38,22] Its prevalence among the elderly may be even lower, since an association exists between mortality from myocardial infarction in younger age cohorts and hyperlipoproteinemia. However, a cause and effect relation has not been established. For this reason, the lipid hypothesis is now the subject of a massive study[40] which will be completed in 1983 at which time analysis of data collected in men with type IIa hyperbetalipoproteinemia half of whom will have received a cholesterol-lowering drug (cholestyramine) for seven years will become available. Although, as noted above, dietary manipulation has little value in elderly persons with hyperlipoproteinemia, it is always important to recognize that hyperlipoproteinemia may be a complication of another primary disease which may be amenable to medical treatment. Among these other primary diseases are diabetes mellitus, hypothyroidism, nephrosis, hepatic disease with biliary obstruction, chronic pancreatitis and diseases associated with increases in plasma globulins, particularly myeloma proteins or macroglobulins.[38,21] Hence, elevated serum concentrations of cholesterol and/or triglycerides always deserve careful scrutiny even though lipoprotein phenotyping is rarely indicated in clinical practice.[38,25,20]

Hereditary angioedema (HAE), a condition known for almost a century,[39] had before the advent of modern diagnosis and therapy a mortality rate of about 30%.[14] The cardinal signs are edema without urticaria of the face and extremities, the submucosa of the gastrointestinal tract and the pharynx and larynx. These may be accompanied by abdominal pain often severe enough to mimic an acute abdominal emergency and respiratory obstruction severe enough to cause death if not relieved by a tracheostomy. While the majority of such patients recall the onset of symptoms in childhood, some become symptomatic after age 50. Patients with this condition often present histories of recent ingestion of a particular food or drug which is known to produce anaphylaxis or anaphylactoid reactions. It is important to inquire into a family history of such symptoms and not to be lulled into a false sense of security by a history of food or drug ingestion. The condition is caused by a deficient or a malfunctioning inhibitor of the esterase activity of the first component of complement (Cl INH).[19] Management of the acute problem consists of treatment with 1:1,000 epinephrine subcutaneously (a precautionary measure of questionable benefit) and, most importantly, the availability of the equipment and personnel to perform a tracheostomy. Danazol, an attenuated androgen which induces increase in Cl esterase inhibitors, and epsilon-aminocaproic acid (EACA) and its analog, tranexamic acid, are agents of choice in preventing but not treating acute attacks. No benefit is derived from dietary modifications in management.

Another hereditary disorder with nutritional connotations is hemochromatosis or bronze diabetes. The heritable form is characterized by excessive deposits of iron in the body and by hepatomegaly with cirrhosis, skin pigmentation, diabetes mellitus and often cardiac failure. Patients with primary, heritable hemochromatosis absorb excessive quantities of iron from a diet of normal iron content. The treatment of the condition is the removal of the excess body iron by venesection of 500 ml of blood once or twice a week with the goal of removing 10 to 20 g of iron annually.

Hemochromatosis may also be secondary to other diseases, to medical therapeutic measures or to environmental nutritional conditions. These include alcoholic cirrhosis of the liver, other forms of chronic liver disease and chronic pancreatitis; refractory anemias treated with transfusions; parenteral therapy with iron dextran; and excessive iron intakes such as that found among the Bantu who use iron cooking pots and drink copious amounts of an alcoholic beverage prepared in iron drums and that found among persons who consume excessive quantities of iron-containing over-the-counter medications.

Hemochromatosis became a highly publicized nutritional issue in the decade of the seventies when lobbies representing patients with hemochromatosis fought state compliance with federal regulations requiring iron fortification in wheat flour and rice moving in interstate commerce. No evidence has shown that iron fortification levels used in wheat flour and rice have any adverse effect on persons with hemochromatosis. In fact, recent reports[3,11,27] clearly demonstrate that primary hemochromatosis is an autosomal-recessive disease. In both homozygotes and heterozygotes, the biochemical expression of the alleles enables such persons to absorb enough iron from the ordinary diet eventually to produce symptomatic disease. Persons with primary hemochromatosis require venesections semiweekly. Hemochromatosis is another example of a disease involving an essential nutrient which in its primary form is best managed by nonnutritional means.

Wilson's disease, also known as hepatolenticular degeneration, is another heritable condition with a nutritional component. The pathogenic nutrient is copper which is absorbed in excess of need and is deposited in the liver and the brain where it accumulates to the point of producing serious functional damage. Management of this condition requires both a low copper diet (no more than 1 mg daily) and therapy with either BAL (British antilewisite) or, preferably, d-penicillamine, both chelating agents which dramatically increase urinary loss of copper from the body.

While some of the conditions mentioned in this section may seem rare, they do enter into many differential diagnoses and, in these, nutritional considerations are often involved. In some, nutritional measures enter into the therapy but in others they have no role. It is as important to recognize when nutrition plays no role as it is to recognize when its contribution is vital.

NUTRITIONAL MANAGEMENT OF ACUTE ILLNESSES

Acute febrile illnesses among older persons are situations in which nutrition and hydration, combined always with appropriate medical and surgical therapy, play truly life-saving roles. The principal reason lies in the fact that older persons have limited reserves of essential nutrients. In addition, acute febrile illnesses usually diminish appetite and debilitate the sense of thirst yielding a patient who may have no desire to eat or to drink. Furthermore, such illnesses, as was shown in the first decade of this century in patients with typhoid fever,[43,13,15] lead to tremendous losses of body protein, minerals, electrolytes and water. Procrastination in the successful administration of nutrients and water to such patients can be fatal or at least can lead to unnecessary trips to emergency rooms and to unnecessary hospitalization. While appropriate therapy with antibiotics and other indicated drugs is also essential, nutrition and hydration must be a part of the overall regimen. In the home of the patient, this can be accomplished by relatives and friends encouraging the patient to take small, frequent servings of food together with small, frequent servings of water or other well-tolerated drinks. The training of family and friends to play this crucial role is the key to success of the venture. Enteral nutrition mixtures may be used in place of standard food but are considerably more expensive. Tube feeding and parenteral alimentation and hydration are possible even in the home environment but do require professional supervision and constant attendance by a trained observer.

Among the infectious diseases afflicting the elderly, pneumonia and influenza are among the most common. Another recently recognized to be a frequent cause of morbidity and mortality is Legionnaires' disease.[24] Infections of the genito-urinary tract and various gastroenteritides, including perforated viscera, are frequent problems. Important to include in the differential diagnoses of acute infectious diseases are hepatitis and pancreatitis. The cardio-vascular system is often host to infectious diseases such as pericarditis, myocarditis, valvulitis and septicemias. Improperly managed traumatic wounds including bites of insects, animals and even other humans may lead to acute soft-tissue infections. The meningitides and the encephalomyelitides may infect the nervous system. Whatever the cause, an acute febrile illness results in a rapid loss of the diminished reserves of the older person. These must be replaced quantitatively as soon as the losses begin if serious consequences for the patient are to be avoided.

Acute illnesses must also include mental aberrations as well as those caused by an infectious agent. From a nutritional point of view, acute depression is capable of producing serious debility largely because its presence is less likely to be noticed than is that of an acute febrile episode. Here again, the hour-by-hour attention of caring friends and relatives as well as indicated medications is the key to successful management. The role of interpersonal relationships in assuring appropriate nutrition and the role of meals with others in preventing emotional crises among the elderly are often overlooked.

Whenever possible, it is wise to allow a sick older person to remain in his or her place of residence during the course of an illness. This is recommended not only in deference to the usual wishes of the patient but also in recognition of the fact that hospitals are notorious harborers of nosocomial infectious agents which usually are far more resistant to antibiotic management than are infectious agents in the community.

NUTRITIONAL MANAGEMENT OF TRAUMA

Trauma to older persons may result from accidents in the home or outside while a pedestrian or a passenger in a motor vehicle. Frequently, trauma is the result of assault by criminals or by vicious animals. It may even result from self-inflicted wounds. Regardless of its cause, trauma is a threat to nutritional status of the elderly for two major reasons. First, the trauma itself may produce physical or emotional conditions which prevent normal ingestion of food. Its immediate consequence is almost always pain. Pain itself is debilitating and, if severe enough, may cause nausea and vomiting. It may be treated with drugs which by themselves are capable of reducing appetite and depressing thirst. Self-feeding may be impossible if the trauma produces dysfunction of the upper extremities or obstruction of the mouth or other parts of the upper gastrointestinal tract. Second, trauma usually results in therapeutic immobilization of the elderly victim. During such therapeutic immobilization, the patient is supposed to heal fractures and reunite severed tissues, all of which processes require adequate nutritional status for good results. However, as is the case with acute infections (v.s.), trauma induces large, immediate losses of protein and minerals which must be replaced by appropriate nutrition and hydration. The therapeutic immobilization in turn results in additional losses of protein, calcium, phosphorus, potassium and magnesium, *inter alia*, which are associated with loss of bone. This bone loss produces osteopenia, the most obvious consequences of which are so-called pathologic fractures, *i.e.*, long bone fractures from stresses which, in the absence of osteopenia, would be of insufficient force to break bones. Less apparent, but often even more life threatening, is the fact that the bone components lost through immobilization may reappear in the form of kidney or bladder stones which induce further pain, morbidity and deterioration of nutritional status. The same ingredients derived from normal bone may appear in muscle as heterotopic bone, *i.e.*, bone forming in locations in which it normally is never found. In addition to producing osteopenia, immobilization also induces weakness and wasting of skeletal muscles and, unless nursing care is excellent, may lead to pressure ulcers of the skin.

Certain varieties of trauma, generally termed catastrophic, result in additional problems. Complications of injuries resulting in dysfunction of the spinal cord (quadriplegia or paraplegia) comprise neurogenic dysfunction of the bladder and bowel, gastrointestinal bleeding, carbohydrate intolerance, refractory anemias and virtually total dependency especially when the trauma results in quadriplegia.[49,46] Apart from these difficulties, the patient is usually a victim of severe anxiety and depression. The impact of such trauma on the patient's nutritional status requires great ingenuity for its restoration if it has already deteriorated and for its maintenance if the patient is the victim of a very recent trauma.

In many respects, the nutritional management of a traumatized older person is comparable to that described under the heading of acute infectious diseases. First, the patient requires reassurance from the professional and subprofessional staffs of institutions and also from relatives and friends whether institutionalized or at home. Second, nutrient intake and fluid intake must be sufficient to compensate for the immediate posttrauma losses and to prevent further deterioration of nutritional status if immobilization is required. Regimens high in protein and high in calcium[34,29,35] accompanied by copious amounts of fluid are needed. Depending on the cardiovascular, hepatic and renal status of the patient, sodium intake should be carefully monitored. Energy intake should be high in the absence of contraindications. Very importantly, active or, if necessary, passive physical exercise should be undertaken or administered. Finally, therapeutic immobilization should be terminated at the earliest possible time, and the patient should return to normal activity through a graded series of steps.

NUTRITIONAL MANAGEMENT OF OLDER PERSONS RECOVERING FROM ACUTE ILLNESS OR TRAUMA

Quantification of the impact of the acute episode, whether characterized by infectious disease or by trauma, at the time the patient is first encountered is essential. Unfortunately, malnutrition is often present when patients are admitted to hospitals, but, even more tragically, nutritional status has been shown to deteriorate during hospital stays.[9,7,10] Furthermore, antibiotics and other drugs may in themselves induce malnutrition.[41] Quantification of nutritional status is possible by observing the patients' weight in relation to their height, inspecting their skin-fold thicknesses, inquiring about their dietary intake before and after their illness or trauma, obtaining an accurate accounting of their drug therapy and reviewing the nutritionally related effects of any surgery that may have been performed. Many physical signs indicative of malnutrition may be readily observed by examining the mouth and by observing the conjunctivae and the skin. If indicated by history and physical findings, and if laboratory measurements of pertinent parameters are available, biochemical measurements of nutrients in the blood and urine may be useful in quantifying the specifics of deficiencies present.

Nutritional rehabilitation along with physical and psychological rehabilitation should begin immediately and simultaneously. When oral alimentation is feasible, nutritional rehabilitation should begin promptly with patients' consumption of small, frequent servings of regular food combined in regimens which provide nutrients and energy indicated by the evaluation of the patients' nutritional status. Small, frequent servings of water and other fluids are required physiologically but also may assist in helping the patients swallow the small servings of food proffered. The consistency or viscosity of the nutrient- and water-containing regimens should be tailored to meet the requirements imposed by pathology and to some extent the tastes of the patients. However, care must be taken to avoid the patients' consumption of only thin broths or gruels which contain inadequate nutrients when accumulated on a daily basis. The proposals of the Committee on Dietary Allowances of the Food and Nutrition Board, National Research Council, National Academy of Sciences, as contained in the Ninth (1980) Edition of *Recommended Dietary Allowances*[17] should represent the absolute minimum nutrient intake unless specific medical contraindications are present. The provision of energy must be given a high priority suggesting that the dietary regimen may require a high percentage of total energy from fat to obtain the highest possible concentration of energy in each of the small feedings administered. In addition to its energy density, fat has the added advantage when properly used of improving the organoleptic qualities of the servings offered.

Care must be taken to assure that the recuperating patients actually consume the dietary regimens offered. This requires constant monitoring and assistance in the alimentation process by institutional staff, relatives and friends. The hours devoted to assuring consumption of the dietary offered will pay off in terms of far more rapid return to self-care status by the patients involved.

NUTRITIONAL MANAGEMENT OF OLDER PERSONS NOT AFFLICTED BY ACUTE ILLNESS OR TRAUMA

Many older persons have neither acute illness nor trauma but still require nutritional management to bring their health status to an optimum level. Among these are those who are demonstrating emaciation and cachexia, who use excessive amounts of alcohol, who smoke excessively and who are addicted to either prescription or over-the-counter drugs. Many other elderly suffer from changes in hearing, vision, taste and smell which relate importantly to their eating and drinking patterns. The condition of the dentition has a major impact on the quantity and quality of food consumption. Social deprivation, problems of income maintenance and housing and transportation difficulties plague many older persons.

Some of these problems are readily managed by applying well-tried solutions until the difficulty is resolved. The emaciation resulting from gross undernutrition can be handled by increasing the food intake, provided disease-related causes of the condition have been ruled out. Often such increases will in themselves eliminate the anorexia and also improve the affect of the emaciated individual so that resocialization takes place greatly abetting the maintenance of improved nutritional status. Alcohol, tobacco and drug abuse are difficult problems at any age but among the elderly the drain on income which they create often provides added leverage which, when combined with education and resocialization, leads to successful resolutions. Improved oral hygiene and dental care for those who still have teeth can eliminate one cause of malnutrition. For those without teeth, prostheses are often difficult to fit as well as being very costly. Such persons are frequently best managed by developing menus which contain foods previously processed to assure easy mastication and swallowing. Social, income maintenance, housing and transportation problems require time and persistence on the part of elderly persons themselves or on the part of their advocates who are knowledgeable in the solutions available in particular communities.

Sensory losses, those of hearing, vision, taste and smell, are more difficult to manage. Impaired hearing prevents elderly persons from enjoying meals in large, noisy places because, even when equipped with hearing aids, they cannot hear what their dinner companions are saying over background noise. Slowing of dark adaptation and a need for much more illumination deprives the elderly of opportunities to enjoy dining in dimly lit or candlelighted places frequented by younger persons. Diminished taste and smell[42,12] in older persons are well documented. They lead to aversions to many foods and may in fact lead to total aversion to the ingredients of an adequate diet. There is no agreement in regard to resolutions of the loss of taste and smell. Surgical procedures and electronic devices have yet to overcome the hearing losss problem. Nutritional measures[5,28,54,6] have had no influence on dark adaptation and the need for more light. Resolution of these problems which are capable of inducing malnutrition is largely a matter of educating the older persons so afflicted about physiologic aging and of guiding them in developing menus and life styles which will to the greatest extent possible compensate for these deficits.

A factor not generally recognized which accounts for poor nutrition among many otherwise healthy elderly persons is the amount of physical exertion which is required to assure an adequate diet day after day. Shopping for food requires trips to and from the food stores. Preparation of the meals, cooking the ingredients, serving the meal and finally cleaning up the utensils and place settings used require the expenditure of much energy. One meal may require two to three hours of time and also the energy which in many elderly is reduced. While convenience foods provide some relief for those who can afford them, many elderly either go without appropriate nutrition on a regular basis or may resort to a constant regimen of a few items which are easy to obtain and to prepare. Solutions to this problem are manifold. They include such items as group meal service,[48] home delivered meals, home delivered groceries, meats and produce, foods with long shelf lives whose preparation is simple and quick and the availability of food processors and microwave ovens which can substantially reduce the time required to have a meal ready for consumption.

CONCLUSION

This chapter has dealt with the nutritional management of older persons with heritable conditions, those who are the victims of acute illness and trauma, those who are recovering from acute illness and trauma, those whose nutritional status is diminished by a variety of other problems not related to acute illness or trauma. It has also pointed to the futility of conventional prevention by nutritional means among the elderly. The nutritional aspects of diseases of the aged have been detailed in Chapter 14.

Throughout this chapter, certain needs have been evident. Foremost of these is the need for education of the elderly themselves, of their relatives and friends and of the professionals and subprofessionals who serve them in institutions and in community agencies. Since nutrition, health and aging form an inseparable triad of constantly interrelating agglomerates,[52] education in nutrition must be combined with education about health and about aging *per se*. The full report of first-wave findings of the longitudinal evaluation of the Nutrition Program authorized by Title VII (now Title IIIc) of the Older Americans Act of 1965, as Amended[1] suggests but does not prove that participation in a Nutrition Program project site activities does result in improved nutrient intake as measured by 24-hour recall, some improvement in life satisfaction and psychological well-being but no differences in self-perception of health. Unfortunately, objective measures of health and nutritional status were not obtained. The very modest accomplishments reported in this evaluation may reflect the failure to give emphasis to education in the first five years of the Nutrition Program's operations. In 1980, a new system of nutrition education for the elderly based on the peer counseling principle was developed.[8] This effort will require many years to implement and even longer to evaluate its results. Nonetheless, it does represent acknowledgement by the Bureau of Health Education, Center for Disease Control, Department of Health and Human Services that education among the elderly and their advocates is presently inadequate.

Another need stressed repeatedly is that of greater involvement by the health industry in the application of nutrition principles to the management of acute illnesses and trauma, of rehabilitation and of certain other generic problems related to nutrition among the elderly. Among professionals and subprofessionals at all levels, nutrition should become a therapeutic tool which is never neglected in the management of all patients but particularly, because of their diminished reserves, of the elderly.

A final need is that of accessibility to available resources. Among these are included the Nutrition Program[48] and the Food Stamp Program.[18] To become fully effective, their accessibility must be increased by far greater emphasis on outreach. Public assistance and the benefits of Social Security must become more readily accessible.

In the health field, Medicare and Medicaid benefits and, for those who are eligible, Veterans Administration benefits must become accessible to larger numbers. Better supervision of long term care institutions with particular emphasis on increasing turn-over by improving nutritional management should assist in making needed facilities more accessible. Finally, with the aid of the food processing industry, foods packaged with consideration of the needs of older persons which have long shelf lives and which require a minimum of time and effort in preparation should become more accessible.

Modern nutrition does play a vital role in optimizing the health status of older persons. It plays this role when the principles on which it stands are applied to the solution of specific problems. No magic is involved. The application of principles requires personnel who have knowledge and who also have the dedication and the time to make the application succeed. Finally, more research in nutrition and aging will be one of the greatest services to be performed in behalf of older persons. While that research is being conducted, those who are already old deserve vigorous application of all the knowledge of modern nutrition available today.

REFERENCES

(1) Administration on Aging, U.S. Department of Health, Education and Welfare. 1980. Longitudinal Evaluation of the National Nutrition Program for the Elderly. Report of First-Wave Findings, by Kirschner Associates, Inc. and Opinion Research Corporation. DHEW Publication No. (OHDS) 80-20249. U.S. Government Printing Office, Washington, DC.
(2) Andres, R. 1971. Aging and diabetes. *Med. Clin. North America* 55: 835-846.
(3) Beaumont, C., Marcel, S., Fauchet, R., Hespel, J.-P., Brissot, P., Genetet, B. and Bourel, M. 1979. Serum ferritin as a possible marker of the hemochromatosis allele. *New England J. Med.* 301:169-174.
(4) Belloc, N.B. and Breslow, L. 1972. Relation of physical health status and health practices. *Prev. Med.* 1:409-421.
(5) Birren, J.E., Bick, M.W. and Fox, C. 1948. Age changes in the light threshold of the dark-adapted eye. *J. Gerontology* 3:267-271.
(6) Birren, J.E., Bick, M.W. and Yiengst, M. 1950. The relation of structural changes of the eye and vitamin A to elevation of the light threshold in later life. *J. Exper. Psychol.* 40:260-266.
(7) Bistrian, B.R., Blackburn, G.L., Hallowell, E. and Hiddle, R. 1974. Protein status of general surgical patients. *J. Amer. Med. Assoc.* 230:858-860.
(8) Bureau of Health Education, Center for Disease Control, U.S. Department of Health, Education and Welfare. 1979. Nutrition Education for the Elderly, by Westinghouse Public Applied Systems. Contract No. HEW 200-79-0923. Center for Disease Control, Atlanta, GA.
(9) Butterworth, C.E., Jr. 1974. The skeleton in the hospital closet. *Nutrition Today* 9:4-8.
(10) Butterworth, C.E., Jr. and Blackburn, G.L. 1975. Hospital malnutrition. *Nutrition Today* 10:8-18.

(11) Cartwright, G.E., Edwards, C.Q., Kravitz, K., Skolnick, M., Amos, D.B., Johnson, A. and Buskjaer, L. 1979. Hereditary hemochromatosis: phenotypic expression of the disease. *New England J. Med.* 301: 175-179.
(12) Cohen, T. and Gitman, L. 1959. Oral complaints and taste perception in the aged. *J. Gerontology* 14:294-298.
(13) Coleman, W. and DuBois, E.F. 1914. The influence of high Calorie diet on the respiratory exchanges in typhoid fever. *Arch. Internal Med.* 14:168-209.
(14) Donaldson, V.H. and Rosen, F.S. 1966. Hereditary angioneurotic edema: a clinical survey. *Pediatrics* 37:1017-1027.
(15) DuBois, E.F. 1922. Metabolism in fever and certain infections. In L.F. Barker, R.G. Hoskins and H.O. Mosenthal (eds.) *Endocrinology and Metabolism, Volume 4*. pp. 95-152. Appleton, New York.
(16) Elliott, J. 1980. Blame it all on brown fat now. *J. Amer. Med. Assoc.* 243:1983-1985.
(17) Food and Nutrition Board, Committee on Dietary Allowances, H.N. Munro, Chairman. 1980. *Recommended Dietary Allowances, Ninth Revised Edition*. National Research Council, National Academy of Sciences. Washington, DC.
(18) Food and Nutrition Service, U.S. Department of Agriculture. 1979. How to Apply For and Use Food Stamps. Program Aid No. 1226. U.S. Government Printing Office, Washington, DC.
(19) Frank, M.M., Gelfand, J.A. and Atkinson, J.P. 1976. Hereditary angioedema: the clinical syndrome and its management. *Ann. Internal Med.* 84:580-593.
(20) Fredrickson, D.S. 1975. It's time to be practical. *Circulation* 51:209-211.
(21) Fredrickson, D.S. and Lees, R.S. 1966. Familial hyperlipoproteinemia. In J.B. Stanbury, J.B. Wyngaarden and D.S. Fredrickson (eds.) *The Metabolic Basis of Inherited Disease, Second Edition*. pp. 429-485. The Blakiston Division, McGraw-Hill Book Company, New York.
(22) Glueck, C.J. 1977. Classification and diagnosis of hyperlipoproteinemia. In B.M. Rifkind and R.I. Levy (eds.) *Hyperlipidemia: Diagnosis and Therapy*. pp. 17-40. Grune and Stratton, New York.
(23) Glueck, C.J., Gartside, P., Fallot, R.W., Sielski, J. and Steiner, P.M. 1976. Longevity Syndromes: Familial hypobeta and familial hyperalphalipoproteinemia. *J. Lab. Clin. Med.* 88:941-957.
(24) Gregory, D.W., Schaffner, W., Alford, R.H., Kaiser, A.B. and Zell, A.M. 1979. Sporadic cases of Legionnaires disease: expanding clinical spectrum. *Ann. Internal Med.* 90:518-521.
(25) Havel, R.J. 1975. Editorial. Hyperlipoproteinemia: problems in diagnosis and challenges posed by the type III disorder. *Ann. Internal Med.* 82:273-274.
(26) Hazzard, W.R. 1976. Aging and atherosclerosis: interactions with diet, heredity and associated risk factors. In M. Rockstein and M.L. Sussman (eds.) *Nutrition, Longevity and Aging*. pp. 143-195. Academic Press, New York.
(27) Kidd, K.K. 1979. Editorial. Genetic linkage and hemochromatosis. *New England J. Med.* 301:209-210.
(28) Kirk, E. and Chieffi, M. 1948. Vitamin studies in middle-aged and old individuals. I. The vitamin A, total carotene and alpha and beta carotene concentrations in plasma. *J. Nutrition* 36:315-322.

(29) Krook, L., Lutwak, L., Henrikson, P.-A., Kallfelz, F., Hirsch, C., Romanus, B., Belanger, L.F., Marier, J.R. and Sheffy, B.E. 1971. Reversibility of nutritional osteoporosis: physicochemical data on bones from an experimental study in dogs. *J. Nutrition* 101:233-246.
(30) Linn, B.S. 1968. Viewpoint: a personal opinion on progress in care of the aged. Elite physical resistance seen as key to longevity. *Geriatrics* 23:No. 4 (Apr.), 48, 55, 58.
(31) Linn, B.S., Linn, M.W. and Gurel, L. 1968. Cumulative illness rating scale. *J. Amer. Geriatrics Soc.* 16:622-626.
(32) Linn, B.S., Linn, M.W. and Gurel, L. 1969. Physical resistance and longevity. *Geront. Clin. (Basel)* 11:362-370.
(33) Linn, M.W., Linn, B.S. and Gurel, L. 1967. Physical resistance of the aged. *Geriatrics* 22:134-138.
(34) Lutwak, L. 1969. Nutritional aspects of osteoporosis. *J. Amer. Geriatrics Soc.* 17:115-119.
(35) Lutwak, L., Krook, L., Henrikson, P.-A., Uris, R., Whalen, J., Coulston, A. and Lesser, G. 1971. Calcium deficiency and human periodontal disease. *Israel J. Med. Sci.* 7:504-505.
(36) McWhirter, N. 1980. *Guinness Book of World Records.* pp. 25-29. Bantam Books, Inc., New York.
(37) Metropolitan Life Insurance Company. 1959. New weight standards for men and women. *Statistical Bull.* 40:1-4.
(38) Motulsky, A.G. 1976. The genetic hyperlipidemias. *New England J. Med.* 294:823-827.
(39) Osler, W. 1882. Hereditary angioneurotic oedema. *Amer. J. Med. Sci.* 95:362-367.
(40) Rifkin, B.M., Chief, Program Office. 1979. The Coronary Primary Prevention Trial: design and implementation. The Lipid Research Clinics Program. *J. Chronic Dis.* 32:609-631.
(41) Roe, D.A. 1976. *Drug-Induced Nutritional Deficiencies.* AVI Publishing. Westport, CT.
(42) Schiffman, S.S. 1981. Effects of age on taste and smell. In D. Harman (ed.) *Proceedings of a Symposium on Aging and Nutrition at the Ninth Annual National Meeting of the American Aging Association (AGE), Washington, DC, September 20-22, 1979.* Raven Press, NY. (In Press.)
(43) Shaffer, P.A. and Coleman W. 1909. Protein metabolism in typhoid fever. *Arch. Internal Med.* 4:538-600.
(44) Stock, M.J. 1981. Diet-induced thermogenesis: a role for brown adipose tissue. In E.J. Bassett and R. Beers, Jr. (eds.) *Nutrition Factors: Modulating Effects on Metabolic Processes.* Proceedings of the 13th Miles International Symposium, Baltimore, June 18-20, 1980. Raven Press, New York. (In Press.)
(45) Vaisrub, S. 1980. Editorial. Beware of the lean and hungry look. *J. Amer. Med. Assoc.* 243:1844.
(46) Watkin, D.M. 1972. Practical solutions to malnutrition in spinal cord dysfunction. In H. Talbot (ed.) *Proceedings of the Joint Meeting of the Eighteenth Spinal Cord Injury Conference of the Department of Medicine and Surgery of the United States Veterans Administration and the International Medical Society of Paraplegia, Boston, October 5-7, 1971.* pp. 231-233. U.S. Government Printing Office, Washington, DC.

(47) Watkin, D.M. 1973. The aged. In J. Mayer (ed.) *U.S. Nutrition Policies for the Seventies.* pp. 53-63. W.H. Freeman and Company, San Francisco.
(48) Watkin, D.M. 1977. The Nutrition Program for Older Americans: a successful application of current knowledge in nutrition and gerontology. *World Rev. Nutr. Diet.* 26:26-40.
(49) Watkin, D.M. 1977. Aging, nutrition and the continuum of health care. *Annals N.Y. Acad. Sci.* 300:299:290-297.
(50) Watkin, D.M. 1979. Nutrition, health and aging. In M. Rechcigl, Jr. (ed.) *Nutrition and the World Food Problem.* pp. 20-62. S. Karger, Basel.
(51) Watkin, D.M. 1979. Mutual relationships among aging, nutrition and health. In R. Hodges (ed.) *Nutrition: Metabolic and Clinical Applications.* pp. 219-240. Volume 4. In R.B. Alfin-Slater and D. Kritchevsky (general eds.) *Human Nutrition: A Comprehensive Treatise.* Plenum Press, New York.
(52) Watkin, D.M. 1981. *The Nutrition-Health-Aging Triad: Integrating the Sciences, Medicine and Public Health.* Raven Press, New York. (In Press.)
(53) Watkin, D.M. 1981. Certain aspects of the effect of age on the acquisition of nutrients. In D. Harman (ed.) *Proceedings of a Symposium of Aging and Nutrition at the Ninth Annual Meeting of the American Aging Association (AGE), Washington, DC, September 20-22, 1979.* Raven Press, New York. (In Press.)
(54) Yiengst, M. and Shock, N.W. 1949. Effect of oral administration of vitamin A on plasma levels of vitamin A and carotene in aged males. *J. Gerontology* 4:205-211.

–18–
Food Facts, Fads, Fallacies and Folklore of the Elderly

Audrey K. Davis and Robert L. Davis

INTRODUCTION

Throughout history, people have dosed themselves with nostrums and potions of foods or mixtures of food to ward off disease, to protect against real or imagined environmental dangers, and to promote physical attributes, fertility, or sexual powers. Man has believed in superstition, witch doctors and medicine men, and today is little different.[48] It has been estimated that Americans are wasting between five and ten billion dollars each year on unneeded, useless, and frequently harmful food supplements, and other forms of nutrition quackery. There seems to be something inherent in man's nature to use food as a panacea for his many problems since a substantial portion of the population continues to place faith and hope in the magic of many "health" foods that aren't healthy, supplements that do nothing, and diets that falsely claim to cure incurable diseases or ward off the aging process. Since this trend is apparently increasing rather than decreasing, it is probably realistic to accept the prospect that there will continue to be those who choose to believe in food magic in whatever form it may take in the future.[1,21,24,48,50,56]

The current interest in health and nutrition has been given impetus by the recognition that either directly or indirectly, nutrition plays a part in practically all of the major chronic health problems associated with aging.[9,14,25,28] For this reason, it is not surprising that nutrition has become a topic of vital and personal interest to older people.[10,31] In addition, improved mass media communication systems have made it possible to rapidly disseminate throughout the world a great volume of nutrition information. Never before has the public had access so quickly to so much health information both good and bad, as is available today.[12] Eating is a very personal activity and one over which the individual has considerable capacity to control or to manipulate as seems appropriate.[8,11] The availability of nutrition information, regardless of reliability, has created a natural desire on the part of the elderly, as well as other age groups, to apply whatever knowledge is available in an effort to promote health and prevent or cure disease.[29] Unfortunately, nutrition misinformation has been more readily available and appealing than reliable information.

SOURCES OF NUTRITION MISINFORMATION

Specialty Food Stores

The local specialty or "health" food store may well be the most accessible source of nutrition information presently available to the public, although a considerable amount of the information dispensed, either verbally or through printed material, may be misleading or unreliable. These business enterprises offer an assortment of "health" literature, and "nutrition" products, but some nutrition-related terms such as "natural," "organic," and "health" food are used in ways that confuse and deceive people by implying that such items are superior to ordinary supermarket products.[46] This is not necessarily correct. The cost is usually the only factor that is superior, and since there are no precise definitions for these terms as applied to food, any interpretation of their meaning is possible.[2] "Organic" and "natural" can be, and are used loosely to promote almost anything from vegetables to shampoo, beer, or pet food.

Organic foods are usually much higher in cost than similar supermarket foods, although there appears to be no objective evidence that the type of fertilizer used to grow "organic" foods will have any effect on the taste[4] or nutritional value. Nevertheless, many people are apparently willing to pay premium prices for natural or organic products in the mistaken belief that they are getting better nutrition or that higher cost insures higher quality.

Publications

There are many pocket-sized books and magazines on the market today which present a unique brand of health information designed to create a desire on the part of the public for "health" products and food supplements that ordinarily would have little market value. In spite of their unreliability, some of these publications have achieved the status of best sellers, and have been and continue to be a major source of nutrition misinformation.[26] Misleading health publications quote scientific journals out of context, distort meaning, and manage to sound highly authentic. Although this type of printed material relies almost entirely upon personal testimonials concerning the benefits to be derived from the use of some vitamin supplement or secret remedy, the selling technique is highly effective in generating a market for expensive food supplements and other products.[2,18,25,39]

"Health" publications serve as a vital link between the producer of food supplements and the public because the Food and Drug Administration (FDA) restricts label information, package inserts and other types of advertising that can be attached to "health" food items. Without these publications, the public would hear very little about many such products, much less buy them. Many subscription-type nutrition magazines devote at least one-half of the pages to recognizable advertising of health products and the other half to articles written to create a desire for the products being advertised. Although it may not always be recognized by the reader, this type of publication is essentially an advertising medium, and the buyer of these magazines pays to read advertising of one kind of another.

Testimonials

Other sources of nutrition misinformation are those persons who sincerely believe that they have been cured of a terminal illness as a result of having taken "miracle" supplements or followed some exotic dietary routine. Much erroneous health information can be broadcast along with the dramatic testimonials made by these people.

Folklore

Misinformation is also spread by individuals who are the innocent advocates of special foods or diets resulting from beliefs passed on from person to person and sometimes generation to generation. These individuals are usually receptive to practical health information provided it is presented in a nonjudgemental atmosphere, with adequate explanations and the individual can be shown that the new information is of personal value.

FOOD FADDISM AND QUACKERY

Definition

The continuing appeal of food faddism is sometimes hard to understand, but according to Schafer,[40] food faddism becomes an outlet for personal expression and a vehicle for meeting psychological needs. Food faddism is defined as "an unusual pattern of behavior enthusiastically adopted by its adherents and may be expressed in one of three ways:

(a) Acceptance of special virtues of a particular food to cure specific diseases;

(b) Elimination of certain foods from the diet in the belief that harmful elements are present; and

(c) Emphasis on "natural foods."[40]

Jarvis[27] defines a food quack as "someone who, because of avoidable ignorance, delusion, misconception, or intent to deceive, makes excessive claims or promises for the value of a nutritional substance or dietary practice to prevent, alleviate, or cure a disease, extend life-span, or improve physical or mental performance."

Characteristics

There are some general characteristics of food faddism that help to identify the more common varieties of nutrition nonsense and misinformation.

Antiestablishment: It is characteristic of food faddists that they are invariably antiestablishment. They are opposed to orthodox ways of farming, food processing, and medical treatment. They usually claim that they are being persecuted or suppressed by some agency or group that is conspiring to hamper their business operations, or prevent them from selling some untested "miracle" formula. Favorite targets of oppression include the FDA, which is the agency that has authority to control product labeling and to remove products from the market, the Federal Trade Commission (FTC), the agency that controls transportation of products across state lines, the United States Department of Agriculture (USDA), drug companies, and the American Medical Association (AMA) which seldom appreciates the faddist's approach to nutrition and methods of treating disease. By magnifying and sensationalizing topics of general concern, such as additives, pesticides, and radiation, faddists promote the idea that federal agencies are not protecting the public and that physicians and other professionals are not to be trusted. This undermines confidence in the orthodox and encourages unorthodox approaches to health care. The purpose of this strategy is to instill the notion that protection of the "little people" depends upon self-diagnosis and treatment with "natural," "organic," "health foods," and supplements. While this is an effective technique for promoting sales of these products, the results can be inadequate health care for older people.

Believe that All Disease Is Caused by Faulty Diet: One of the basic premises of food faddists is that improper diet is the cause of all disease and food has unlimited power to cure. Proper diet is important from the standpoint of preventing some abnormalities and also in the treatment and control of some conditions, but important as nutrition is, it is but one aspect of health and disease. Heredity, infections, and many other factors play a role. Food is not a cure-all, and food supplements should not be used as medicine.[31] Everyone has a right to make personal decisions about food, but neither fad diets nor food supplements are appropriate substitutes for accurate diagnosis and medical treatment.[32,47] Some people would have us believe that the old ways of treating disease are better than what is now available. For some people, especially the elderly, anything from long ago is better than anything modern. This back-to-nature philosophy effectively promotes self medication which can delay or take the place of medical treatment.

The food faddist characteristically endorses the notion that health is the natural state and disease is unnatural. The number of diet books, food treatment plans and food supplements promoted as cures for the "unnatural" state represented by disease is staggering. Many of these cures are unrealistically based on the idea that the body can be "detoxified" and returned to its natural healthy state by periods of fasting, the use of liquid diets, herbs, laxatives, and enemas, along with inevitable large doses of food supplements. Older people deserve better care than herbs, vitamins, and enemas for their problems.

Believe that the Soil Is Depleted: Another fundamental characteristic of food faddists and vitamin merchants is that they claim that the use of commercial fertilizer has depleted soil nutrients which, in turn, produces food that is of low quality. The notion that Americans are no longer healthy because their food is "devitalized" is a fallacy, but this line of reasoning is remarkably successful in convincing people to purchase vitamin supplements.[45] Concerns about nutritionally "depleted" food and the possible harmful effects of food additives have created a profitable market for almost anything that is touted as being "natural" or "organic." Anything advertised as natural apparently sells better and has more sales appeal than other products.

Believe that Processed Foods Lack Nutrients: Food faddists also erroneously claim that processed or refined foods are low in food value. Actually, food enrichment programs that replace minerals and vitamins removed in milling or other processing methods and the fortification of many foods, along with scientific farming methods, have made available the most abundant and nutritious food supply in history. Americans do not always make wise food choices, but the food itself has not deteriorated.

Believe that Organically Grown Food Provides Superior Nutrition: It is characteristic of food faddists to erroneously claim that foods produced with organic fertilizer are superior nutritionally to foods produced with other fertilizers. This can not be true because the chemical composition of each species of plant is determined genetically and will not vary significantly regardless of what fertilizer is used. Organic and commercial fertilizers function in identical ways to promote plant growth. Organic fertilizer simply takes longer to become available to plants since it must first be broken down in the soil by worms and bacteria to the same chemicals found in commercial fertilizers. However, the use of organic fertilizers poses some health risk because manure and moldy organic matter may infect the crops with "fungal toxins," e.g., aflatoxin, which is an extremely potent carcinogen. The nature of organically grown foods is such that there may be some risk of contamination with salmonella bacteria, which may cause a serious type of food poisoning, particularly when it occurs in older people. Anyone who uses organically grown food should exercise caution to avoid infection by thoroughly washing and cleaning these foods before placing them on kitchen counters, in the refrigerator, or consuming them.

It is frequently assumed that so-called "organic" food has been grown with organic matter and natural rock minerals, and has not been exposed to fungicides, herbicides, or pesticides, and that no preservatives, additives, fresheners, or coloring agents have been added. This is not necessarily so, and in actual practice there is no way for the consumer to determine whether or not a food has been "organically" grown. The FDA does prohibit the use of the term "natural" on products that have artificial flavors or colors, or those that have highly purified food additives. The USDA also prohibits use of "natural" and "organic" descriptions on most poultry or egg product labels.

WHY MISINFORMATION IS DANGEROUS

Misinformation is dangerous because it encourages self-diagnosis and self-medication, and may delay or prevent needed treatment of serious medical problems. In addition, the use of unneeded or useless food supplements and special foods can be very costly. Unfortunately older people on limited income provide one of the most lucrative markets for these expensive products.

VULNERABILITY OF THE ELDERLY

The elderly are particularly vulnerable to nutritional misinformation for several reasons. First, this generation of older adults did not generally have the benefit of nutrition education or physiology as part of their formal schooling. Those who lack this basic knowledge become easy targets for authentic-sounding health information that is not based on fact.[25] Second, most older people have several chronic medical problems for which there are no cures at this time. The possibility that an unknown miracle product may be able to correct a supposedly incurable condition has understandable appeal. A third reason that many older people fall victim to the persuasive arguments and misinformation of the food faddists and profiteers is the general attitude of the public concerning age itself.[44] In our culture, growing old is perceived as undesirable, not quite socially acceptable, and a symbol of reduced economic and social status.[43] On the practical side, age is likely to carry with it some realities of discrimination related to employment, economic potential, and community influence. It is understandable that our youth-oriented culture seeks to avoid the stigma of graying hair, baldness, wrinkles, skin changes, sagging tissues and reduced physical capacity associated with growing old.[3] People see these changes happening to themselves and are willing to try almost any product that seems to offer some promise of reprieve. People are apparently willing to spend huge sums of money to dose themselves with all sorts of potions in hope of retarding the aging process or increasing the possibility of rejuvenation.[7]

Opportunists and profiteers are fully aware of these circumstances and the desire of the public for health information and products that might offer some hope of prevention or retardation of aging and the problems associated with it. The older population provides a large and receptive target for "health" publications and products designed to cater to the chronic problems of aging. It is no accident that product outlets and advertising campaigns are concentrated in the retirement states and communities.

NUTRITION FACTS AND FALLACIES

Megavitamin Therapy

Vitamins are tremendously popular with the public and account for a large share of self-prescribed medications taken by older adults. It is a fallacy that everyone needs to take vitamin supplements.[36] The medical community generally accepts the recommendation that unless there is a condition that interferes with eating or absorption of food, the normal diet of 1500 calories or more that includes the foods recommended in the basic food groups, will usually provide adequate nutrients without the need for supplementation. However, it is also recognized that some older people may have a need for additional vitamins. When eating habits deteriorate, meals are skipped, variety is limited, and essential foods are omitted from the diet, supplemental vitamins may be advisable. There is also the possibility that older people may need some vitamins in larger quantities than the amounts presently recommended, but it is not expected that megadose levels would be suggested. Megadoses of vitamins cannot serve vitamin functions; they can only serve chemical functions, some of which may be harmful.[22,23]

The AMA has recommended that when daily vitamin supplements are indicated, the dosage should contain amounts no greater than the Recommended Dietary Allowance (RDA). Therapeutic dosage at levels of 3 to 5 times the RDA may be justified when there is prolonged illness, chronic disease, and poor food intake over a long period of time.

There is much misinformation about vitamins and how they function. Some of these notions began long ago and continue to survive because reliable information has not been readily available and because scientific facts are comparatively uninteresting compared to the sensationalism of some types of advertising.

Vitamins do not give a person quick pep and energy, although people often take them for this reason. In a survey of almost 3,000 adults, nearly three-fourths agreed that feeling tired and run-down indicates a need for more vitamins and minerals. While 86% of the population feel that they can get all the vitamins they need from a well-balanced diet, over 50% take vitamins and other dietary supplements.[41] The survey shows many Americans grasping for cures, concerned about diet, and taking supplements "just in case" they might be needed. Forty-three percent of adults surveyed never used supplemental vitamins.

The quantity of vitamins that the body can utilize is believed to be close to the levels of the RDA.[37] Megavitamin dosage is considered to be ten times or more in excess of the recommended amounts, and at this level symptoms of toxicity may develop. Megavitamins are not completely harmless and while many people experience no problems, other individuals may not be able to tolerate excessive amounts.[6,19] Unfortunately, vitamin toxicity is not always easy to diagnose. Symptoms are not clear cut and problems related to overdose could be overlooked or attributed to other pathological conditions. Physicians may not ask about self-prescribed supplements and even if they do, many older people do not know the dosage they are taking or how it compares with the RDA.

Excessive intake of vitamin A can cause loss of appetite, enlarged liver, spleen, dry cracking skin, aches, bone pain, irritability, and loss of hair. Symptoms of vitamin A toxicity may be similar to those of a brain tumor because of increased pressure within the skull.[34] Toxicity resulting from excess dosage must be considered when symptoms exist.[13]

Overdoses of nicotinic acid or niacin can result in severe flushing, skin disorders, itching, liver damage, gout, blood sugar problems and ulcers. Studies have shown that megadoses of niacin can produce increased gastro-intestinal problems, cardiac arrhythmias, and abnormal blood chemistry. Megavitamin therapy for treatment of schizophrenia has not been accepted as effective therapy according to the American Psychiatric Association Task Force on Vitamin Therapy in Psychiatry. However, niacin has been of benefit in cases of mental confusion caused by severe nutritional deficiency states and pellagra.

Pangamic acid has been widely promoted and is also referred to as vitamin B_{15}. The FDA has stated that this product is not a vitamin, has no known value for humans, and may be harmful.[25]

Vitamin C is one of the vitamins most frequently consumed in megadose quantities. The RDA for vitamin C is 60 mg; one-half cup of orange juice daily would supply approximately 55 mg which would largely meet the requirement for this vitamin. The average diet would also contribute additional quantities of vitamin C from other foods. However, it is common to find older people taking self-prescribed vitamin C at doses twenty or more times the RDA.

Excess vitamin C can destroy 50 to 90% of the vitamin B_{12} in a regular diet and could create serious vitamin B_{12} deficiency symptoms. Megadoses of vitamin C may also interfere with normal utilization of vitamin A and have been reported to raise the uric acid level and precipitate gout in some people.[26] Stopping megadoses of vitamin C suddenly has resulted in a rebound action with the development of symptoms similar to scurvy: bleeding gums, skin hemorrhages, and loosened teeth. Large doses of vitamin C have produced severe hemolysis and death. Some ethnic groups including blacks, orientals, and Sephardic Jews are particularly susceptible to this type of disorder.

Megadoses of vitamin C may be dangerous for persons who have diabetes since this vitamin in large quantities can interfere with accurate urine testing. In addition, laboratory tests for blood in the stool may be inaccurate when large doses of this vitamin are used.

Claims that vitamin C will prevent respiratory infections have not been substantiated by scientific evidence. Vitamin C may have some mild antihistaminic effect, but personal testimonials, rather than research, have been largely responsible for the claims that vitamin C will cure "colds" as well as a great variety of other ailments.

Vitamin E enjoys tremendous popularity due chiefly to testimonial type advertising which implies that this vitamin has almost unlimited potential to cure almost anything.[21] Implications that this vitamin could be a panacea for many age-related problems, particularly impotency and heart disease, will probably insure its continued popularity among the elderly.[38]

Numerous research studies have failed to substantiate advertising that vitamin E will perform the miracles claimed for it.[21] The Food and Nutrition Board of the National Research Council has stated that the claims concerning the effectiveness of vitamin E to cure heart disease, sterility, impotence, cancer, ulcers, muscle weakness, skin problems, burns, and delay the aging process, are the result of improper interpretation of animal research data or to outright misrepresentation designed to promote sales of this vitamin.

Vitamin E is essential and the RDA is listed as 10 mg for adult men over 51 years of age and 8 mg for women of the same age. Vitamin E is widely distributed in unsaturated fats and oils, grains, egg yolks, and meats, and it is almost impossible to not get enough of this vitamin if any food at all is eaten. Excessive amounts of vitamin E may interfere with vitamin K and blood clotting function.

Some elderly people can not tolerate large doses of vitamin E. Excessive dosage has been reported to cause fatigue, headaches, nausea, chapped lips, inflammation of the mouth, gastro-intestinal problems, bleeding tendencies, degenerative changes, hypoglycemia, and blurred vision. Instead of increasing sexual function, as many people have been led to believe, vitamin E may actually have the opposite effect and suppress function.[21] However, experimental evidence indicates vitamin E may be of value in the treatment of some cases of intermittent claudication.[16]

Natural Versus Synthetic Vitamins

Advertising designed to sell expensive vitamin products has induced many people to believe that "natural" vitamins are superior to synthetic vitamins made in the laboratory. However, synthetic vitamins are identical in chemical formula to "natural" vitamins and function identically in the body. Many natural vitamin products have synthetic vitamins added to them such as "vitamin C with rose hips."

Food Supplements

(1) Acidophilus supplements are widely advertised and sold for the purpose of supposedly creating a "healthy, clean digestive tract." In reality the digestive tract does not ordinarily need cleaning and is usually better left to function normally.

(2) Bioflavonoids are substances sometimes sold as "vitamin P." These are popular and profitable items. There have been claims that these are essential vitamins, but since no deficiency has been found in humans and it has not been possible to induce a deficiency condition in laboratory animals, it is highly questionable that they have a vital place in human nutrition or are really vitamins.

(3) Garlic is sold as a juice, extract, or capsule and is falsely claimed to be a "miracle" medicine that can prevent or cure diabetes, tuberculosis, hypertension, allergies, hypoglycemia, pneumonia, colds, dysentery, cancer, and many other diseases.

(4) Kelp is advertised as a source of necessary iodine but there is no practical advantage to using this product. A much less expensive source of this mineral is iodized salt and iodine is also present in seafood and many other foods. The amount of iodine required appears to decrease with age.

(5) Lecithin is a phospholipid present in many foods including liver, meats, egg yolk, whole grains, and soy beans. The human body also manufactures large quantities of lecithin. It has been claimed that lecithin is capable of reducing blood cholesterol levels, preventing heart disease, curing arthritis, mental disorders, and performing other spectacular cures. Unfortunately, there has been no significant scientific evidence that this substance has been effective in treating these conditions. Since lecithin has not been shown to be an essential nutrient, there would seem to be no reason to take it as a supplement.[15]

(6) Pangamic acid was falsely called vitamin B_{15} and has been advertised as a cure for a variety of ailments.[28] The FDA has stated that this substance does not meet the criteria used to establish a substance as a vitamin, and there is no evidence that pangamic acid is either effective or safe to use as a food additive or supplement and considers sale of it illegal.[25] Pangamic acid is known to contain mutagenic agents which are associated with ability to cause cancer.

(7) Papain is a protein digestive enzyme derived from the papaya fruit. There is ordinarily very little need for such supplemental

enzymes and conditions which indicate a need for them should be medically supervised.

(8) Protein supplements have been widely advertised, but use of protein tablets, liquids, and powders for weight reduction and other health benefits prompted the FTC to propose disclosure information on protein supplement labels. The FTC would prohibit false claims that excess protein promotes muscle development, intellectual performance or delays the aging process.

(9) Potassium chloride, sold as a diet supplement, is potentially dangerous for some people. Overdose has been the cause of death. Healthy people have little need for extra potassium, but potassium is sometimes prescribed to offset depletion due to the action of some medications. In these cases the dosage should be carefully adjusted according to laboratory tests that monitor body potassium levels. The FDA believes that no one should take potassium without medical supervision.[26]

(10) Rose hips and acerola berries are advertised as a source of "natural" vitamin C and are expensive forms of this vitamin that are not superior in any way to lower cost synthetic brands.

(11) Sea salt is advertised as being superior to regular table salt because it "comes from the sea and contains essential minerals." The chief mineral referred to is iodine, but most of the iodine is lost during the process of evaporation.

(12) Some substances found in foods such as inositol, para-aminobenzoic acid (PABA), citrus bioflavonoids, hesperidin, and rutin, have not been shown to be of any particular importance in human nutrition. However, this does not seem to affect the sale of these supplements.

Medicinal Herbs and Teas

The promotion of herbal teas as "natural" beverages or as "organic" old-time remedies for various ailments are examples of the popular belief that if something is old and contains no additives or chemicals it must be good for you. Herbal teas may be sold as a single ingredient or as complex mixtures of as many as 20 different plant parts consisting of leaves, flowers, stems, bark, roots, and seeds. A mixture of herbs is actually a mixture of chemicals and can have quite active pharmacological properties.[19,23] The public is not generally aware of the chemical effects of some of these plants, but like any drug, these "natural" products can have side effects and should be used with caution.

It is estimated that there are presently as many as 5 to 6 million ginseng users in the United States. Ginseng contains estrogen and has been reported to cause painful and swollen breasts, diarrhea, nervousness, skin eruptions, insomnia, and elevation of blood pressure. Mandrake root contains scopolamine and may be sold as ginseng. Some other products sold as ginseng that contain no ginseng are Texas pieplant, wild pieplant, sorrel, wild rhubarb, and dock.[14]

Weight Reduction Plans

Low Carbohydrate-High Protein Diets: Each year or so another version of this ketogenic type diet surfaces under a new name. Although there are minor differences unique to each particular diet, the general characteristics are that the carbohydrate content is very low while the protein and sometimes the fat content is high, but total calories remain unrestricted.[48]

Some diets of this type that have run their course from time to time include the Air Force Diet, the Calories Don't Count Diet, the Drinking Man's Diet, the Stillman Diet, Doctors Quick Weight Loss Diet, and Dr. Atkin's Diet. The impact of these diets has had no effect on the general problem of obesity and there is no evidence that they offer any metabolic advantage over conventional calorie reduction diets.

Very low carbohydrate diets result in a loss of body electrolytes and dehydration. Unrestricted fat intake can elevate blood cholesterol and triglyceride levels which are risk factors associated with heart disease. These diets place extra stress on the kidneys and may be dangerous if used when there is impaired renal function, especially in the elderly.

Success resulting from this type of dietary manipulation is almost always of a temporary nature. Any weight control diet plan is almost surely doomed to failure unless it results in a permanent change in eating habits that balances energy intake with energy expenditure.

Liquid Protein Modified Fast Diets: The concept advanced by advocates of the liquid (or powdered) protein formulas for weight reduction was that protein without carbohydrate would spare muscle tissue and lean body mass while burning excess body fat to meet energy needs. Such modified fasting diets put considerable strain on the body and should be used only under close medical supervision. Protein products enjoyed tremendous popularity until the FDA reported at least 31 sudden deaths associated with their use. There were also other reports of dehydration, potassium deficiencies, fatigue and muscle weakness, constipation, diarrhea, nausea, and vomiting.

Appetite Suppressants: Appetite suppressants are widely advertised as an aid to weight loss and are frequently taken by people who are unaware of the possible dangers involved in their use. Over-the-counter diet pills are readily available and patients often take them after a physician has recommended weight reduction. Such drugs are effective for only a short period of time, usually one to three weeks, and are not the solution to long-term management of caloric intake. Appetite suppressant medications have many side effects including rapid pulse, dizziness and palpitations. The people who need to lose weight because they have hypertension, heart disease, kidney or thyroid problems, are the very people (many elderly) who should not use these drugs.

The Fructose Diet: Fructose is widely advertised as a "new and natural" sugar, but there is really nothing new about fructose. Common table sugar (sucrose) breaks down to fructose and glucose in the body. Advertising would have us believe that the consumption of fructose will cause weight loss, and testimonials have claimed that fructose solves problems of obesity, alcoholism, mental disease, poor concentration, hypoglycemia, and "that tired feeling."

This is an example of misinformation that promotes the idea that one particular food can cure a wide variety of physical and mental problems. There is a basic fallacy in the notion that adding any kind of sugar to the diet without reducing calorie intake or increasing energy expenditure can result in weight loss. Two tablespoons of either table sugar or fructose will provide about 96 calories. Simply changing the kind of sugar will not reduce weight. The American Dietetic Association (ADA) has noted that there is no advantage in using fructose in a weight reduction diet because there are as many calories in this form of sugar as there are in ordinary table sugar.[33]

The Scarsdale Diet: This diet is another ketogenic, high protein, low carbohydrate weight reduction diet and is a little less rigid than some other diets of this type. The plan includes the recommendation that it be used only by healthy adults and be continued no longer than two weeks. The ketogenic state causes a loss of body fluids, dehydration, and loss of appetite. Side effects may include weakness, dizziness, and irritability. This diet does not limit the quantity of protein foods that may be consumed and, for this reason, some people who follow it may not lose weight. People who do lose weight on this diet plan will usually regain a large portion as soon as the diet is stopped and the body returns to its normal state of hydration unless energy intake is adjusted to a higher level than before the diet was started. This diet is nutritionally inadequate and results are likely to be temporary.

The Pritikin Diet: This program is based on the hypothesis that excess dietary fat "poisons the blood" and is said to be at least partially responsible for a wide variety of conditions including arthritis, hypertension, atherosclerosis, gallstones, gout, and other degenerative diseases associated with aging. The Pritikin thesis maintains that these ordinarily separate diseases are symptoms of a single disease. The diet regimen is low in protein, fat and cholesterol (less than 100 milligrams per day). Meat and fish intake is restricted to about 3½ ounces not over three times a week. Refined carbohydrates are restricted, but complex forms of carbohydrate such as found in whole grains, vegetables and fruits may be used. There are no restrictions in the quantity of "permissible" foods that can be consumed, but salt is limited to 2 to 4 grams per day. Coffee, tea, alcohol, and smoking are prohibited. Mineral and vitamin supplements are not recommended. A daily program of walking or jogging is an important part of this plan. The Pritikin diet book contains many incorrect statements such as grains, vegetables and fruits are better sources of protein than meats, fish, eggs, or milk. The AMA considers this diet plan experimental since neither the claim for prevention nor claims for reversal of disease have been established.[35]

Disease Cures

Cardiac Cures: Although dietary supplements of vitamin E, vitamin C and lecithin have been promoted vigorously as a means of preventing or curing heart disease, claims so far have not been substantiated by recognized research.

Arthritis Cures: Copper bracelets as a cure for arthritis are still around and have been joined by brass tubes filled with barium chloride, various electronic devices, vitamins, "natural" foods, alfalfa tea, sea brine, and cod liver oil to "lubricate the joints." People who have arthritis turn to unproven remedies in desperation, but copper, brass, vitamins, minerals, and special diets do not relieve the pain or alter the course of arthritis. Many so-called cures have occurred during periods when the disease happened to go into remission and the temporary improvement had nothing to do with the diet or drugs being taken.[5,6]

Food Beliefs

Misinformation about food and how the body metabolizes it causes much unnecessary concern and sometimes inadvisable food restrictions. Reliable information helps to combat misinformation and to alleviate unwarranted concern about possible harmful effects of some foods.

Citrus Fruit: Citrus fruits are frequently omitted from the diet by older people because it is falsely believed that oranges and grapefruit "make the stomach acid." This is a misunderstanding because the stomach is normally acid due to the presence of hydrochloric acid which is essential for proper digestion of food. Problems of heartburn and reflux attributed to "acid stomach" are often associated with a hiatal hernia and other conditions common among older people.

Butter Versus Margarine: Butter is probably a "status" food among older adults, some of whom insist that butter is the "real thing" and, therefore, is a superior product to margarine. Butter and margarine are both fats, each contains about 135 calories per tablespoon, and each has about the same vitamin A content because margarine is fortified with this vitamin. Since butter is of animal origin, it is high in cholesterol and saturated fats. Margarine is made from vegetable oils, contains no cholesterol, and is a comparatively unsaturated fat. However, unsaturated vegetable oil that has been hardened by adding hydrogen to make it more like butter in consistency is not as unsaturated as the liquid oil. Persons who wish to emphasize unsaturated fats in their diets should use the special soft or liquid margarines.

Breads: White bread receives its share of criticism because many people think that it has little food value simply because it is white. The fact is that white flour and breads are not necessarily low in nutritional value. Although the bran and wheat germ are removed in the process of milling, most breads and flours are now enriched and white bread may be identical in food value to brown and wheat breads. The replacement of thiamin, niacin, riboflavin, and iron may in some cases even bring the vitamin and iron content of white bread above that of original whole grain. In spite of what the back-to-nature people claim, bleached flour is not harmful and will not "poison" the digestive tract.

Wheat Germ: Wheat germ and wheat germ oil are very popular food supplements because they are a source of vitamin E which is probably the world's favorite vitamin. Also, wheat germ has been advertised in ways that create an association with the "germ of life" with "prolongation of life," "cell rejuvenation," "stamina building," "fertility," "virility," and sexual power.

Sugars: Refined sugar will not poison the body as some people believe. Certainly sugar is an item that can be omitted when there is a need for weight control or other problems of carbohydrate metabolism. Sugar is also a food of low nutrient density, meaning that it provides little or no nutrients in proportion to the calories it contains. However, there is nothing in sugar itself or the refining process that turns it into a harmful poison. Sugar metabolizes to glucose the same as other carbohydrate foods. Raw sugar is not superior to regular refined sugar except as a matter of personal preference.

Molasses: Molasses has been promoted as a treatment for rheumatism and anemia and to correct that "run-down feeling" because it contains some iron. Actually, one tablespoon of blackstrap molasses contains about the same amount of iron as two-thirds cup of dried beans or 3 ounces of cooked beef.

Honey: Honey has been advertised as beneficial in treating high blood pressure, sleeplessness, muscle cramps, and hay fever, but no actual therapeutic effect on these problems has been demonstrated. Most people who have diabetes should not use honey because it contains about the same calories and carbohydrates as table sugar. Honey contains fructose and glucose. There are traces of vitamins, minerals, and protein present in honey, but the amounts are so small that they are not nutritionally significant.[17]

Milk and Cheese: Milk and cheese are believed to be constipating by many people. This idea has been prevalent for a long time and is responsible for large numbers of older people eliminating milk from their diets. Inquiry into the eating habits of those who complain of this chronic problem will usually show that the diet has been very low in fiber content and high in soft, high carbohydrate food. Frequently older people will intentionally avoid nutritionally essential foods because they believe they may contribute to constipation. In these cases not only milk and cheese may be omitted from the diet, but also whole grain cereals and breads, fruits and vegetables may be purposely omitted in an effort to improve the situation. But decreasing dietary fiber only perpetuates the problem.

Some people are opposed to pasteurization of milk because the heating process destroys vitamin C. It is true that vitamin C is destroyed by heat, but milk provides no significant quantity of vitamin C and the protection afforded by pasteurization of milk far outweighs any loss of this vitamin.

Yogurt: Yogurt is milk that has been treated with a bacterial culture. This product became popular in England where it was erroneously thought to have special therapeutic effects. Long ago it was believed that yogurt had special powers to promote longevity by being able to "reduce putrefaction in the bowel and to eradicate autointoxication."

Eggs: Brown eggs are sometimes said to be nutritionally superior to white eggs, but the nutritive value of brown and white eggs is the same. Neither are fertile eggs nutritionally superior to nonfertile eggs, however, they will spoil much more rapidly than eggs that are not fertile.

Vegetables: Some people believe that fresh vegetables cooked at home will always have a higher nutritive value than vegetables that are canned or frozen. Actually, the rapid movement of vegetables from the field to commercial plants may preserve more nutrients than vegetables that have been slowly moved from the field, to market, to home. Fresh vegetables may deteriorate considerably during the holding time necessary for distribution and sale of the product in addition to the time stored in the home refrigerator before being used. It is not dangerous to store food in the original can after it has been opened but any prepared food should be kept covered and refrigerated.

Foods Containing Iron: Beets, tomato juice, and grapefruit are sometimes said to "build blood," but these vegetables are not really good sources of iron. Protein foods such as meats, egg yolk, and some leafy vegetables are much better sources of this mineral.

Food for Sex: Oysters, olives, raw eggs, rare or raw meat have not been proven to have special properties to increase sexual powers.

Food Combinations: Food combinations have been of needless concern to many people. Misinformed persons claim that carbohydrates should not be eaten with proteins, or that starches should not be combined with proteins. The notion that it is harmful to consume combinations of foods such as breads and milk or potatoes is unrealistic. Practically all foods are mixtures of carbohydrates and proteins, and it would be impossible to eat a balanced diet and not combine foods that contain these nutrients in natural combinations.

Diet Foods: "Diet" and "dietetic" are terms applied to food and the meaning is frequently misunderstood. These terms may be brand names not intended to have specific meaning and may refer to products that contain no added sugar, salt, or fat, or fewer calories than foods of a similar nature. "Diet" foods are not necessarily beneficial for any particular type of diet, and may be high in fat or cholesterol or carbohydrates. The only way for the shopper to determine the composition of any product is to read the labels. Currently, regulations do not require that labels show the quantity of sodium contained in a product.

Weight Watchers Foods: Weight Watchers food items are products sold under the brand name of the weight reduction organization. The labels on these items show the caloric content of the product and this can be useful information for those who are counting calories. Some Weight Watchers packaged foods are high in sodium. The Weight Watchers program is concerned with weight reduction; it is not concerned with sodium, fat, or cholesterol control and, for this reason, these products which bear this label may not be suitable for persons who need dietary control other than reduction of calories.

Food Equipment: Food processors, food mills, and blenders are useful in preparing various kinds of food, but they have no special properties to make food more healthful or nutritious than it was before it was chopped, ground, sliced, or liquified. Changing the form or consistency of food does not increase nutritional value in spite of the implications and claims of advertising campaigns designed to sell these appliances.

TYPES OF NUTRITION NONSENSE

Some forms of misinformation and food faddism are easily recognized. Other forms are subtle, and professionals including physicians as well as the public may have difficulty separating facts from fiction. Some examples of nutrition nonsense, profiteering, and unreliable sources of information include:

(1) Any food supplement, diet plan, special food, or treatment that is claimed to prevent or cure a large number of unrelated conditions or chronic diseases.

(2) Any "secret" formula, cure or "miracle" substance offered for sale that the medical profession refuses to prescribe or "does not know about."

(3) Plans, products, or course of treatment that claim to "balance body chemistry," "detoxify the body," or "flush" or "cleanse" the digestive tract.

(4) Publications or individuals that offer nutrition advice and have "health" products or equipment of their own for sale.

(5) Sources that claim that all disease is related to faulty diet.

(6) Sources that claim depleted soil results in food of poor nutritional value, therefore everyone should take vitamin and mineral supplements to be healthy.

(7) Sources that claim organically grown food is nutritionally superior to food grown with commercial fertilizer.

(8) Sources that claim "natural" or "organic" vitamins are superior to synthetic products.

(9) Any proposal to cure diabetes with "herb insulin" or advice to use mixtures of herbs to prevent or cure cancer, heart disease, hypertension, arthritis, kidney, or gastro-intestinal disease.

(10) Books, magazines, TV or radio programs, and newspaper articles or advertising that promote weight reduction plans, equipment, or treatment programs through testimonials by people who say they have lost weight by using the plan.

(11) Exotic diets or health books that are promoted by an entertainment personality on TV or radio talk shows.

(12) Sources that recommend taking supplements such as kelp, fiber, enzymes or massive daily doses of vitamins to protect against background radiation and pollution.

(13) Anti-aging pills, potions, lotions, and mixtures of minerals and vitamins, or secret formulas advertised as being able to restore youth to tired skin, remove wrinkles, or reverse the aging process.

(14) Laboratory testing services that analyze hair, nails, and saliva and report results in elaborate computer printout form that claim to identify nutritional deficiencies or "unbalanced body chemistry." The printout is typically interpreted by the firm's pseudo-nutritionist who prescribes long-term treatment with house brand food supplements to cure the multitude of problems inevitably found.

(15) Anti-sugar zealots who claim that sugar is a "poison" and elimination of it from the diet will provide protection against disease. Also those who insist that sugar should be eliminated to protect the teeth, but advocate the use of honey and dried fruits which may be more cariogenic than sugar itself.

(16) Professionals who scare patients into having dietary intake records analyzed and then recommend a particular brand of supplements to correct deficiencies.

(17) Spot reducing gadgets and sweat-inducing garments including belts, girdles, weights, and body wraps that are claimed to cause weight loss in specific parts of the body or to cause fat to move from one part of the anatomy to another.

(18) Pills or creams that promise to "melt" body fat.

(19) Exercisers, home gymnasiums, and spa programs that promise to take weight off "without effort" on the part of the user.

(20) Any weight reduction plan that claims "calories don't count."

(21) Foundations or institutes that welcome tax deductible contributions to "carry on their goals" which are identified as being able to sell more products and to "educate professionals and lay people" to understand the advantage of using their advertised products.

REALITIES OF ADVERTISING

The question is often raised as to why the public is not better protected from exposure to false or misleading nutrition information. Why do newspapers carry advertising known to be questionable or false? The answer is simple. First, misleading, slanted articles and false claims are shielded by freedom of the press. Second, the purpose of any advertising media including television, newspapers, radio, and magazines is to sell advertising space or time, and the sponsors of food supplements and so-called "health" foods are good customers.

Advertising for many questionable nutrition products depends to a great extent on personal accounts of benefits, cures, and testimonials extolling the virtues of a particular product. Since it is a constitutional right that anyone may speak out or write about almost anything, one does not necessarily have to be a trained and qualified person to publish accounts of what are believed to be benefits derived from use of a food or product. Such testimonials can be very persuasive and are widely used by the sponsors of products with full knowledge that this is an effective marketing strategy. The scientific process of controlled research receives little consideration in this type of selling. The FDA regulations prohibit false labeling information on packaged products, and neither printed advertising for supplements, nor the labels on the containers in which they are sold are permitted to carry claims concerning directions for use or benefits to be derived. The package labels simply identify the contents, and it is indeed a buyer beware market. Purchasers of food supplements are free to dose themselves for whatever condition may exist, with whatever quantity of a substance is felt to be appropriate.

The FDA is limited in capacity to regulate food production and sale of products. At the same time, the rights of all people to freedom of choice must be preserved. This situation emphasizes the desirability of providing basic nutrition education for all age groups. Sound nutrition knowledge and a basic understanding of body functions appears to offer the most promise for responsible food consumption behavior.

PERSONAL AND PUBLIC RESPONSIBILITY

Americans, including the older population, have become increasingly nutrition conscious and the concept of personal involvement in preventive health measures to protect or to enhance physical and emotional well-being has been rapidly adopted by the older population.

With the emphasis being shifted from crisis treatment to personal involvement in disease prevention, there is a great need to separate reliable nutrition information from the confusion of misinformation, faddism, and quackery that presently bombards the public. People have a right to choose a lifestyle and to eat as they wish, but there is also some obligation on the part of the government, professionals, and the food industry to make available dependable information on which informed choices can be made.

Public interest in food and nutrition should be encouraged since reliable information is the best defense against misinformation and exploitation. It is desirable that people have sufficient knowledge to be able to select diets that promote health and well being, but in the absence of sound nutrition knowledge, useless or dangerous misinformation abounds and charlatans are able to reap huge profits from an uninformed public.

The views expressed in this paper are not necessarily those of the Veterans Administration.

REFERENCES

(1) Arje, S.L., and Smith L. 1975. The cruelest killers. In Barrett, Stephen, and Knight (eds.), *The Health Robbers*, pp. 2-13. George F. Stickley Co., Philadelphia.
(2) Barrett, S., Stephen, A., and Knight, G. (eds.). 1976. *The Health Robbers*, George F. Stickley Co., Philadelphia.
(3) Bender, A.E. 1971. Nutrition of the elderly. *Royal Soc. Health J.* 91: 115-124.
(4) Bender, A.E. 1979. Health foods. *Proc. Nutr. Soc.* 38:163-171.
(5) Benzaia, D. 1975. The misery merchants. In Barrett, Stephen, and Knight (eds.). *The Health Robbers*, pp. 34-43. George F. Stickley Co., Philadelphia.
(6) Brin, M., and Roe, D. 1979. Drug-diet interactions. *J. Fla. Med. Assn.* 66:424-428.
(7) Brown, E.L. 1976. Factors influencing food choices and intake. *Geriatrics* 31:89-92.
(8) Busse, E.W. 1978. How mind, body and environment influence nutrition in the elderly. *Postgrad. Med.* 63:118-122.
(9) Butler, R.N. 1968. Why are the older consumers so susceptible? *Geriatrics* 2:83-88.
(10) Clancy, K.L. 1975. Preliminary observations on media use and food habits of elderly. *Gerontologist* 15:529-532.
(11) daCosta, F., and Moorhouse, J.A. 1969. Protein nutrition in aged individuals on self-selected diets. *Amer. J. Clin. Nutr.* 22:1618-1624.
(12) Davies, L. 1976. Nutrition education for the elderly. *Proc. Nutr. Soc.* 35:125-130.
(13) DeLuca, L.M. 1978. Vitamin A. In L.M. DeLuca (ed.), *Handbook of Lipid Research* Fat Soluble Vitamins. Plenum Press, New York.
(14) Deutsch, R. 1977. *New Nuts Among the Berries: How Nutrition Nonsense Captured America*, pp. 281-292. Bull Publishing Co., Palo Alto.
(15) Fletcher, D.C. 1978. Lecithin for hyperlipemia: Harmless but uesless. Questions and Answers, *JAMA* 238:64.
(16) Haeger, K. 1976. Long-time treatment of intermittent claudication with vitamin E. *Amer. J. Clin. Nutr.* 27:1179-1181.
(17) Hamilton, E.M., and Whitney, E. 1979. *Nutrition Concepts and Controversies*, pp. 60-61. West Publishing Co., St. Paul.

(18) Harper, A.E., 1978. Dietary goals-a skeptical view. *Amer. J. Clin. Nutr.* 31:310-321.
(19) Hathcock, N.J., and Coon, J. 1977. Nutrition and drug interactions. *A Nutrition Foundation Monograph.* Academic Press, New York.
(20) Herbert, V. 1976. The health hustlers. In Barrett, Stephen, and Knight (eds.), *The Health Robbers,* pp. 100-119. George F. Stickley Co., Philadelphia.
(21) Herbert, V. 1977. Toxicity of vitamin E. *Nutr. Rev.* 35:158.
(22) Herbert, V. 1979. Facts and fiction about megavitamin therapy. *J. Fla. Med. Assn.* 66:475-581.
(23) Herbert, V. 1979. Megavitamin therapy. *N.Y. State J. Med.* 79:278-279.
(24) Hopkins, H. 1978. Debunking Pangamic Acids. FDA Consumer. HEW Pub. No. (FDA) 79-1050.
(25) Howell, S.C., and Loeb, M.D. 1969. Nutrition and aging. *Gerontologist* 9:1-122.
(26) Ibrahim, I.K., Ritch, A.E.S., MacLennan, F., and May, T. 1978. Are potassium supplements for the elderly necessary. *Age and Ageing* 7:165-170.
(27) Jarvis, W.T. 1980. Food quackery is dangerous business. *Nutr. News.* 43:1-2.
(28) Kelly, J.T. 1978. Nutritional problems of the elderly. *Geriatrics* 33:41-49.
(29) Kippel, R., and Sweeney, T. 1974. The use of information sources by the aged consumer. *Gerontologist* 14:163-166.
(30) Krehl, W.A. 1974. The influence of nutritional environment on aging. *Geriatrics* 29:65-76.
(31) Lane, M.M. 1967. Stereotype ideas about food for the elderly. *Nursing Homes* 16:27-32.
(32) Lane, M.M. 1968. Foods, fads, fallacies and dietary quackery. *Nursing Homes* 17:17-28.
(33) Lecos, C. 1980. Fructose: Questionable Diet Aid. FDA Consumer. HEW Pub. No. (FDA) 80-2130.
(34) Moore, T. 1957. *Vitamin A.* Elsevier, Amsterdam.
(35) Mondeika, T. 1980. The Pritikin diet plan. Questions and Answers. *JAMA* 243:67.
(36) Position paper on food and nutrition misinformation on selected topics (anonymous). *J. Amer. Diet. Assn.* 66:277-279.
(37) *Recommended Dietary Allowances,* 9th ed. 1980. The National Research Council, National Academy of Sciences. Washington, D.C.
(38) Riccitelli, L. 1972. Vitamin C therapy in geriatric practice. *J. Amer. Geriatrics Soc.* 20:34-39.
(39) Rynearson, E.H. 1974. Americans love hogwash. *Nutr. Res. Suppl.* 1:1-14.
(40) Schafer, R., and Yetley, E.A. 1975. Social psychology of food faddism. *J. Amer. Diet. Assn.* 66:129-133.
(41) Schneider, H.A., and Hesia, J.T. 1973. The way it is. *Nutr. Rev.* 31:233-237.
(42) Shifflett, A. 1976. Folklore and food habits. *J. Amer. Diet. Assn.* 68:347-349.
(43) Slesinger, D.C., McDiuitt, M., and O'Donnell, F.M. 1980. Food patterns in an urban population: Age and sociodemographic correlates. *J. Geront.* 15:432-441.
(44) Stare, F.J., and Whelan, E.M. 1979. Nutrition facts and fallacies. *Long-Term Health Svc. Adm. Qtr.* 3:48-57.

(45) Stare, F. 1980. Nutrition–sense nonsense. *Postgrad. Med.* 67:147-153.
(46) Stephenson, M. 1978. The Confusing World of Health Foods, FDA Consumer. HEW Pub. No. (FDA) 79-2108.
(47) Watkin, D.M. 1965. New findings in nutrition of older people. *Amer. J. Publ. Health* 55:548-555.
(48) White, P.L. 1974. Nutrition misinformation and food faddism. *Nutr. Rev. Suppl.* 1:32.
(49) White, P., and Mondeika, T.D. 1980. Food fads and faddism. In R.S. Goodhart and M.E. Shils (eds.), *Modern Nutrition in Health and Disease*, 6th ed., pp. 456-462. Lea and Febiger, Philadelphia.

−19−
Research Needs

Charles H. Barrows and Gertrude C. Kokkonen

GENERAL CONSIDERATIONS

Future research needs are dictated by the goals individuals or organizations establish. The goals of gerontological research have varied considerably over the past twenty years. One notable example among these was a lack of desire to increase the life span of man. Therefore gerontologists had suggested that efforts should be made to add life to years rather than years to life. Unfortunately there is no way of knowing whether successful aging research will provide either or both of these goals. Therefore it would be unwise to exclude studies on the effects of experimental variables on life span. Another practical aspect should be considered. To the purist the life span of an organism may not reflect the rate of aging; on the other hand, except for specific disease prone strains of animals, there must be a relationship between the rate of aging and the life span of a species. When true age parameters are considered, very few have been described within a given species. This lack of credible parameters of aging is even greater when one considers agreement among species. Thus careful consideration should be made before life span studies are excluded from gerontological research.

The dietary manipulation usually employed in studies concerning life extension has been that of feeding animals levels of nutrients which do not support maximal growth. This concept seems to be contradictory to that employed in the vast majority of past nutritional studies, namely, "the bigger the better." To what extent does the term "the bigger," express the well documented redundancy found in biological systems? For example, it is known that nutritionally restricted animals in a sheltered environment can survive with some enzyme levels as low as 50% of that seen in animals fed diets which support maximal growth.[17,25] The major questions are: Can such restricted animals without this apparent redundancy respond adequately to stresses in an open environment? Are organisms operating at maximal levels overly designed systems? Is the cost of this redundancy an inability to reach maximal life span? The answers to these fundamental questions determine the applicability of the concept of under-nutrition to human populations.

At present, there are two discrete areas of research, the successful completion of which would result in an understanding of the relationship between nutrition and aging. The first research area is the effect of age on adult nutritional requirements. The second research area is an understanding of the basic biological mechanism which, in response to dietary restriction, expresses itself in at least one aspect by an extension of life span.

EFFECT OF AGE ON ADULT NUTRITIONAL REQUIREMENTS

Nutritional requirements have been defined as the minimal intakes of nutrients necessary to maintain normal function and health. These requirements are therefore directly related to disease and well-being. Part of the difficulty in establishing nutritional requirements on the basis of diseases and health is the lack of working criteria. The first relationship, that between nutritional requirements and diseases, has been emphasized in the past and is easy to recognize since diseases are relatively well defined. On the other hand, well-being or health requires definition. In order to proceed with these problems, health will have to be arbitrarily defined, perhaps on the basis of the level of those physiological functions which are considered most necessary for normal performance.

Nutritional status is an expression of how an individual rates in terms of fulfilling his nutritional requirements. Since the latter is not well established it follows that data available on nutritional status must be somewhat compromised. Nevertheless, efforts have been made to establish the effects of age on nutritional status and have recently been reviewed.[3] These efforts are made difficult due to the marked difference in genetic background, social environment, and economic status of the populations studied. These variables have a marked impact on the nutrition of an individual. Furthermore, the great selection of foods available complicates the problem. Other difficulties arise because of the various ways of assessing nutritional status. These include records of dietary intake, plasma, blood, or tissue contents of the nutrient, urinary or fecal excretion under various intakes, and the measurement of a biochemical system in which the nutrient plays a role. At present, nutritionists do not agree on the adequacy of any one method for the assessment of nutritional status. Although there seem to be many reports on the subject of nutrition and aging, many are difficult to interpret due to a small population, lack of adequate distribution and questionable techniques.

One approach to the problem of establishing nutritional requirements and status may well be the utilization of data available through the many longitudinal studies presently being carried out. For example, in many of these studies there are dietary records as well as serum nutrient levels on hundreds of individuals, some for as long as twenty years. In addition, there are physiological and biochemical data, medical records, the incidences of various diseases, and mortality statistics. Thus, valuable information regarding criteria for nutritional needs may result from a correlation of these nutritional variables with the other variables measured in these longitudinal studies.

Although there is little or no data presently available to discriminate between adults of different ages as to nutritional requirements, there are data from various nutritional surveys and laboratory studies[3] which can be drawn upon to gain some knowledge of the nutritional status of the aged. Nutritional surveys fail to consistently provide evidence that a poor nutritional state exists generally among members of the aged population in the United States. However, significant numbers of many of the groups studied consumed less than the recommended RDA of certain nutrients. These included protein, calcium, ascorbic acid, and vitamin A. One of the most consistent findings in the national surveys was that low dietary intakes were associated with poor health and low income. Laboratory studies in which the economic variable was minimized indicated that age was accompanied by a decrease in the intake of all nutrients. The decrement in total caloric intake was approximated by: (1) an age-associated decrease in basal metabolism, primarily due to cellular loss, and (2) a decline in physical activity. A number of carefully conducted laboratory studies in which only small numbers of subjects participated substantiated the findings of the national surveys regarding the poor nutritional status of limited numbers of old individuals. In general, little correlation has been observed between low plasma levels of various vitamins and physiological impairments associated with their deficiencies. The low plasma levels of various vitamins, often referred to as marginal deficiency states, may be due to malabsorption as well as reduced intake. In man marginal deficiency states can be overcome by increasing the intake of a particular vitamin. Thus far, there is no evidence to indicate malabsorption exists in normal elderly human subjects. However, the administration of normal dietary levels of various radioactive labeled vitamins has indicated that malabsorption may occur in certain animal species.[12,13]

The limited amount of data available as well as the time required to accrue such data for the establishment of nutritional requirements and nutritional status of the aged makes the task ahead almost staggering. However, there are data presently available which may begin to solve the problem within the next five to ten years. For example, dietary history records were obtained in 1948-49 on 577 individuals over 55 years of age, essentially equally distributed by sex, in San Mateo County, California.[9] Fifty-two percent of the men and 46% of the women were older than 65 years. These data were compared with that obtained in 1952 and the findings were related to the mortality statistics of the group. The data suggest that a relationship exists between mortality and the intake of vitamins A, C, and niacin. For example, among the individuals consuming less than 5,000 I.U. of vitamin A per day, a mortality rate of 13.9% was observed. For those subjects whose intake was above 5,000 I.U. per day, the mortality was 5.4%. For people who consumed diets that provided less than 50 mg vitamin C per day the mortality was 18.5%. On the other hand, for those subjects whose intake was greater than 50 mg per day the mortality was 4.5%. A reduction in mortality with increasing niacin intake was apparent but not as marked as that for vitamins A and C. Similar data for vitamin C were obtained in 1948 on 100 women in Michigan.[29] Between 1948 and 1972, 60 of the women died. The mean intake of vitamin C in 1948 was 51 and 73 mg for those who died and those who survived until 1972, respectively. Definitive conclusions of this study is made difficult by the fact that in 1948, the ages of the women who died prior to 1972 averaged 67.4 years while those who survived averaged 52.1 years. In addition, the dietary records did not include estimates of supplementary vitamins. Both of these studies suggest that beneficial effects on longevity may result when vitamin intake is above the Recommended Daily Allowance. Nevertheless, the retrospective nature of these reports necessitates further investigation. Such studies should include dietary records as well as the supplementation of varying, but known, amounts of vitamins. Owing to the positive response of the elderly to attention, it is important that the effects of placebo also be evaluated.

DIETARY RESTRICTION AND LIFE EXTENSION

Life extension due to dietary restriction has been observed in a variety of species and may likely represent a very basic biological process.[5] Unfortunately, the mechanisms responsible remain unknown. There are a number of key questions which must be answered, however, in order to understand this relationship. For example, the dietary nutrient responsible for an increase in life span must be identified in order to gain some insight into how the nutrient can affect biological processes. Any increase in life span associated with dietary manipulations is generally believed to be due to a restriction of dietary calories. However, most studies in an attempt to accomplish caloric restriction have restricted the intake of a nutritionally adequate diet so that not only has the caloric intake been reduced but also the protein and other dietary components as well. It must be recognized that it is experimentally difficult to hold all dietary components constant and reduce only calories. In order to achieve only caloric restriction under *ad libitum* conditions, there must be adjustments in the diets according to an animal's intake which changes markedly with growth and is dependent upon dietary composition. Thus far, caloric restriction has been successfully accomplished only once and the data indicated the restriction of calories indeed increased the life span of C_3H mice.[33] The *ad libitum* feeding of a diet containing insufficient amounts of protein to support maximal growth has been shown to increase the life span of animals.[3] However, it is not clear the degree to which caloric restriction occurs under these experimental conditions. For example, it has been reported that reduced dietary protein did not affect the caloric intake of adult rats.[12] However, Ross[26] reported that the caloric intake of growing rats fed a synthetic diet containing 8% casein was reduced when compared to that of animals fed a commercial diet. Barrows et al[6] reported an 11% decrease in the intake of young rats fed a 10% as compared to a 20% protein diet. Therefore, on the basis of data presently available, it is not possible to conclude that calories are the sole dietary component which influence life span.

Another important question that must be answered is the age at which dietary restriction will be effective. For example, it has been generally believed that nutritional manipulations which increase life span had to be imposed during growth. This concept originated as a result of the early work of Minot[23,24] postulating that senescence follows the cessation of growth. In addition, McCay et al[20,21,22] showed that increased life span of rats was associated with growth retardation. Furthermore, Lansing[16] indicated that aging in the rotifer involves a cytoplasmic factor the appearance of which coincided with the cessation of growth. However, more recently, studies have indicated that dietary restriction imposed in adult life was effective in increasing life span.[5] This would imply that the mechanism is not solely related to growth.

It is of interest to establish whether the various dietary manipulations which increase life span do so by a common biological mechanism. In general, dietary restriction has been brought about by: 1) reducing the daily intake of a nutritionally adequate diet (one which supports maximal growth); 2) intermittently feeding a nutritionally adequate diet (e.g., feeding every second, third, or fourth day); and 3) feeding *ad libitum* a diet containing insufficient amounts of protein to support maximal growth. A recent series of studies measuring various biochemical, immunological, and morphological variables was carried out in mice fed two of these dietary regimes reported to increase life span; namely, low protein and intermittent feeding.[7,8,32] Unfortunately the data of these experiments are not in agreement. For example, the activities of succinoxidase, cholinesterase, and malic dehydrogenase based on DNA were decreased in the 4% and intermittent-fasted animals and increased in the intermittent-fed animals. Furthermore, the mean values of the enzymatic activities of the intermittent-fed and intermittent-fasted animals were essentially the same as that of the 24% *ad libitum* controls.[4] Therefore these data did not indicate the existence of a common biochemical alteration to explain the phenomenon of increased life span due to dietary restriction. However, the morphological data showed that microvilli present on the surface of podocytes and their processes of the kidney, varied according to age and dietary restriction. The results indicated that both types of dietary restriction produced the same effect; i.e., a lower amount of microvilli as compared to normal animals of the same age.[14] Similarly, the mitogenic activity of three week old mice placed on these diets for a period of one month were lower in both animals fed the 4% protein diet and those intermittently-fed as compared to control animals.[19] Therefore, these morphological and immunological data may represent evidence of a common biological alteration found in these dietarily restricted animals fed dietary regimes reported to increase life span.

Another major difficulty in establishing the relationship between dietary restriction and life extension, is the lack of aging parameters. So that although it is possible to increase life span by dietary manipulations, there are no variables that one can compare in order to determine whether and how one is affecting the aging process. Recognizing these shortcomings, however, there are a number of studies which have measured various physiological and biochemical variables in normal animals as well as those whose life span has been increased by dietary restriction. From such data, working models may evolve which would be useful in proposing various testable hypotheses related to this phenomenon.

Leto et al has shown that animals whose life span has been increased by low protein feeding (4%) have a lower rectal temperature than do those fed the control diet (24% protein). In addition, the low body temperature of these mice was associated with an increased oxygen consumption.[18] Unfortunately, little information is available on the effect of body temperature on the life span of homeothermic animals. Nevertheless the life span of poikilothermic animals increases with decreased environmental temperature.[31] It is generally assumed that this latter finding is a result of a decreased metabolic rate due to the lowering of the rates of biochemical reactions at the reduced temperatures. However, the low body temperatures of the mice described by Leto et al[18] were associated with an increased oxygen consumption. Furthermore, studies of Clark and Kidwell,[10] and Clarke and Smith,[11] in which poikilotherms had been exposed to different temperatures at various times in the life cycle, suggest a more complicated mechanism which may be independent of basal metabolic rate. Complete agreement on the effects of oxygen uptake on the life span of animals is not found in data presently available. For many years an inverse correlation has been described among various species of mammals, i.e., the higher the oxygen uptake per unit of body weight, the shorter the life span.[27] Indeed Kibler and Johnson[15] showed that rats exposed to cold temperatures throughout their life experienced a marked decrease in longevity and a 40% increase in oxygen consumption. However, Weiss[35,36] reported that although the life span of the F-1 generations was longer and the BMR lower than either parental strains (AXC and Fisher), the BMR of the parents was essentially the same in spite of marked differences in longevity. Finally, Storer[30] has reported a direct relationship between oxygen consumption and life span among 18 strains of mice. Should the longevity of individuals within a strain vary inversely with basal metabolic rate, the increased oxygen consumption due to dietary restriction would shorten life span, whereas should the converse relationship exist, the increased oxygen consumption of these mice would result in an increased life span. Therefore, at the present time, the interrelationship among life span, body temperature, and basal metabolic rate provides no information toward an understanding of the mechanism responsible for the increased life span associated with dietary restriction.

It has been proposed that the fertilized egg contains all the genetic information necessary for the orderly sequence of events occurring during the life of an organism. The coded information of the DNA is transcribed to messenger RNA and translated into proteins. It is obvious that new information must be transmitted at different ages in order to account for the various changes seen in an organism throughout its life.[1] It is possible that the rate of occurrence of these programmed events is influenced by the rate of synthesis of specific regulatory RNAs and proteins. Thus, it may be proposed that a possible way to retard the rate of aging would be to decrease the rates of synthesis of specific RNAs and proteins during the life of an organism. Studies measuring the changes in enzymatic activities throughout the life span of rotifers and rats may support this concept. For example, in these studies the typical pattern of changes observed in most enzymatic activities throughout the life span, i.e., an increase in concentration during early life, a period of relatively stable values during adulthood, and decreased activity during senescence were found regardless of the life span. However, dietary restriction which increased life span delayed the time of occurrence of these changes.[5] This concept may also be supported by the immunological studies of Walford et al.[34] who proposed that the initial suppression of the immunological responses of restricted animals was followed by a maturation of a normal response later in life. This peak response was then carried on into later life whereas control animals showed their age related decline in immunological responses. If diseases are an integral part of aging and are genetically controlled, then it seems to follow that dietary restriction should delay the onset of specific diseases as well as increased life span. Indeed, it has been previously pointed out that the onset of many diseases found in rats and mice is delayed in animals whose life span has been increased by dietary restriction.[5] Therefore there are many studies which support the concept that dietary restriction which increases life span also delays the occurrence of specific programmed events at least during early life.

At present there is no evidence to support the concept that aging results from the expression of deleterious genes in late life. In addition the ability to increase the life span of adult organisms indicates that another mechanism must exist independent of alterations in the readout of the genetic code. It has been proposed that dietary restriction results in limited amounts of amino acids which result in decreased protein synthesis and consequent decreased use of the genetic code. It was further proposed that the age changes observed in tissue proteins are the result of selective errors caused by the use of the genetic code.[4] In addition, studies suggest a reduced use of the genetic code in the tissues of restricted animals. For example, feeding *ad libitum* a diet containing 4% protein increased the life span of female C57BL/6J mice. The activities of various enzymes calculated on the basis of DNA were low in these restricted animals.[17] In addition, Ross[25] has reported that hepatocytic catalase, alkaline phosphatase, histidase, and adenosine triphosphatase activities were markedly lower in animals whose life span was increased by dietary restriction as compared to control rats. The work of Schimke[28] supports the premise that reduced enzymatic activity per DNA or per cell represents a reduced enzyme synthesis. For example, when rats were fed diets in which the protein was decreased from 70 to 8%, the animals reduced their total liver arginase from approximately 9 mg to 2 mg and the rates of both synthesis and degradation decreased from 0.7 mg arginase per gram protein per 3 days to 0.2 mg arginase per gram protein per 3 days. Therefore, a reduction in cellular enzymatic activity due to a reduction in dietary protein is the result of a decreased rate of protein synthesis when steady state conditions are reached. A decreased rate of protein synthesis is most likely to result in a reduced use of the genetic code.

The question is always posed as to whether there is evidence that dietary restriction would increase the life span of man. There is no scientifically acceptable documentation to support this. In general undernutrition that exists in underdeveloped countries is most frequently characterized by imbalances of nutrients and usually results in high mortality. The more pertinent question may be: has the time come to conduct clinical studies of dietary restriction on human populations? The large number of studies carried out in animal model systems has shown that nutritional influences are not simply laboratory curiosities. In addition, it has been shown that reducing dietary proteins by 50% in sixteen month old female rats resulted in an increase in life span without a change in the body weight of the animals.[2] Although changes in body weight may be considered among the most general estimates of well-being, nevertheless this variable is also the most frequently used. Therefore, serious consideration should be given to limited studies on the effects of low protein diets in adult humans. It would, of course, be necessary to define the criteria for such an investigation. The health status of the individual would be a key factor—in other words, the person's total state of health related to dietary intake and particularly to protein intake. Under proper medical supervision, it may be possible to administer low protein diets for a specific number of years and to compare health and life span. It is also possible to work retrospectively from mortality statistics that are presently available if the data include sufficient information on nutritional status. From an epidemiologic standpoint, it would be preferable to begin a study of middle-aged individuals who could be followed closely for a prolonged period.

REFERENCES

(1) Barrows, C.H. 1972. Nutrition, aging and genetic program. *Am. J. Clin. Nutr.* 25:829-833.
(2) Barrows, C.H. and Kokkonen, G.C. 1975. Protein synthesis, development, growth and life span. *Growth* 39:525-533.
(3) Barrows, C.H. and Kokkonen, G.C. 1977. Relationship between nutrition and Aging. In H. Draper (ed.), *Advances in Nutritional Research*, pp. 253-298. New York.
(4) Barrows, C.H. and Kokkonen, G.C. 1978. The effect of various dietary restricted regimes on biochemical variables in the mouse. *Growth* 42: 71-85.
(5) Barrows, C.H. and Kokkonen, G.C. 1978. Diet and life extension in animal model systems. *Age* 1:131-143.
(6) Barrows, C.H., Roeder, L.M. and Fanestil, D.D. 1965. The effects of restriction of total dietary intake and protein intake, and of fasting interval on the biochemical composition of rat tissues. *J. Geront.* 20:374-378.
(7) Beauchene, R.E., Bales, C.W., Smith, C.A., Tucker, S.M. and Mason, R.L. 1979. The effect of food restriction on body composition and longevity of rats. *Physiologist* 22:8.
(8) Carlson, A.J. and Hoelzel, F. 1946. Apparent prolongation of the life span of rats by intermittent fasting. *J. Nutr.* 31:363-375.
(9) Chope, H.D. 1954. Relation to nutrition of health in aging persons. A four year follow-up of a study in San Mateo County, Calif. *State J. Med.* 81:335-338.
(10) Clark, A.M. and Kidwell, R.N. 1967. Effects of developmental temperature on the adult life span of *Mormoniella vitripenuis* females. *Exptl. Geront.* 2:79-84.
(11) Clark, J.M. and Smith, J.M. 1961. Independence of temperature on the rate of aging in *Drosophila subobscura*. *Nature* 190:1027-1028.

(12) Draper, H.H. 1958. Physiological aspects of aging. I. Efficiency of absorption and phosphorylation of radiothiamin. *Proc. Soc. Expt. Biol. Med.* 97:121-127.
(13) Draper, H.H. and Lowe, C. 1958. Physiologic aspects of aging. II. Radiocyanocobalamin absorption and protein digestion by the rat. *J. Geront.* 13:252-254.
(14) Johnson, J.E. and Barrows, C.H. 1980. Effects of age and dietary restriction on the kidney glomeruli of mice: observations by scanning electron microscopy. *Anat. Rec.* 196:145-151.
(15) Kibler, H.H. and Johnson, H.D. 1961. Metabolic rate and aging in rats during exposure to cold. *J. Geront.* 16:13-16.
(16) Lansing, A. 1948. Evidence of aging as a consequence of growth cessation. *Proc. Natl. Acad. Sci.* USA 34:304-310.
(17) Leto, S., Kokkonen, G.C. and Barrows, C.H. 1976. Dietary protein, life-span, and biochemical variables in female mice. *J. Gerontol.* 31:144-148.
(18) Leto, S., Kokkonen, G.C. and Barrows, C.H. 1976. Dietary protein, life-span, and physiological variables in female mice. *J. Gerontol.* 31:149-154.
(19) Mann, P.L. 1978. The effect of various dietary restricted regimes on some immunological parameters of mice. *Growth* 42:87-103.
(20) McCay, C.M., Crowell, M.F. and Maynard, L.A. 1935. The effect of retarded growth upon the length of life span and upon the ultimate body size. *J. Nutr.* 10:63-79.
(21) McCay, C.M., Maynard, L.A., Sperling, G. and Barnes, L.L. 1939. Retarded growth, life span, ultimate body size and age changes in the albino rat after feeding diets restricted in calories. *J. Nutr.* 18:1-13.
(22) McCay, C.M., Sperling, G. and Barnes, L.L. 1943. Growth, ageing, chronic diseases, and life span in rats, *Arch. Biochem.* 2:469-479.
(23) Minot, C.S. 1908. The problem of age, growth, and death; a study of cytomorphis based on lectures at the Levell Institute, March, 1907. London, 1908.
(24) Minot, C.S. 1913. *Moderne Probleme der Biologie*, Jena.
(25) Ross, M.H. 1969. Aging, nutrition and hepatic enzyme activity patterns in the rat. *J. Nutr.* 97:565-602.
(26) Ross, M.H. 1961. Length of life and nutrition in the rat. *J. Nutr.* 75:197-210.
(27) Rubner, M. 1908. *Das Problem der Lebensdauer und seine beziehungen zu Wachstum und Ernahrung*, R. Oldenbourg, Munchen.
(28) Schimke, R.R. 1964. The importance of both synthesis and degradation in the control of arginase levels in rat liver. *J. Biol. Chem.* 239:3808-3817.
(29) Schlenker, E.D. 1976. The nutritional status of older women, Ph.D. thesis, Michigan State University, East Lansing.
(30) Storer, J.B. 1967. Relation of lifespan to brain weight, body weight and metabolic rates among inbred mouse strains. *Exptl. Geront.* 2:173-182.
(31) Strehler, B.L. 1961. Studies on the comparative physiology of aging. II. On the mechanism of temperature life-shortening in *Drosophila melanogaster. J. Geront.* 16:2-12.
(32) Tucker, S.M., Mason, R.L. and Beauchene, R.E. 1976. Influence of diet and feed restriction on kidney function of aging male rats. *J. Geront.* 31:364-370.

(33) Visscher, M.B., Ball, Z.B., Barnes, R. and Sivertsen, I. 1942. The influence of caloric restriction upon the incidence of spontaneous mammary carcinoma in mice. *Surgery* 11:48-55.
(34) Walford, R.L., Lui, R.K., Gerbase-DeLima, M., Mathies, M. and Smith, G.S. 1974. Longterm dietary restriction and immune function in mice: response to sheep red blood cells and to mitogenic agents. *Mech. Ageing and Develop.* 2:447-454.
(35) Weiss, A.K. 1962. Metabolism during aging in highly inbred and F_1 hybrid rats. *Fed. Proc.* 21:219.
(36) Weiss, A.K. 1966. A lifespan study of rat metabolic rates. In *Proceedings of the 7th International Congress of Gerontology*, Vienna, Viennese Medical Academy, 1:215-217.

Index

Abdominal muscle, 59
Acerola berries, 338
Acetylcholine, 271
Acetyl Coenzyme-A, 97
Achlorhydria, 145
Achromotrichia, 196
Acidophilus supplements, 337
Acrodermatitis enteropathica (AE), 193, 206
Actin, 57
Adipose tissue, 180
Administration on Aging, The
 programs, 46
 report, 45
Adrenal gland, 131, 144, 179, 181
Adrianzen, T.B., 13
Aflatoxin, 332
Africa, 280, 282, 284, 291, 292
Agriculture, Department of
 studies, 45
Air Force Diet, 339
Albright, –, 242
Albumin, 57, 74, 190, 192, 197
Alcohol, 270, 275, 321
 effect on cardiovascular mortality, 272
 prenatal effects, 12
 rum, 122
 wine, 122
Alcoholism, 193, 220
Alfin-Slater, R.B., 183
Alkaline phosphatase, 164, 357
Alopecia, 189, 194, 201
Aluminum, 251, 261
Aluminum hydroxide, 223, 227, 257-258, 261
American Association of Retired Persons, 50
American Dietetic Association (ADA), 340
American Medical Association (AMA), 331, 334
Amine oxidase, 196, 198
Amino acids, 197 (Also see individual amino
  acids)
 in brain function, 70, 71
 in food protein, 56
 in growth of fetus, 12
 nonessential, 76
 sulfur-containing, 176, 178
para-Aminobenzoic acid (PABA), 338
epsilon-Aminocaproic acid, 316
delta-Aminolevulinic acid, 133
Amnesia, 70
Anderson, J.W., 280, 289, 292
Andres, R., 4
Anemia, 196, 268, 271, 298, 303
Anorexia, 35, 223, 274
Antacids, 223, 227, 303

Antibiotics, 274, 320
Antibody formation, 133
Anticoagulant treatment, 274
Anticonvulsant therapy, 167
Antioxidants, 181-182, 201
Anxiety, 27, 28, 29, 35-36, 92, 227, 314, 319
Aorta, 272
 amine oxidase in, 198
 insoluble collagen content of, 59
 protein synthesis and turnover in, 61-62
 smooth muscle cell necrosis of, 103
 soluble collagen content of, 59
Appendicitis, 284
Appetite, 24, 25, 26, 30, 121, 189, 302
 suppressants, 339
Arginine, 67
Armstrong, W.D., 252
Arrhythmia, 272, 335
Arterial adventitia, 89-90
Arterial intima, 67, 89, 90, 91, 92
Arterial media, 89-90, 91
Arteries
 necrosis of, 103
 structure, 89-90
Arteriosclerosis
 protein alterations in, 67
 in pyridoxine deficient animals, 136
 in senile dementia, 69
Arthritis, 26
Aschoff, L., 92
Ascorbate sulfate, 129
 radioactive, 130
Ascorbic acid. See Vitamin C
Asp, N., 120
Atherogenesis, 67, 88, 90-109
Atheroma, 67, 69, 94, 97
Atherosclerosis, 3, 10, 268, 271, 272, 289-291
 in Caucasus, 122
 in diabetes, 268
 effect of cholesterol metabolism on, 88-109
 protein alterations in, 67
 serum copper concentration in, 199
 in swine, 118
Atkin's Diet, 339
Australia, 196

Bahlburg, H.F., 12
Baker, E.M., 130, 132
BAL (British antilewisite), 317
Baltimore Longitudinal Study, 4
Barnes, K., 180
Barns, P.M., 193
Barragry, J.M., 165

Barrows, C.H., 8, 354
Basal metabolic rate (BMR), 356
Bass, M.A., 44, 46
Batata, M., 139
Baugh, C.M., 136
BBPS. See Build and Blood Pressure Study
Belgium, study in, 168
Belloc, N.B., 2, 4, 313
Berdanier, C.D., 114, 115
Berger, R., 30
Berkeley, U. of California
　study on protein needs, 73
Bertelsen, S., 90
Berlyne, G.M., 166
Bhattacharya, S.K., 100
Bile, 138, 197, 231
Bile acids, 96, 97, 101, 282, 286, 289-291
Bile salts, 96, 97, 159, 177, 231
Biliary tract disease, 268
Binge-eating, 28
Bioflavonoids. See Vitamin P
Blakely, S.R., 117
Blindness, 201
Blood
　concentration of pyridoxine phosphate in, 133
　lipids, 67
　magnesium in, 219, 220
　phosphorus in, 221
Blood vessel, 179
Blurred vision, 267, 336
Body mass index, 26
Bogert, L.J., 21
Bone, 159, 313
　calcification, 183, 232
　calcium in, 223
　collagen metabolism in, 64-65
　demineralization, 196
　effect of fluoride on, 250-260
　formation, 223
　magnesium in, 220
　metabolism in osteoporosis, 68
　phosphorus in, 220
　resorption, 168, 169, 260
Bone loss, sketetal. See Osteoporosis
Bone marrow, 143, 180
Booth, D.A., 25
Bowel function, 29
　effect of fiber on, 119, 120
Bowers, E.F., 132
Boxer, G.E., 135
Boykin, L.S., 22
Brain
　copper accumulation in, 199
　development, 11
　fatty acid metabolism in, 137
　folacin in, 137
　pigmentation in, 179
　protein, 11, 65
　in senile dementia, 69
　vitamin E concentration in, 181
Brammer, L.M., 19
Bran, 286, 293
Bras, G., 111
Breads, 342
Breslow, L., 2, 313
Briggs, G.M., 21
Briggs, M.H., 195, 200
Brook, M., 130
Brosin, H.W., 35
Brown fat, 4

Brown, M.L., 9
Buell, S.J., 69
Build and Blood Pressure Study (BBPS), 4
Burch, R.E., 193
Burkitt, D.P., 280
Burns, J.J., 130
Busse, E.W., 31
Butler, R.N., 45
Buttermilk, 122
Butylated hydroxytoluene (BHT), 181

Cachexia, 269, 274
Cadmium, 190, 202
Caffein, 272
Calcitonin, 68, 166-167, 233
Calcium, 161, 181, 190, 219-248, 251, 272, 273, 292, 304, 305, 320, 352
　absorption, 119-120, 128, 163, 165, 166, 168-169
　in atherosclerosis, 90, 91
　in bone metabolism, 68
　in cholesterol metabolism, 96
　deficiency, 169
　effect of fluoride on, 257, 260
　in high-fiber diets, 120
　homeostasis, 157, 159
　intake by elderly, 123, 130
　in serum, 19, 167
　sources, 298
Calcium:phosphorus ratio, 229, 231, 234, 236
Caldwell, D.F., 13
California
　Alemeda County study, 4
　San Mateo County study, 353
Calloway, D.H., 21, 76
Cameron, N., 22
Canada
　report on protein intake, 74
Cancer, 32, 182, 270, 273, 274, 297
　endometrial, 26
　colon, 120, 286-287
　protein alterations in, 67
Carbohydrate
　intake by Hunzukuts, 122
　intake in U.S. senior citizens, 122
　metabolism, 267-268
　nutrition, 110-124
Carbonated beverages, 190
Cardiovascular disease, 26, 30, 272, 273, 318
Carroll, K.K., 67
Cartwright, G.E., 200
Casein, 67, 111, 123, 354
Castle, W.B., 140
Cataracts, 201, 274
Catecholamine, 70, 71
Catechol-o-methyltransferase, 70
Central nervous system, 12, 13
　folate depletion in, 137
　lesions in, 196
Cereal grains, 177, 220
Cerebrovascular accident, 271
Cerebrum, 70, 179
Ceruloplasmin, 196, 197, 198, 199, 200
Chase, H.P., 11
Chelation therapy, 189
Chen, L.H., 180
Chen, W.L., 280, 289, 292
Chenney, M.C., 135
Chenodeoxycholic acid, 291
Chicago Peoples Gas Company study, 4

Index

Chieffi, M., 130
China, 201
Chio, K.S., 180
Cholecalciferol, 231
Cholelithiasis, 291
Cholic acid, 291
Cholesterol
   in carbohydrate metabolism, 115, 117
   metabolism, 88-109
   radioactive, 101
   in serum. See serum cholesterol
   synthesis and absorption in small intestine, 97, 98
   in tissue, 272
Cholestyramine, 3, 315
Choline, 271
Chromium, 188, 194, 203-206, 268, 272, 275
Chromosomes, 179
Chylomicrons, 98, 177
Circulatory system, 30
   disorders, 201
Cirrhosis, 134, 144, 178, 193, 268, 270-271, 316
Citrus fruit, 341
Clancy, K.L., 45
Clark, A.M., 355
Clark, M., 45, 50
Clarke, J.M., 355
Cleave, T.L., 288, 292
Cobalamin. See Vitamin B_{12}
Cobalt, 140, 275
Cobamides, 140-141
Coenzyme B_{12}, 140
Coffee, 189, 272, 275
Cohen, A.M., 116
Cohen, T., 119
Colavita, F., 30
Coleman, P.D., 69
Cole, T.G., 4
Collagen, 57, 58, 129, 176, 206
   in connective tissue, 77
   in human arterial walls, 67, 90, 91
   insoluble, 59-61
   metabolism in bone, 64-65
   in osteoporosis, 68
   oxidative alteration in, 179
   soluble, 59-61
Collagenase
   activity in osteoporosis, 68
Colon, 282-283
Colorado
   study on protein intake, 74
Comfort, A., 181
Committee on Public Information, 182
Compulsive eating, 299
Comstock, G.W., 12
Congenital abnormalities, 196
Connective tissue, 62
Constipation, 29, 35, 120, 270, 282, 304
Copper, 120, 180, 188, 190, 194, 195-200, 271, 272, 275
   metalloproteins, 198
   radioactive, 197
   in Wilson's disease, 317
Corless, D., 164, 165
Cornstarch
   in diet of rats, 114, 116
   effect on longevity of rats, 111
   in swine diet, 118
Coronary heart disease (CHD), 88, 93, 94, 95, 96, 99, 103, 272

Coronary occlusion, 88
Coronary Primary Prevention Trial (CPPT), 3, 4
Corrick, J.A., 183
Corticosteroid therapy, 242-244, 273
Cortisol, 12
Cousins, R.J., 190
CPPT. See Coronary Primary Prevention Trial
Creatine phosphokinase, 90
Creatinine, 166
Creatinuria, 183
Cultural taboos, 22
Cysteine, 190
Cystic fibrosis, 178
Cytochrome oxidase, 196, 198, 199

Dairy products, 122, 177, 197, 223, 224, 237, 239, 298, 303, 341, 342-343
Dalderup, L.M., 111
Daly, M.M., 94
Danazol, 316
Das, B.C., 100
Davie, M., 164
Davis, I.J., 192
Davis, R.L., 113, 145
Deafness, 275
Death, 35
   of spouse, 48, 49
Decker, T.N., 33
Dehydroascorbic acid, 129
Dehydrocholesterol, 157
Denmark, studies in 164, 168
Dental care, 322
Dental caries, 202
Dentition, 298-299
Dentures, 118-119, 123, 246, 299
Deoxycholic acid, 286, 291
Deoxyuridine suppression test, 139
Depression, mental, 29, 33, 34, 35, 71, 271, 318, 319
Desai, I.D., 179, 180
Deutsch, R., 50
Dextrimaltose, 76
DHEW-USDA *Dietary Guidelines for Americans*, 5
Diabetes mellitus, 10, 26, 30, 32, 266-272, 275, 297, 315
   in atherogenesis, 93, 94-95
   fiber intake in, 292
   protein alterations in, 67
   serum copper concentrations in, 199
Diabetic macroangiopathy, 95
Diabetic microangiopathy, 95
Diarrhea, 35, 194, 196, 270
Dietary Guidelines for Americans, 5
Diets, weight reduction, 339-341
Digestive enzymes, 29
Digestive juice, 128
Digestive system, 29
Digitoxin, 121
Diphenylhydantoin, 167, 222
Disability
   in rural aged, 51
   of spouse, 48, 49
Diuretics, 121, 227, 273, 275
Diverticulitis, 282
Diverticulosis, 120
Dixon, D.G., 289
DNA, 143, 176, 355, 356, 357
   synthesis, 191

total organ, in protein and energy restricted diets, 9
DNA polymerase, 191
Dopamine, 70
Dowager's hump. See Kyphosis
Draper, H.H., 180
Driskell, J.A., 136
Drug abuse, 321
Drug treatment, 121
Durand, A.M.A., 111
dU suppression test, 143
Dvorsky, R., 145
Dwarfism, 189
Dyspnea, 275

EAA/NEAA (essential amino acid/nonessential amino acid) score, 74
Eastwood, M.A., 291
Eating behavior, 26, 27
Eating patterns, 29, 44
Ecuador
 longevity study, 121-122
Eczematous dermatitis, 197
Eggs, 122, 177, 343
Egg yolk, 103, 336
Egypt
 zinc deficiency in, 189
Eichhorn, G.L., 188
Elastin, 57, 58, 61, 198, 206
 in human arterial walls, 67, 90
 oxidative alteration in, 179
Electrocardiographic abnormalities, 268
Elftman, H., 179
Elwood, P.C., 139
Emanuel, I., 13
Emanuel, N.M., 181
Encephalomalacia, 178
Endocrine system, 12
Engen, T., 30
Enzymatic activity, 357
Eppright, E., 31
Epstein, F.H., 95
Erikson, E., 19
Ernst, P., 33
Erythrocytes, 134, 135, 136, 138
 hemolysis of, 178
Estrogen, 62
 effect in osteoporosis, 68, 167
 therapy, 242
Ethoxyquin, 181
2-Ethyl-6-methyl-3-hydroxypyridine-HCl, 181
Evers, W.D., 114
Extrinsic factor, 140
Exudative diathesis, 178
Eyre, D.R., 57

Fad foods, 48
Family traditions, 22
FAO/WHO Expert Group, 145
Farrell, P.M., 182
Fat
 brown, 4, 315
 effect on calcium absorption, 231
 excretion with fiber intake, 289
 intake by Hunzukuts, 122
 metabolism, 268
 in vitamin E absorption, 177
Fatigue, 29
Fatty streak, 91-92
Federal Trade Commission (FTC), 331, 338

Fehling, C., 137
Fernstrom, J.D., 71
Feurig, J.S., 26
Fiber intake, 190, 279-294
 effect on bowel function, 52, 120
 effect on calcium absorption, 222, 232-233
 effect on glucose tolerance, 52
 effect on serum cholesterol levels, 103
 by Hunzukuts, 122
 relationship to bowel diseases, 120
Fibroblasts, 91
Fibrous plaque, 92-93
Finicky eaters, 45
Fish, 224
Fiske, C.H., 252
Flatulence, 292
Fleck, H., 24, 33
Flexner, J.B., 70
Florida
 study in St. Petersburg, 45
Fluoride, 233
 isotopic, 260
 metabolism, 250-261
5-Fluorouracil
 effect on taste, 32
Folacin. See folic acid
Folic acid, 136-140, 141
Food
 behavior, 53
 preferences, 44, 46, 50
 programs, 46
Food and Drug Administration (FDA), 330, 331, 333, 335, 337, 339, 346
Food and Nutrition Board, *Toward Healthful Diets*, 5
Food poisoning, 189
Food Stamp Program, 323
Food symbolism, 23-24
Formiminoglutamic acid (FIGLU), 138-139
Fractures, 236, 238, 241, 297, 299, 319
Framingham Study, 4, 103
Freed, S.C., 27
Freeland, J.H., 190
Free radical reactions, 179, 180, 181
Freud, Sigmund, 20
Fried foods, 177
Friedman, M., 25
Frisch, R.E., 10
Fructose
 Diet, 340
 in human diet, 116, 117
 in rat diet, 114-115, 116-117
Fructose-1,6-diphosphate aldolase, 115
Fructose-1-phosphate aldolase, 115
Fruits, 122, 190

Gaffney, G.W., 144, 145
Galactose, 116
Gallagher, J.C., 163, 164, 169
Gall bladder disease, 120, 297
Gallery, E.D.M., 189
Gallstones, 26, 291
Gander, J., 132
Gangrene, 88
Ganther, H.E., 202
Garlic, 122, 337
Garza, C., 76
Gastrointestinal problems, 35, 189, 297, 335, 336
Gastrointestinal system, 29, 30

Index

Gault, M.E., 200
George, R., 119
Gingivitis, 197
Ginsburg, S.W., 20
Ginseng, 338
Gitman, L., 119
Glass, G.J.B., 145
Glickman, 246
Globulin, 190
Glucocorticoids, 159, 168
Glucose, 25, 268
  in ascorbic acid synthesis, 129
  in atheroma formation, 94
  blood level in mice, 123
  effect on longevity in rats, 111
  in rat diet, 114-115
Glucose-6-phosphatase activity, 116
Glucose-6-phosphate dehydrogenase (G6PD)
  activity, 114, 117-118
  deficiency, 132
Glucose tolerance, 116, 117, 119, 120, 123, 203, 267, 292, 315
Glucose tolerance factor (GTF), 204-206, 268
Glutamic-oxaloacetic transaminase (GOT), 135, 136
Glutamic-pyruvic transaminase (GPT), 135, 136
gamma-Glutamylcarboxypeptidase, 137
Glutathione peroxidase, 176, 202, 203, 206
α-Glycerol phosphate dehydrogenase, 115
Glycine, radioactive, 65
Glycogen phosphorylase, 90, 133
Glycoprotein, 57, 61, 70
Golden, M.H.N., 66
Gonadal steroids, 167
Gordon, J.B., 29
Gormican, A., 190
Gout, 272, 273, 336
Graham, G.C., 13
Great Britain. See United Kingdom
Greek Orthodox Church, religious influence on diet of, 47
Greenberg, L.D., 134
Greene, H.L., 194
Greger, J.L., 191
Grief, 35
Grimshaw, J.J., 130
Grinna, L.S., 181
Grossman, S.P., 24
Grotkowski, M.L., 121
Growth hormone, 10, 12, 159, 167
Growth retardation, 11, 12, 189, 196, 354
L-Gulonolactone oxidase, 129
Guthrie, H.A., 9

Haeflein, K.A., 190
Hair
  growth, 189
  zinc in, 192
Halbrook, T., 194
Halsted, J.A., 200
Hambidge, K.M., 13, 192
Hamburger, W.W., 28
Hamfelt, A., 134, 135
Hamilton, J.C., 165
Harman, D., 181, 199
Harper, L.J., 77
HDL. See high density lipoprotein
Health food, 329, 331

Heart
  disease, 297
  pigmentation in, 179, 180
  vitamin E content in, 181
Heart muscle
  soluble protein content of, 58
Hegsted, D.M., 71
Hellman, I., 130
Hematologic disorders, 199
Hemochromatosis (bronze diabetes), 316-317
Hemoglobin, 57, 177, 275
Hemolytic anemia, 178, 182
Hemorrhoids, 284
Heparin, 227
Hepatic disease, 270, 315, 318
Hepatic lipogenesis, 114, 117, 123
Hepatic necrosis, 134, 196
Herbert, V., 132, 143
Herbs, 338, 345
Hereditary angioedema (HAE), 316
Hernias, 287-289
Herraiz, M.L., 179
Herring, B.W., 192, 198
High density lipoprotein (HDL), 96, 313
Hill, C.H., 198
Hirai, S., 181
Histidine, 138
HMG-CoA reductase. See hydroxymethylglutaryl Coenzyme-A reductase
Hodgkin, D.C., 140
Hodgkin's disease, 199
Holland
  pregnant women in, 13
Honey, 342
Hormones
  calcitonin, 68
  parathyroid (parathormone), 68
  receptor levels, 116
Horning, D., 129
Hsu, J.M., 144
Huennekens, F.M., 137
Human Aging Study. See Libow, L.S.
Hunger, 24, 25
Hurdle, A.D.F., 139
Hurley, L.S., 190
7-alpha Hydroxylase, 97
Hydroxylysine, 57
Hydroxymethylglutaryl Coenzyme-A reductase (HMG-CoA reductase), 97
Hydroxyproline, 68, 69, 273
Hypercalcemia, 169, 229, 270
Hypercalciuria, 69, 270, 273
Hypercholesterolemia, 93-109
Hyperglycemia, 93, 94, 267
Hyperirritability, 220
Hyperlipidemia, 93, 94, 289
Hyperlipoproteinemia, 315
Hyperparathyroidism, 166, 220, 229, 234, 273
Hypertension, 26, 92, 93-94, 95, 99, 199, 271, 297
Hyperthyroidism, 199
Hypertriglyceridemia, 95
Hyperuricemia, 272-273
Hyperzincuria, 193
Hypocalcemia, 159, 164
Hypochlorhydria, 145
Hypocholesterolemia, 289
Hypochondria, 36
Hypoglycemia, 336

Hypogonadism, 189
Hypoparathyroidism, 157
Hypophosphatemia, 159, 164, 222, 223
Hypotension, 275
Hypothalamus, 25, 26

Ileitis, 143
Immobilization, 273
Income, low
  effects on nutrition, 49, 297
  Incontinence, urinary, 30
India
  cholesterol levels study, 100
  osteomalacia study, 168
  protein-energy interaction study, 76
Indigestion, 35
Indoleamine, 70, 71
Infection, 133
Infertility, 182
Influenza, 275, 318
Inositol, 338
Inoue, G., 76
Insomnia, 71
Institute of Food Technologist's Expert Panel on Food Safety and Nutrition, 182
Institutionalization
  negative effects of, 48
  as solutions for social problems, 308
Insulin, 10, 12, 32, 203, 268, 292, 315
  in carbohydrate metabolism, 115, 117
  in cholesterol metabolism, 102
  in increasing age, 95, 116
  in obesity, 95
Intermittent claudication, 182, 336
International Atherosclerosis Project, 91
Intestinal distress, 183
Intestinal enzymes
  glycolytic, 137
Intestine, large, 252, 259
Intestine, small, 159, 191
  calcium absorption in, 225
  cholesterol absorption in, 98
  cholesterol synthesis in, 97, 98
  chromium absorption in, 204
  copper absorption in, 197
  fluoride absorption in, 260
  folate absorption in, 137
  vitamin B_6 absorption in, 133
  vitamin B_{12} absorption in, 141
  vitamin D absorption in, 159, 168
  zinc absorption in, 190
Intrinsic factor (IF), 141, 143
Involutional osteopenia, 161, 166, 167, 168-169
Iodine, 338
IQ, 11, 14
Iran
  zinc deficiency in, 188
Iron, 120, 130, 141, 221, 275, 292, 298, 303, 316-317, 343
Irwin, M.I., 71
Isotopic studies
  B_6, 134
  B_{12}, 143
  cholecalciferol, 165
Israel
  stature of people, 8

Jacob, E., 132
Jacobs, A., 135
Jagerstad, M., 144

Japan
  oldest man, 312
  stature of people, 8
  studies of PTH levels, 166
Jarvik, L.F., 35
Jarvis, W.T., 331
Jenkins, D.J.A., 289
Johnson, H.D., 356
Johnson, L.C., 276
Jolly, J.S., 235
Jordan, 205
Judaism
  religious influences on diet by, 47
Junk food, 50, 51

Kalish, R.A., 44
Kao, K.Y.T., 61
Kashmir
  longevity study, 121-122
Kelp, 337
Kelsay, J.L., 120
Kent, S., 182
α-Keratin, 57
Keshan disease, 201
Keto acids, 76
Ketone bodies, 76
Kheim, T., 136
Kibler, H.H., 356
Kidney, 227, 297
  changes in diabetes, 95
  copper accumulation in, 199
  effect of dieting on, 339
  fluoride excretion by, 252, 259
  function in aging, 121, 181
  insoluble protein content of, 59
  pigmentation in, 179
  soluble protein content of, 58
  synthesis of 1,25(OH)$_2$D$_3$ in, 159
  zinc excretion by, 191
Kidney stone disease, 235, 319
Kidwell, R.N., 355
Kipshidze, N.N., 100
Kirchmann, L.L., 132
Kirk, J.E., 130, 136
Koch, J., 137
Kohrs, M.B., 74
Kosher food, 47
Kritchevsky, D., 67
Krumdieck, C.L., 137
Kubik, M.M., 132
Küstner, R., 179
Kwashiorkor, 178
Kyphosis (dowager's hump), 236

Labadarios, D., 134
Lactase deficiency, 232
Lactose, 231-232
Lahey, M.E., 200
Lanner, E., 194
Lansing, A., 354
Larkins, R.G., 167
Lawoyin, S., 168
Laxatives, 270, 303
  abuse, 168
  effect on calcium absorption, 222, 227
LDL. See low density lipoproteins
Leaf, A., 122
LeBovit, C., 130
Lecithin, 271, 337
Lee, C.J., 10

Index

Moser, P.B., 115
Moynahan, E.J., 193
Mucopolysaccharides, 179
Multiple myeloma, 273
Murphy, E.W., 190
Murphy, W.P., 140
Mursell, S.L., 25
Muscle
  magnesium in, 219
  pigmentation in, 179
  pyridoxine phosphate concentration in, 134
  weakness, 223
Muscular dystrophy, 178, 182
Myelin, 96
Myocardial infarction, 88, 97, 99, 269, 272, 313
Myopathy, 168, 201
Myosin, 57
Mysoline, 167

Naege, R.L., 9
Nageswara Rao, C., 76
National Advisory Council on Aging, 44, 46
National Center for Health Statistics
  studies of, 45
National Heart, Lung and Blood Institute (NHLBI), 3
National Research Council, 74, 145
National Retired Teachers Association, 50
Natural foods, 329, 330, 331, 332, 338
Nayal, A.S., 165
Neoplastic disease, 274
Nephrolithiasis, 165
Nephrons, 121
Nephrosis, 199, 315
Nerve cells, 179, 180
Nerve tissue, 134
Nervous system disease, 271
Neurosis, 35-36
Neurotransmitters, 70, 71, 271
Neutropenia, 196
Newman, B., 132
New York State
  study of diet of elderly, 130
New Zealand, 201, 203
NFDM. See non-fat dry milk
NHLBI. See National Heart, Lung and Blood Institute
Niacin, 123, 335
Nibbling, 28
Nicotine, during gestation and lactation, 12
Niederberger, W., 132
Nielsen, A.L., 200
Nigeria, 286
"Night-eating" syndrome, 28
Nisbett, R.E., 26
Nitrogen balance, 76
Nitrogen protein utilization (NPU), 76
Noncholesterol lipids, 117
Non-fat dry milk (NFDM), 225
Nordin, B.E.C., 164, 165, 169
Norepinephrine, 4, 70
Norman, J.R., 12
Northern Ireland. See United Kingdom
Nursing homes, 23
  diet in, 50
  survey of calorie intake of subjects, 122
Nutrition Foundation, Inc.
  report, 46
Nutrition Program
  Older Americans Act of 1965, 323
Nuts, 197, 220

Nyctalopia, 273

Oberleder, M., 36
Obesity, 26-29, 99, 122, 268, 269, 271, 272, 297, 301, 304-305, 307, 309, 315
  diets for, 339-341
  effect of fiber intake on, 292
  effect on atherosclerosis, 93, 95
  effect on CHD, 95
Obstructive jaundice, 178
O'Dell, B.L., 196, 198
O'Hanlon, P., 74
O'Leary, J.A., 195
$25OHD_3$, 159-161, 164-166
$1,25(OH)_2D_3$, 166-167, 169, 170
Ohlson, M.A., 77
Oils, vegetable, 177, 178, 231
Older Americans Act,
  nutrition program, 48, 51
Olfaction, 32
Organic foods, 329, 331, 332-333, 344
Osaki, S., 198
Osteomalacia, 157, 163-164, 168-170, 222, 228, 236, 273
Osteopenia, 319
Osteoporosis, 121, 157, 161-163, 167, 170, 235-248, 268, 270, 273, 297, 305
  effect of calcium on, 120
  effect of estrogen on, 68, 167
  effect of fluoride on, 250, 255-257, 259-260
  protein alterations in, 67-69
  radiographic survey of, 236-237
  role of $1,25(OH)_2D_3$ in, 169
Ovarian dysfunction, 26
Overeating, 35
Overweight. See Obesity
Oxalic acid, 233

Paget's disease, 229, 257, 273
Palmer, G.H., 289
Pancreatic
  insufficiency, 168, 178
Paneth cells, 194
Pangamic acid, 335, 337
Pao, E.M., 123
Papain, 337
Parakeratosis, 189
Paralysis, 275
Paranoia, 33, 36
Parathormone. See Parathyroid hormone
Parathyroid hormone (PTH), 68, 159, 166, 167, 227-228, 233-234
Parkinson's disease, 71
Pectin, 282, 291, 293
Pelcovits, J., 34
d-Penicillamine, 317
Periodontal disease, 119, 244-246
Pernicious anemia, 140, 143, 145, 268, 274
Petering, H.G., 192
Phenobarbital, 167, 222
Phobias, 36
Phosphate, 190
  absorption, 159
  serum, 167, 169
Phosphoenolpyruvate carboxykinase activity, 116
Phospholipids, 90
Phosphorus, 219-224
Physicial activity, 122, 146, 181, 182, 227, 305-306
Physical impairment. See Disability

Legionnaire's disease, 203, 318
Leg muscle, 59
Legumes, 122, 220
Leibovitz, B.E., 176
Leisure, 300
Leitner, Z.A., 180
Leon, G., 26, 27
Lester, E., 164
Lethargy, 189
Leto, S., 355
Leukemia, 199
Leukocytes, 129, 130, 131, 132, 133, 134, 135, 189
Levine, S., 12
Lewis, J.S., 180
Libow, L.S.
  Human Aging Study, 4
Lichen planus, 197
Life span
  in animals, 8-9
  in man, 312-313
Lindeman, R.D., 192
Linolein, 133
Lipid
  in arteries, 90, 91, 272
  malabsorption, 182
  peroxidation, 179, 180, 199
  polyunsaturated, 180
Lipodosis, 91
Lipofuscin pigment (senile pigment), 176, 179 180, 183
Lipoprotein, 67, 70, 99
  in artery, 90, 94
  in blood, 177, 180
  pre-beta, 113
  serum, 57, 113
Lithium carbonate, 32
Liver, 177, 183, 190, 197, 222, 227
  cholesterol synthesis in, 97, 98
  concentration of pyridoxine phosphate in, 133
  conversion of vitamin D in, 159
  copper levels in, 199
  folate stores in, 138
  insoluble collagen content of, 59
  necrosis, 178, 201
  pigmentation in, 179
  soluble protein content of, 58
  vitamin B_{12} storage in, 143-144
  vitamin E level in, 180
Liver function, 26, 165
Liver therapy, 140
Loh, H.S., 131
Lombeck, I., 193
Loneliness, 34
Longevity, 121-122
Lott, I.T., 200
Low density lipoproteins (LDL), 99
Lund, B., 164
Lung
  insoluble content of, 59
  soluble protein content of, 58
Lurie, O.R., 20
Lymphocytes, 143
Lysine, 57, 67
  radioactive, 61
Lysosomal enzyme activity, 94

McCay, C.M., 8, 9, 205, 354
Macdonald, I., 115
McGill, H.C., 87

McKenzie, J.M., 195
MacLennan, W.J., 165
α_2-Macroglobulin, 190
Magnesium, 120, 219-221, 233, 268, 272, 292
Maine
  study on protein intake, 74
Malaise, 223
Mania, 271
Manner, P., 65
March, B.E., 183
Margarine, 341
Marshall, V., 235
Martin, J.C., 12
Massachusetts Institute of Technology (MIT)
  study on protein needs, 73, 77
Mathur, K.S., 289
Maturity onset diabetes, 205, 206
Meal programs, 46
Meat, 122, 190, 298, 336
Medicaid, 324
Medicare, 324
Megaloblastic anemia, 137, 138, 139
Megavitamin therapy, 334-336
  vitamin C, 132
  vitamin E, 182-183
Meindok, H., 145
Melanin formation, 197
Menkes disease, 200
Menopause
  bone loss after, 167, 236-242
  total body calcium in, 161
Mental depression, 194
Mental illness, 275
Mental impairment, 51
Mental retardation, 11
2-Mercaptoethylamine, 181
Mercury, 190, 202
Mertz, W., 204
Methamphetamine, 12, 13
Methionine, 181
Methylcobalamin (methyl-B_{12}), 140, 141
Methyl folate, 138, 141
3-Methylhistidine, 66
Methylmalonate, 140
Methylmalonic acid (MMA), 142
Metropolitan Life Insurance Company,
  desirable weight tables, 4
Meyers, M.B., 12
Michigan
  study of undernourished females, 45, 353
Mickelson, O., 26
Middaugh, L.D., 13
Migraine headaches, 71
Milhaud, G., 68
Milne, J.S., 131
Mineral balance, 120
Minot, C.S., 354
Minot, G.R., 140
Mitochondrial fraction, 199
Mitochondrial metabolism, 177, 183
Miyagishi, T., 179
Monckeberg, F., 11
Mokrasch, L.C., 65
Molasses, 342
Mollin, D.L., 144
Monoamine oxidase, 70
Moon, H.D., 67
Moore, H.B., 23
Morita, R., 166
Morse, E.H., 132

Index

Phytic acid, 190, 222, 233
Picou, D., 66
Piez, K.A., 57
Pigmentation, 179
Pituitary gland, 131, 179
Plasma, 134, 135
Plasma amino acid response, 73
Plasma cholesterol, 102
Pneumonia, 318
Pories, W.H., 194
Porter, H., 199
Potassium, 268, 275
  chloride, 338
  loss, 121
  radioactive, 66
Prasad, A.S., 189
Presbyopia, 32
Pritikin Diet, 340
Processed foods, 332
Prolactin, 159
Protein, 320, 352, 357
  in brain, 10
  in diet of rats, 111
  effect on calcium absorption, 231
  metabolism, 56-71, 177
  nutrition, 56-87
  soluble (nonsclero), 58-61
  soy, 67
  supplements, 338
  synthesis, 191, 356, 357
Protein restriction, effects of
  in animals
    prenatal, 9, 113
    rats, 358
  in humans
    childhood, 14
    early infancy, 9, 11
    maternal diets, 9
    on total serum protein and albumin, 74
Prothrombin, 183
Protocollagen, 61
Protocollagen hydroxylase, 61-62
Pruritus vulvae, 267
Psychosis, 36
Psychosocial stresses, 93
Pulmonary disease, chronic, 199, 273
Pulmonary distress, 189
Pumpian-Mindlin, E., 19
Puromycin, 70
Pyridoxal, 133, 134
  phosphate, 133, 135
Pyridoxamine, 133, 134
  phosphate, 133
Pyridoxic acid, 133, 134, 136
Pyridoxine. See vitamin B_6
Pyruvic kinase, 115

Quadriplegia, 313, 319
Qureshi, R.U., 117

Rader, J.I., 137
Radice, J.C., 179
Radioisotopic assays
  for serum folate, 138
  for vitamin B_{12}, 142, 144-145
  for zinc absorption, 192
Radiotherapy
  effect on taste, 32
Rafsky, H.A., 132
Ranke, E., 135

Rao, D.B., 49
Rasmussen, A.L., 190
Raymunt, J., 200
RDA. See Recommended Dietary Allowances
Read, A.E., 139
Recommended Dietary Allowances (RDA), 4, 74, 77, 123, 130, 133, 321, 334-335, 352, 353
  for ascorbic acid, 132
  for calcium, 224, 237
  for copper, 200
  for folate, 139
  for selenium, 203
  for vitamin B_6, 136
  for vitamin B_{12}, 145
  for vitamin D, 165, 170
  for vitamin E, 177-178, 336
  for zinc, 190, 195
Reddy, K., 179
Reinhold, J.G., 192
Religion
  influence on dietary factors, 47
  as subsystem, 52
Renal disease, 220, 273
Renal function, 165, 166, 168
Renal hydroxylase, 159, 167
Renal tubular acidosis, 273
Reproduction disorders, 196, 201
Reproductive organs, 179, 189
Retina
  changes in diabetes, 95
Reverse transcriptase, 191
Rhead, W.J., 131
Riboflavin. See Vitamin B_2
Rickets, 159, 163, 228, 336
Ridit, 2
Riggs, B.L., 166
Robinson, M.F., 203
Roderuck, C., 130, 132
Rodin, J., 27
Rodriguez, M.S., 137
Roman Catholics
  religious influence on diet of, 47
Roof, B.S., 166
Rose hips, 338
Ross, G.I.M., 144
Ross, M.H., 9, 111, 354, 357
Roth, L., 26, 27
Rotruck, J.T., 202
Running exercises
  effect on HDL, 99, 100
Rural aged, 51
Rushton, C., 165
Russ, E.M., 200
Russia
  cholesterol level study, 100
  longevity study, 121-122
  nutrition study of pregnant women, 13

St. Clair, R.W., 118
Sakurai, H., 203
Savitsky, E., 20
Scarsdale Diet, 340
Schachter, S., 26
Schafer, R., 330

Schenker, J.G., 195, 200
Schiffman, S., 31, 32, 119
Schilling, R.J., 101
Schilling test, 143
Schimke, R.R., 357
Schizophrenia, 71, 335
Schlenker, E.D., 26, 45
Schrauzer, G.N., 131
Scurvy, 129, 130, 131, 132, 273
Seafood, 190
Sea salt, 338
Sedentary life-style, 93
   as risk factor of CHD, 96
Selenium, 176, 178, 181, 188, 194, 201-203, 275
Senile dementia, 69-71, 269
   protein alterations in, 67
Senile pigment. See lipofucsin pigment
Sensory losses, 322
Serine, radioactive, 139
Serotonin, 70, 71, 271
Serum, 135
   folate concentration in, 138
Seruim cholesterol 3, 98, 282
   in CHD, 97
   in diabetes, 95
   with fiber intake, 289
   interrelationship with atherogenesis, 102
   in pectin-fed rabbits, 291
   in progeny of sucrose-fed mice, 113
   in protein metabolism, 67
   study of levels in America, 100
   in sucrose-fed rats, 114
   in sucrose-fed swine, 118
Serum protein
   total, 74
Serum triglycerides, 114, 115, 118
Sever, L.E., 13
Sex glands, 179, 180
Sex hormones
   influence on atherosclerosis, 67
Shapiro, S., 12
Sherwood, S., 28, 43
Shore, H., 32
Shostrum, E.L., 19
Siegel, B.V., 176
Siegel, P.S., 28
Silica, 275
Sims, L.S., 121
Singer, L., 252
Skeletal muscle
   protein metabolism of, 65
   soluble protein content of, 58
Skin
   concentration of pyridoxine phosphate in, 134
   insoluble collagen content of, 59
   protein synthesis and turnover in, 61, 64
   roughness, 189
   soluble collagen content of, 59
   soluble protein content of, 58
Skin lesions, 189, 193
Smell, 119, 322
Smith, J.M., 355
Smoking, 275
   by Abkhazians, 122
   as atherogenic stimulus, 92, 93
   cardiac risks from, 272
   effect on HDL, 99
   effects on perinatal mortality, 12
   effect on vitamin C requirement, 133
   by Vilacabambans, 122
   weight gain with, 304-305

Snack foods, 300-301
Social Security, 323
Sodium, 268
Soul foods, 22
Soy protein, 67
Spellacy, W.N., 195
Sphingomyelin, 90
Sphingomyelinase activity, 90
Spleen
   insoluble collagen content of, 59
   pigment deposit in, 180
   soluble protein content of, 58
Sporty, L., 34, 37
Sprue, 143, 178
Srivanij, D., 179, 180
Starch
   effect on longevity in rats, 111
   in rat diet, 115-118, 123
Stare, F.J., 23
Steatorrhoea, 178
Sterility, 178
Steroid hormones, 96, 159
Steinkamp, R., 130
Steward, R.D., 203
Stiedemann, M., 123
Stillman Diet, 339
Stokstad, E.L.R., 137
Stone, L.H., 26
Storer, J.B., 356
Strain, W.H., 192
Straus, B., 29
Stress, 28, 30, 36, 227, 301, 302
Stricker, E., 25
Stroke, 88, 244, 271, 272, 313
Stunkard, A.J., 25, 28
SubbaRow, Y.T., 252
Sucrose
   effect on longevity of rats, 111, 112
   in mouse diet, 10, 113
   in rat diet, 114-118, 123
   replacement with dextrimaltose, 76
   in swine diet, 118
Sudden death, 272
Sugars, 342
Sulkin, N.W., 179, 180
Sullivan, J.F., 193
Sunlight exposure, 165, 228
Superoxide dismutase, 176
Surgery
   gastric, 168
Swayback, 196
Sweat, 191
Sweden
   dental problems of aged, 119
   report on protein intake, 74
Swendseid, M.E., 155
Systemic conditioning, 132

Tagasaki, E., 235
Taiwan
   children of undernourished mothers, 14
Tappel, A.L., 180, 181
Tardive dyskenesia, 271
Taste, 30, 31, 32, 119, 322
Tauber, S.A., 145
Taylor, D.D., 117
Taylor-Roberts, T., 66
Tea, 189, 275
   herbal, 338
Teeth, 223
   formation, 244-246
Temperature, environmental, 133

Index

Tendon
  insoluble collagen content of, 59
  protein synthesis and turnover in, 61-63
  soluble collagen content of, 59
  soluble protein content of, 58
Ten-State Nutrition Survey, 74, 298
Testosterone, 167
Tetracyclines, 227
Thiamin, 123
Thompson, C.D., 202
Threonine
  requirement, 73
Thyrocalcitonin, 229
Thyroid hormone, 12
Thyroxine, 12, 102
Tobacco, 321
Tobias, A.L., 29
Tocopherol. See Vitamin E
Tocopherol-p-chlorophenoxyacetate, 181
Todd, W.R., 188
Total parenteral nutrition (TPN), 194
Trace elements, 188
Tranexamic acid, 316
Transaminase activity, 134, 135, 136
Transcobalamin, 141
Trauma, 318-321, 323
Triglyceride concentrations
  in adiposity, 95
  in blood of aging, 117
  in mice, 123
Troll, L.E., 22
Tryptophan, 70, 71, 133, 271
  hydroxylase, 70
  load test, 134, 135
  requirement, 73
Tsuchya, K., 203
Turkey, 205
Turner, M.R., 10
Typhoid fever, 317
Tyrosinase, 197-198
Tyrosine, 70
  decarboxylase, 134
  hydroxylase, 70

Uganda, 282
Ulcerative colitis, 284
Ulcers, 268, 275
Ultraviolet radiation, 165, 167
Underwood, B.A., 180, 181
Underwood, E.J., 189
United Kingdom
  appendicitis in England, 284
  fiber intake in England, 282, 286
  study in Derbyshire, 4
  studies in Great Britain, 74, 121, 164, 165, 168
  studies in Northern Ireland, 74
  study in Scotland, 168
  studies in South Wales, 139
  vitamin B_{12} isolation research, 140
United States
  appendicitis in, 284
  bile acid excretion study, 291
  cancer-selenium relationship in, 275
  Ca:P ratio in diet, 229
  chromium in diet, 205
  colon cancer in, 286
  diverticulitis in, 283
  effect of diet deficiencies, 53
  food behavior, 53
  gallstones in, 291
  hemorrhoids in, 284
  religious influences on diet, 47
  senior citizen caloric intake, 123
  serum cholesterol level studies, 100, 289
  studies on osteoporosis, 168, 169
  survey on zinc in diet, 191
  vitamin B_{12} isolation research, 140
U.S. Bureau of Census 1974-1975 report, 296, 297
USDA 1977-1978 Nationwide Food Consumption Survey, 122, 224
Unsaturated fat, 336
  effect on serum cholesterol levels, 103
Uremic osteodystrophy, 157
Uric acid, 272, 273
Urinary amino acid excretion, 73
Urinary disease, 30, 32
Urinary sulfate excretion, 73
Urinary tract stones, 270, 273
Urocanic acid, 138-139
Urolithiasis, 235
U.S.S.R. See Russia
Uterus
  insoluble collagen content of, 59
  pigment deposits in, 179
  protein synthesis and turnover in, 61-63
  soluble collagen content of, 59
  soluble protein content of, 58

Valine, radioactive, 65, 66
Vallee, B.L., 193
Van Reen, R., 189
Van Rij, A.M., 201
Varicose veins, 287-289
Vasodilation, 220
Vatassery, G.T., 180
Vavrik, M., 94
Vegetables, 122, 177, 190, 224, 343
Verzar, F., 57
Very low density lipoproteins (VLDL), 99
Veterans Administration benefits, 324
Vision, 32
Visser, W., 111
Vitamin A, 123, 130, 273, 335, 336, 352, 353
Vitamin B_2 deficiency, 130, 134
Vitamin B_6 (pyridoxine), 133-136, 141
  deficiency in animals, 136
  phosphate, 133, 134, 135
Vitamin B_{12} (cyanocobalamin), 120, 132, 140-145, 274, 275, 336
Vitamin B_{12}a (hydroxycobalamin), 140
Vitamin B_{12}b (aquocobalamin), 140
Vitamin B_{12}c (nitrocobalamin), 140
Vitamin B_{15}. See Pangamic acid
Vitamin C (ascorbic acid), 123, 129-133, 176, 181, 273, 303, 335-336, 352
  effect on mortality, 353
  natural, 338
  radioactive, 130
Vitamin D, 157-170, 224, 227, 233, 236, 240, 273
  in cholesterol metabolism, 96
  1,25-dihydroxy, 68
  effect of reduced kidney function, 121
  effect of vitamin E on, 183
  in osteoporosis treatment, 261
Vitamin E (tocopherol), 176-183, 202, 336
Vitamin K, 183, 274, 336
Vitamin P (bioflavonoids), 337

Vitamins, 334-337, 344
  natural, 337, 344
VLDL. See Very low density lipoproteins
Vomiting, 196

Wakefield, L.M., 45, 50
Walford, R.L., 356
Walravens, P.A., 192
Water, 272
Waterlow, J.C., 65, 66
Weakness, 274, 275
Weglicki, W.B., 179, 181
Wei, C.K., 180
Weight loss, 35, 274
Weight Watchers, 344
Weinberg, J., 34
Weiner, M.F., 37
Weiss, A.K., 356
Weiss, E., 28
Werner, M., 100
Whagner, P.D., 202
Wheat germ, 342
White House Conference on Food and Nutrition, 50
White muscle disease, 178
Widowhood
  reduced income in, 49
Wilner, M., 34, 37

Wilson, C.W.M., 131
Wilson's disease, 199-200, 271, 317
Winick, M., 120
Wintrobe, M.M., 200
Wolinsky, H., 94
Wound healing, 133, 194
Wünscher, W., 179
Wurtman, R.J., 71

Xanthurenic acid, 134, 135
Xylose, 119

Yamamoto, M., 98
Yamamura, Y., 98
Yogurt, 343
  calcium in, 225
  in diet of Abkhazians, 122
  effect on serum cholesterol levels, 103
Yoshikawa, M., 181
Young, P.T., 25, 26
Young, V.R., 73
Yucca root starch, 122
Yunice, A.A., 199

Zemp, J.W., 13
Zinc, 13, 120, 188-195, 270, 275
Zinc-binding ligand (ZBL), 190